汉英-英汉
建模与仿真术语集
（第二版）

Chinese-English English-Chinese
Modeling and Simulation Dictionary
（Second Edition）

李伯虎　　Tuncer Ören　　赵沁平
　　　　　　　　　　　　　　　　编著
吴启迪　陈宗基　肖田元　龚光红

科学出版社
北　京

内 容 简 介

全书共两大部分：第一部分是汉英建模与仿真术语集，收集一万四千余条汉英对照词汇，涵盖仿真建模技术、仿真系统与支撑技术、仿真应用工程技术等研究领域，基本覆盖我国建模与仿真学术界和工程领域的常用名词术语；第二部分是英汉建模与仿真术语集，收集一万三千余条英汉对照词汇，涵盖建模与仿真、控制科学与工程、计算机科学与技术等相关学科领域，覆盖国际建模与仿真学术界和工程领域的常用名词术语。

本书是一本简明、实用、面广的普及型建模与仿真工具书，可为广大从事建模与仿真理论研究和工程技术工作的科技人员和高等院校学生提供术语参考，也可供高等院校的教师教学使用。

（Tuncer Ören 保留本书外文版的专有出版权及其专有信息网络传播权。）

图书在版编目（CIP）数据

汉英-英汉建模与仿真术语集/李伯虎等编著. —2版. —北京：科学出版社，2019.12
ISBN 978-7-03-062150-4

Ⅰ. ①汉… Ⅱ. ①李… Ⅲ. ①建立模型-名词术语-汉、英 ②系统仿真-名词术语-汉、英 Ⅳ. ①O22-61 ②TP391.9-61

中国版本图书馆 CIP 数据核字（2019）第 181913 号

责任编辑：余 丁 / 责任校对：郭瑞芝
责任印制：师艳茹 / 封面设计：蓝 正

科学出版社 出版
北京东黄城根北街 16 号
邮政编码：100717
http://www.sciencep.com

河北鹏润印刷有限公司 印刷
科学出版社发行 各地新华书店经销

*

2019 年 12 月第 一 版　开本：720×1000　1/16
2019 年 12 月第一次印刷　印张：31
字数：620 000

定价：198.00 元
（如有印装质量问题，我社负责调换）

《汉英-英汉建模与仿真术语集（第二版）》编撰组

李伯虎	院士	北京航空航天大学
Tuncer Ören	教授	渥太华大学（加拿大）
赵沁平	院士	北京航空航天大学
吴启迪	教授	同济大学
陈宗基	教授	北京航空航天大学
肖田元	教授	清华大学
龚光红	教授	北京航空航天大学
王行仁	教授	北京航空航天大学
张最良	教授	军事科学院
黄柯棣	教授	国防科学技术大学
刘藻珍	教授	北京理工大学
康凤举	教授	西北工业大学
杨 明	教授	哈尔滨工业大学
张 霖	教授	北京航空航天大学
沈旭昆	教授	北京航空航天大学
范文慧	教授	清华大学
李 革	教授	国防科学技术大学
马 萍	教授	哈尔滨工业大学
郭百巍	副教授	北京理工大学
杨惠珍	副教授	西北工业大学
王中杰	教授	同济大学
吴云洁	教授	北京航空航天大学
王江云	副教授	北京航空航天大学
李 妮	教授	北京航空航天大学
柴旭东	研究员	北京仿真中心
侯宝存	研究员	北京仿真中心
叶必鹏	博士生	北京航空航天大学

前　言

计算科学正成为与理论研究、实验研究并列的第三种科学研究手段。"仿真科学与技术"作为计算科学的核心组成部分，可以研究和观察已发生、尚未发生或设想的各类现象；可以研究和探索难以到达的微观、中观或宏观世界。"仿真科学与技术"具有求解高度复杂问题的能力和普适性，具有综合、协同、集成和共享的特性，是现代科学研究中求解高度复杂问题不可或缺的科学技术；已成为国民经济、国防建设、自然科学、社会科学等各个领域的系统论证、试验、设计、分析、运行、维护、评估及人员培训等的重要科学技术。"仿真科学与技术"对于实现我国创新型国家战略具有重要意义，正朝着以"数字化、虚拟化、网络化、智能化、服务化、普适化"为特征的现代化方向发展，正成为我国走"科技含量高、经济效益好、资源消耗低、环境污染少、人力资源优势得到充分发挥"的新型工业化道路的通用性、战略性科学技术。

目前我国有大量的科技工作者从事建模与仿真理论与方法、仿真系统与技术、仿真应用工程技术等方面的研究，全国高校中约有15%的研究生开展建模与仿真的相关研究。因此有必要集众多学者的智慧编撰并不断更新建模与仿真术语集，为仿真科学与技术领域的广大科研、教学、工程技术和管理人员提供用词参考。

2012年，中国仿真学会与加拿大Tuncer Ören教授合作出版了《汉英-英汉建模与仿真术语集》。在此基础上，Tuncer Ören教授在2016年提出了修订再版该术语集、补充近些年建模与仿真领域新近词汇的想法，并于2017年与中国仿真学会确定了第二版的合作细节。2018年初，Tuncer Ören教授提供了第二版的英文条目。中国仿真学会为促进我国仿真领域的科学研究、国际化学术交流和高水平创新人才培养，由北京航空航天大学牵头，组织了清华大学、国防科学技术大学、北京理工大学、哈尔滨工业大学、同济大学、西北工业大学等国内高校与研究院所的专家学者，历时一年，共同完成了本书《汉英-英汉建模与仿真术语集（第二版）》的编撰工作。Tuncer Ören教授是国际仿真技术与信息技术领域久负盛名的学者与专家，一直是中国仿真学会的好朋友。本书是Tuncer Ören教授与我国仿真科技工作者长期友好合作的结晶和见证。

在编撰本书的工作中，得到了很多学术前辈、专家学者的大力支持，他们为本书的定位、词条的增删、词语的释义等提供了宝贵的意见和建议，在此表达深深的谢意。

中国仿真学会
2019年3月

Preface
(For the second and enlarged edition-2019)

We are fortunate to witness the Renaissance of simulation due to tremendous advancements in all its fronts.

Synergies of simulation with several disciplines contributed to their enhancements [1]. For example, artificial (or computational) intelligence started by simulation of human cognitive abilities. Conversely, higher-order synergies increased the power of simulation drastically. The synergy of artificial intelligence and simulation is well-known since a long time [2]. The synergy of simulation with intelligent software agents leads to agent-directed simulation [3] with three categories of possibilities: agent simulation, (run-time) agent-monitored simulation, and agent-supported simulation for front-end and back-end interfaces.

Advances in hardware allow peta-simulation, exa-simulation, and quantum simulation. Big data era which was foreseeing yotto (10^{24}) byte, created the recent concepts of ronna (20^{27}) byte and even quecca (10^{30}) byte. Simulation systems engineering is already well established.

In addition to hundreds of application areas in science, engineering, social sciences, and humanities, simulation has a vital role for complex systems including cyber-physical systems [4] and safety-critical large-scale systems. Consider for example the control system of a new aircraft or a nuclear plant. Simulation allows testing of the software in normal as well as in extreme operating conditions. Without simulation-based comprehensive testing, the only way to realize any malfunction could be at the occurrence of an unfortunate disaster. Since simulation can be used for experimentation or for gaining experience to enhance three types of skills, the users of such complex systems can be properly trained by appropriate simulators.

It is a distinct pleasure to witness the publication of the second and enlarged edition of the *Chinese-English English-Chinese Modeling and Simulation Dictionary* at a time when simulation-based approach is becoming a vital intellectual tool for many disciplines [5-7]. It is refreshing to notice that Chinese simulationists can converse in their own language in any aspect of application of simulation as well as in advancing its theory and methodology.

The first edition of the dictionary published in 2012 had about 9 000 English terms. This second and enlarged version contains over 13 000 English terms and their Chinese equivalents.

My appreciation goes to my long-time colleague and friend Professor BoHu Li, to Professor GuangHong Gong and to several Chinese colleagues who painstakingly revised the first edition and generated Chinese equivalents of the additional about 4 000 English terms.

May this linguistic tool enhance the works of Chinese simulationists as they continue contributing to many disciplines.

2019-03-14, Ottawa, Ontario, Canada

Tuncer Ören, Ph.D.
- Professor Emeritus, http://www.site.uottawa.ca/~oren/
University of Ottawa, Ottawa, Ontario, Canada
- SCS Fellow (2016) & Modeling and Simulation Hall of Fame (lifetime achievement award) (2011)
- SCS McLeod Founder's Award for Distinguished Service to the Profession (2017)
- Golden Award of Excellence - from the International Institute for Advanced Studies in Systems Studies and Cybernetics (2018)
- Selected by IBM Canada as one of the pioneers of computing in Canada (2005)

[1] Ören T. Powerful higher-order synergies of cybernetics, systems thinking, and agent-directed simulation for cyber-physical systems. Acta Systemica, 2018, 18(2): 1-5.
(Also presented as a keynote at: InterSymp 2018-The 30th International Conference on Systems Research, Informatics and Cybernetics, and the 38th Annual Meeting of the International Institute for Advanced Studies, July 30-August 3, 2018, Baden-Baden, Germany.)
[2] Ören T. Artificial intelligence and simulation: A typology// Proceedings of the 3rd Conference on Computer Simulation, Mexico City, 1995: 1-5.
[3] Yilmaz L, Ören T. Intelligent agent technologies for advancing simulation-based systems engineering via agent-directed simulation. SCS M&S Magazine, 2010.
[4] Ören T. Simulation of cyber-physical systems of systems: Some research areas-computational understanding, awareness and wisdom. The Journal of Chinese Simulation Systems, 2018, (2).
[5] Ören T, Mittal S, Durak U. A shift from model-based to simulation-based paradigm: Timeliness and usefulness for many disciplines. International Journal of Computer & Software Engineering, 2018, 3(1).
[6] Ören T, Mittal S, Durak U. The evolution of simulation and its contributions to many disciplines// Mittal S, Durak U, Ören T. Guide to Simulation-Based Disciplines: Advancing Our Computational Future. Berlin; Springer, 2017: 3-24.
[7] Mittal S, Durak U, Ören T. Guide to Simulation-Based Disciplines: Advancing Our Computational Future. Berlin; Springer, 2017.

Preface
(For the first edition-2011)

Words are labels to represent concepts. For a person who doesn't know a word, the concept it represents does not exist. The lack of knowledge of words that may hinder the quality of thinking and communication in non-technical situations may simply be a personal matter. However, in technical fields, there are two important consequences of the above where practitioners and/or researchers are concerned. For example, there are about 9 000 technical terms in modeling and simulation (M&S) which provide very valuable and sometimes vital infrastructure to understand and offer better solutions in hundreds of application areas and especially in challenging complex problems. Therefore, it is imperative that professionals know the concepts represented by many terms, and use them to benefit from the many possibilities offered by modeling and simulation to develop products and/or services to find solutions to complex problems of scientific, technical, and social nature. Awareness of the existence of many concepts and hence knowing many technical terms are indeed vital for scientists and methodologists who are dedicated to advance science, engineering, technology, and software tools and environments for modeling and simulation.

In 2005, Professor BoHu Li and I agreed to prepare a *Chinese-English English-Chinese Modeling and Simulation Dictionary*. Later, I provided over 9 000 English terms that I had compiled over many years. Professors BoHu Li and GuangHong Gong provided the leadership necessary to form the Chinese work group and coordinate the contributions of thirty Chinese individuals from senior professors to undergraduate students. My Chinese colleagues and future colleagues (currently students) deserve acclamation, since they worked and reworked on each term with great care. We are equally grateful to the contributions of the prestigious institutions they are associated with.

I hope that my long-term personal efforts as well as the dedicated and meticulous efforts of the thirty Chinese contributors in preparing this dictionary will have an effect in cultivating and advancing the already rich Chinese language and culture.

2011-11-21, Ottawa, Ontario, Canada

Tuncer Ören, Ph.D.
- Professor Emeritus, http://www.site.uottawa.ca/~oren/
University of Ottawa, Ottawa, Ontario, Canada
- SCS Modeling and Simulation Hall of Fame, Inducted 2011

使 用 说 明

1. 本术语集分"汉英"和"英汉"两部分。

2. 汉英部分，按中文词汇的第一个汉字拼音音序排序；若第一个汉字拼音相同，则以第二个汉字拼音音序排序；以此类推。

3. 汉英部分，对应于同一个中文词汇，常用的英文词汇间用";"隔开。

4. 汉英部分，以英文字母为中文词汇开头时，将其置于各部分的最前，并按其汉字拼音音序排序；以希腊字母为中文词汇开头时，将其置于各部分的最后，并按其汉字拼音音序排序。

5. 英汉部分，对应于同一个英文词汇，常用的意义相近的中文词汇间用","隔开；不同意义的中文词汇间用";"隔开。

6. 中文词汇中"（ ）"内的文字，表示其在使用时可以省略。

7. 英文词汇后"（ ）"内的"$v.$"表征其为动词形式。

目　　录

前言
Preface(For the second and enlarged edition-2019)
Preface(For the first edition-2011)
使用说明

第一部分　汉英建模与仿真术语集

Aa	3	Nn	139
Bb	3	Oo	141
Cc	13	Pp	142
Dd	23	Qq	145
Ee	39	Rr	151
Ff	39	Ss	155
Gg	56	Tt	177
Hh	68	Ww	184
Ji	73	Xx	191
Kk	106	Yy	206
Ll	115	Zz	222
Mm	125		

第二部分　英汉建模与仿真术语集

Aa	243	Nn	387
Ba	260	Oo	392
Cc	267	Pp	398
Dd	291	Qq	412
Ee	308	Rr	415
Ff	323	Ss	424
Gg	331	Tt	457
Hh	337	Uu	467
Ii	343	Vv	471
Jj	357	Ww	475
Kk	357	Yy	477
Ll	359	Zz	477
Mm	365		

参考文献 ··· 478

第一部分　汉英建模与仿真术语集

Aa

A-稳定 A-stable
阿贝尔沙堆模型 Abelian sandpile model
安全测试 security testing; safety testing
安全层 security layer; safety layer
安全关键系统 safety-critical system; security-critical system
安全机制 security mechanism; safety mechanism
安全壳 containment
安全模型 security model; safety model
安全模型库 secure model repository; safety model repository
安全认证 security certification; safety certification
安全属性 security property; safety property
安全系统 secure system; safety system
安全性 safety; security
安全性分析 security analysis; safety analysis
安全预测 security prediction; safety prediction
安全组件系统 secure component system; safety component system
安置 allocation
安装 installation
安装程序错误 installer error
安装误差 installation error
按表格计算的 tabular
按词典排序 lexicographic ordering
按需仿真 on-demand simulation; simulation on-demand
暗含依存关系 implicit dependency
奥卡姆剃刀理论 Occam's razor
α测试 alpha test
α值 alpha value

Bb

Bouchaud-Mézard 模型 Bouchaud-Mézard model
白板的理解 tabula rasa understanding
白盒 white box
白盒测试 white box testing
白盒法 white box approach
白盒模型 white box model
白盒验证 white box validation
白箱 white box
白箱测试 white box testing
白箱法 white box approach
白箱模型 white box model
百分比误差 percentage error
柏拉图式的现实 Platonic reality
拜占庭故障 Byzantine fault
拜占庭故障模型 Byzantine fault model
拜占庭行为 Byzantine behavior
拜占庭建模 Byzantine modeling
拜占庭模型 Byzantine model
版本 version
版本控制 version control; versioning
半代码生成 semicode generation
半混沌 semichaotic
半混沌状态 semichaotic state
半解析算法 semianalytic algorithm

半经验模型　semiempirical model
半马尔可夫过程　semi-Markov process
半马尔可夫模型　semi-Markov model
半实物　hardware-in-the-loop
半实物仿真　hardware-in-the-loop simulation
半线性去模糊化　semilinear defuzzification
半隐式方法　semi-implicit method
半隐式算法　semi-implicit algorithm
半隐式梯形公式　semi-implicit trapezoidal formula
半正式的　semiformal
半正式定义　semiformal definition
半正式语言　semiformal language
半自动半代码生成　semiautomatic semicode generation
半自动代码生成　semiautomatic code generation
伴随变量　concomitant variable
帮助　help
包　package; wrap
包含关系　subsumption relation
包容性　containment
包图　package diagram
包装的　wrapped
包装技术　wrapping technique
保持验证　holdout validation
保护机制　protection mechanism
保留的数据　retained data
保守并行仿真　conservative parallel simulation
保守的假设　conservative assumption
保守方法　conservative method
保守仿真　conservative simulation
保守时间管理　conservative time management
保守事件仿真　conservative event simulation
保守同步协议　conservative synchronization protocol
保障性　supportability
保真度　fidelity
保真度管理　fidelity management
保真度和分辨率　fidelity and resolution
保真度建模　fidelity modeling
保证　assurance
报偿　payoff
报告　report
报关　declaration
曝光　exposure
暴露　exposure
暴露变量　exposure variable
贝叶斯　Bayesian
贝叶斯规则　Bayesian rule
贝叶斯建模　Bayesian modeling
贝叶斯克里格模型　Bayesian Kriging model
贝叶斯模型　Bayesian model
贝叶斯模型融合　Bayesian model merging
贝叶斯时空模型　Bayesian spatio-temporal model
备份　back up (*v.*); backup
备选的解决方案　alternative; alternative solution
背景　background
背景变量　background variable

背景知识　background knowledge
被辨识的　identified
被辨识的参数　identified parameter
被辨识的模型　identified model
被测变量　measured variable
被测量　measured quantity
被动多模型　passive multimodel
被动接受的输入　passively accepted input
被动式群体激发工作　passive stigmergy
被动式群体激发工作现象的系统　passive stigmergic system
被动系统　passive system
被动状态　passive state
被发现的错误　discovered error
被交换的模型　exchanged model
被交换的数据　exchanged data
被交换的信息　exchanged information
被禁止的情境　forbidden situation
被考虑系统　considered system
被控变量　controlled variable
被描述的　be described
被认证的　certified
被认证的建模仿真专家　certified M&S professional
被认证的模型　certified model
被认证的数据　certified data
被识别的结构　identified structure
被实验的　experimented
被推荐的　recommended
被修正的误差　corrected error
被压缩的　compressed
被异化的　dissimilated
被证明的　certified
被智能体化的　agentified
被转换感知数据　converted sensory data
被转换值　converted value
本地备份　onsite backup
本地化　localizability
本体　ontology
本体标记语言　ontological markup language
本体表达　ontological representation
本体词典　ontological dictionary
本体仿真　ontomimetic simulation
本体论　ontology
本体论视角　ontological perspective
本体模型　ontology model; ontomimetic model
本体偏移　ontological shift
本体融合　ontology fusion
本体涌现　ontological emergence
本体语言　ontology language
本征模拟　intrinsic simulation
本质级联耦合　essentially cascade coupling
本质模型　intrinsic model
逼近　approximation
逼近阶次　approximation order
逼近精度　approximation accuracy
逼近器　approximator
逼真的　verisimilar
逼真的仿真　coersing simulation
逼真的虚拟现实娱乐　realistic virtual reality entertainment
逼真度等级　level of fidelity
逼真性　verisimilitude
比较　alternative; comparison

比较测试　comparison testing
比较分析法　comparative analysis
比例误差　proportional error
比率　ratio
比率变量　ratio variable
比率尺度　ratio scale
比率尺度变量　ratio scale variable
比率仿真　ratio simulation
比率值　ratio value
比赛　playing
比赛者　player
比喻比较　figurative comparison
笔控输入　pen input
笔纸仿真　pen and paper simulation
必需的　required
必需条件　required condition
必要的　imperative
闭包　closure
闭环　closed-loop
闭环仿真　closed-loop simulation
闭环仿真稳定性　closed-loop simulation stability
闭环连接　closed-loop connection
闭环实时仿真　closed-loop real-time simulation
闭环系统　closed-loop system
闭群　closed cluster
避免死锁　deadlock avoidance
避障范式　failure avoidance paradigm
边迹稳定的　marginally stable
边迹稳定结果　marginally stable result
边迹稳定性　marginal stability
边界　boundary
边界变量　bound variable
边界对称条件　boundary symmetry condition
边界分析　boundary analysis
边界条件　boundary condition
边界验证　bound validation
边界元法　boundary element method
边界值　boundary value
边界值测试　boundary value testing
边界值条件　boundary value condition
边界值问题　boundary value problem
边缘仿真　edge simulation
边缘化输入　marginal input
边缘计算　edge computing
编程　program (v.); programming
编程错误　programming error
编程技术　programming technique
编程技术评估　assessment of programming technique
编程语言　programming language
编队　formation
编辑器　editor
编码错误　encoding error
编码阶段　coding phase
编写摘要　abstracting
编译　compile (v.)
编译错误　compilation error
编译模型　compiled model
编译器　compiler
扁平化保守并行仿真器　flattened conservative parallel simulator
扁平化仿真　flattened simulation
扁平化仿真结构　flattened simulation structure

扁平化广义离散事件系统规范仿真 flattened G-DEVS simulation
扁平顺序仿真器 flat sequential simulator
扁平顺序映射 flat sequential mapping
变保真度仿真 variable fidelity simulation
变步长法 variable-step method
变步长积分算法 variable-step integration algorithm
变差 variation
变度 variance
变分 variation
变分辨率仿真 variable resolution simulation
变分辨率建模 variable resolution modeling
变更率 volatility
变构仿真 allostelic simulation
变构系统 allostelic system
变构系统仿真 allostelic system simulation
变构游戏 allostelic game
变构转变 allosteric transition
变化 change; changing
变化率 rate of change
变换 transformation
变换链 transformation chain
变换驱动的 transformation-driven
变换驱动的复制 transformation-driven replication
变换算法 transformation algorithm
变换引擎 transformation engine
变结构 variable structure
变结构多模型 variable-structure multimodel
变结构建模 variable-structure modeling
变结构模型 variable-structure model
变结构目标 goal with variable structure
变结构耦合模型 variable-structure coupled model; variable-structure coupling model
变结构系统 variable-structure system
变结构自动机 variable-structure automaton
变量 variate; variable; variant
变量的轨迹 trajectory of a variable
变量的含义 meaning of a variable
变量的可导性 derivability of a variable
变量范围 range of a variable
变量范围集 range set of a variable
变量分离 separation of variables
变量可微性 variable derivability
变量类型 variable type
变量耦合 variable coupling
变量耦合建模 variable coupling modeling
变量声明 declaration of a variable
变量有效性 variable validity
变量转换 variable transformation
变迁 transition
变速器 transmission
变体 variant
变拓扑结构的仿真 variable topology simulation
变形 metamorph; metamorphosis; warp

变异测试　mutation testing
变异测试工具　mutation tool for
变异的　variant
变异模型　variant model
变异算子　mutation operator
变异性　variability
变异有效性　effectiveness of mutation
变质　metamorph
变质的　metamorphic
变质的多模型　metamorphic multimodel
变质仿真　metamorphic simulation
变质模型　metamorph model; metamorphic model
变质现实　metamorphic reality
变质作用　metamorphism
便携式　portable
遍历的　ergodic
遍历行为　ergodic behavior
遍历假设　ergodic hypothesis
遍历理论　ergodic theory
遍历系统　ergodic system
遍历性　ergodicity
辨识　identification
辨识的状态　identified state
辨识的状态变量　identified state-variable
辨识激励　discriminative stimulus
辨识模型　discriminative model
辨识网模型　discrimination net model
辨识误差　identification error
辨证的系统　dialectical system
标称变量　nominal variable
标称步长　nominal step size
标称尺度　nominal scale

标称尺度变量　nominal-scale variable
标称的有意涌现　nominal intentional emergence
标称涌现　nominal emergence
标称值　nominal value
标定　standardization
标度共轭梯度算法　scaled conjugate gradient algorithm
标杆　benchmarking
标记　markup; marking; stamp; token
标记法　notation
标记耦合　stamp coupling
标记迁移　labeled transition
标记迁移系统　labeled-transition system
标记数值属性　token numeric attribute
标记语言　markup language
标记知识属性　token knowledge attribute
标记值　tagged value; token value
标记状态转移系统　labeled state-transition system
标记字符串属性　token string attribute
标量变量　scalar variable
标示图　marked graph
标志性事件　sentinel event
标准　standard; criterion; criteria
标准变量　criterion variable
标准博弈　rigid game
标准测试　standards testing
标准差　standard error
标准的　normative
标准化　standardization
标准化变量　standardized variable

标准化值　standardized value
标准假设　standard assumption
标准开发　criteria development
标准模型　normative model
标准评估　normative assessment
标准时间　slow time
标准矢量　criteria vector
标准视图　standard view
标准输入　standard input
表　tableau
表达　expression
表达式　expression
表格查阅　table look up
表格错误　table error
表格剪切　tableau Butcher
表格式的　tabular
表格式的模型　tabular model
表观模型　appearance model
表面　surface
表面的模型　superficial model
表面建模　surface modeling
表面模型　surface model
表面效度　face validity
表面真实　apparent reality
表示（法）　representation
表示层　presentation layer
表示错误　representation error
表示的通用性　universality of representation
表示的唯一性　uniqueness of representation
表示技术　representation technology
表示一致性　representation consistency
表现的　expressive

表现力　expressive power
表现力形式主义　expressiveness formalism
表型　phenotype
表型变化　phenotypical change
表型的　phenotypic；phenotypical
表型地　phenotypically
表型建模　phenotypic modeling
表征　characterization；characterize (*v.*)
表征精度　representational accuracy
表征模型　representational model
表征现实　representational reality
并发仿真　concurrent simulation
并发仿真器　concurrent simulator
并发任务　concurrent task
并发系统　concurrent system
并行　parallel
并行处理　concurrent processing
并行处理分析　concurrent process analysis
并行处理器　parallel processor
并行仿真　parallelized simulation；parallel simulation
并行仿真器　parallel simulator
并行仿真引擎　parallel simulation engine
并行分布式仿真　parallel distributed simulation
并行工程　concurrent engineering
并行行为　parallel behavior
并行化　parallelization
并行化模式　parallelization pattern
并行计算　parallel computing
并行加速　parallelization speedup
并行可视化　parallel visualization

并行离散事件仿真　parallel discrete-event simulation
并行离散事件系统规范 P-DEVS；parallel DEVS
并行连接　parallel connection
并行逻辑变量　concurrent-logic variable
并行模型　concurrent model
并行配置　parallel configuration
并行执行　parallel execution
并置变量　collocated variable
并置法　collocation method
并置式仿真　collocated simulation
并置式克里格仿真　collocated Cokriging simulation
并置式协同仿真　collocated cosimulation
波动方程　wave equation
波动性　volatility
波段　band
波普尔伪造试验　Popperian falsification test
玻璃箱　glass box
玻璃箱测试　glass box testing
玻璃箱模型　glass box model
播放　play
伯努利变量　Bernoulli variable
伯努利随机变量　Bernoulli random variable
泊松分布　Poisson distribution
泊松随机变量　Poisson random variable
泊松稳定性　Poisson stability
博弈　game; gaming
博弈安全　game security
博弈参数　game parameter
博弈策略　game-playing strategy
博弈场景　game scenario
博弈场景管理　game scenario management
博弈多仿真　gaming multisimulation
博弈范式　gaming paradigm
博弈仿真　game simulation; gaming simulation
博弈管理　game management
博弈过程　gaming process
博弈机制　game mechanics
博弈技术　gaming technique
博弈经验　game experience
博弈可靠性　game reliability
博弈控制空间　game control room
博弈论　game theory
博弈论的　game-theoretical
博弈论方法　game-theoretic approach
博弈论仿真　game-theoretic simulation
博弈论模型　game-theoretical model
博弈论学者　gamist
博弈群　gaming group
博弈式仿真　game-like simulation
博弈水平　level of game
博弈图　game graph
博弈研究　research game
博弈周期　game cycle
博弈组织　gaming organization
补偿模型　compensatory model
补偿误差　compensating error; compensated error
补偿系统　compansated system
补偿状态　competence state

捕获误差　capture error
不变的　invariable; invariant; stable; stationary
不变化的　unvariable
不变量　invariant
不变嵌入　invariant embedding
不变条件　invariant
不变性　invariance; invariance property
不成熟的模型　immature model
不充分激励　inadequate stimulus
不当数据　inadequate data
不断演化的现实　evolving reality
不对称　asymmetry
不对称的　asymmetric
不对称仿真　asymmetric simulation
不对称分布　asymmetrical distribution
不对称信息　asymmetric information
不符合高层体系结构的仿真　non-HLA-compliant simulation
不复杂的　uncomplicated
不规则变量　irregular variable
不规则论域　irregular domain
不规则曲面细分　irregular tessellation
不规则误差　erratic error; irregular error; nonregular error
不合理假设　unjustified assumption
不合理信任　unjustified reliance
不合需要的系统　undesirable system
不活动状态　inactive state
不活跃的行为　inactive behavior
不计时的　untimed
不计时的离散事件系统模型　untimed discrete-event system model
不计时形式　untimed formalism
不兼容接口　incompatible interface
不兼容属性　incompatible property
不兼容形式　incompatible formalism
不精确　imprecision
不可持续的应急行为　unsustainable emergent behavior
不可仿真的　unsimulatable
不可仿真的模型　nonsimulatable model
不可分的系统　indecomposable system
不可改变的特性　unmodifiable property
不可更正的错误　irrecoverable error
不可观变量　nonobservable variable
不可观事件　nonobservable event
不可恢复的错误　nonrecoverable error; unrecoverable error
不可检测的突现行为　untestable emergent behavior
不可减少的不确定性　irreducible uncertainty
不可控变量　uncontrollable variable
不可控的组件　uncontrollable component
不可控假设　noncontrollable assumption
不可控因素　uncontrollable factor
不可信解　incredible solution
不可预见的行为　unpredictable behavior
不可预见的数据　unpredictable data
不可预见的突现　unpredictable emergence

不可预见的突现行为　unpredictable emergent behavior
不连续　discontinuity
不连续变量　discontinuous variable
不连续的　discontinuous
不连续函数　discontinuous function
不连续双曲型偏微分方程　discontinuous hyperbolic PDE
不连续状态变量　discontinuous state-variable
不良事件　adverse event
不明确　ambiguity
不明确的　ambiguous
不明确的输入　ambiguous input
不平衡误差　unbalanced error
不确定的　uncertain
不确定的假设　dubious assumption
不确定的状态　indeterminate state
不确定数据　uncertain data
不确定条件　contingent condition
不确定信息　uncertain information
不确定性　uncertainty
不确定性仿真　uncertainty simulation
不确定性建模　uncertainty modeling
不确定性量化　uncertainty quantification
不确定性推理　reasoning under uncertainty
不确定性问题　nondeterministic problem
不确实的数值　metadiscursive value
不适合的仿真　unsuitable simulation
不同　dissimilarity
不透明的思维实验　opaque thought experiment
不完全模型　inadequate model
不完整数据　incomplete data
不完整信息　incomplete information
不稳定　instability
不稳定的　unstable
不稳定结果　unstable result
不稳定模型　unstable model
不稳定误差　unstable error
不稳定性　volatility
不相干的变量　irrelevant variable
不相干激励　irrelevant stimulus
不相干模型　irrelevant model
不相干输入　irrelevant input
不相关的　irrelevant
不相关误差　uncorrelated error
不相容性原理　incompatibility principle
不一致的模型　inconsistent model
不一致的涌现行为　inconsistent emergent behavior
不一致公式误差　inconsistent formula error
不一致涌现　inconsistent emergence
不易移动的　stiff
不应期　refractory period
不由自主的情绪行为　involuntary emotional behavior
不正确的　erroneous
不正确地　erroneously
不正确结果　incorrect result
不重要的状态事件　unimportant state event
不重要事件　insignificant event
不准确　inaccuracy
不准确的　inaccurate

不准确的模型　inaccurate model
不准确地　inaccurately
布尔模型　Boolean model
布尔型变量　Boolean variable
布尔值　Boolean value
布局　layout
布朗模型　Brownian model
步　step
步进的　step-by-step
步长　step-size
步长可控的算法　step-size controlled algorithm
步长控制　step-size control
步长控制算法　step-size control algorithm
步长调整　step-size adjustment
步骤　stage
部分　part; section
部分错误的　partially false
部分规范　partial specification
部分可分解性　partial decomposability
部分求值　partial evaluation
部分求值程序　partial evaluator
部分误差　fractional error
部分析因设计　fractional factorial design
部分因果代数方程系统　partially-causalized algebraic equation system
部分因果方程系统　partially-causalized equation system
部分因果关系的　partially causalized
部分有效性　partial validity
部分有序集合　partially ordered set
部分着色结构有向图　partially-colored structure digraph
部分真实　partial truth
部分整体关系　part-whole relationship
部分正确的　partially true
部分状态　partial state
部署　deployment
部署的模型　deployed model
部署模型　deployment model
部署图　deployment diagram
β变量　beta variable
β测试　beta test; beta testing
β分布　beta distribution

Cc

C4ISR 仿真的互操作性　simulation-C4ISR interoperability
采办　acquisition
采办仿真　simulation for acquisition
采办模型　acquisition model
采办视图　acquisition view
采购　sourcing
采集　capture
采样　sampling
采样分布　sampling distribution
采样数据控制系统　sampled-data control system
采样数据系统　sampled-data system
采样误差　sampling error
彩色的标记　colored token
菜单驱动的工具　menu-driven tool
参考　reference
参考版本　reference version
参考本体　referential ontology
参考变量　reference variable;

referenced variable
参考架构　reference architecture
参考模型　reference model
参考输入　reference input
参考数据　reference data
参考约束　referential constraint
参考值　reference value; referential value
参数　parameter
参数变量　parametric variable
参数辨识　parameter identification
参数辨识方法学　parameter identification methodology
参数不确定性　parametric uncertainty
参数错误　parameter error
参数的　parametric
参数估计　parameter estimation
参数管理器　parameter manager
参数化　parameterization; parametric
参数化的　parameterized
参数化鲁棒性　parameterized robustness
参数化模型　parametric model
参数化实验　parametrized experimentation
参数化试验框架　parameterized experimental frame; parametric experimental frame
参数化误差　parametrization error
参数集　parameter set
参数结构　parameter structure
参数可变性　parameter variability
参数可接受性　parameter acceptability
参数控制　parametric control
参数库　parameter base
参数库管理器　parameter base manager
参数灵敏度　parameter sensitivity
参数灵敏度分析方法学　parameter sensitivity analysis methodology
参数敏感度分析　parameter sensitivity analysis
参数平滑　parameter smoothing
参数随机变量　parametric random variate
参数调整　parameter tuning
参数同态　parameter morphism
参数校准　parameter calibration
参数一致性　parameter consistency
参数有效性　parameter validity
参数值　parameter value
参与的　participative
参与培训　engagement training
参与式　participative
参与式仿真　participative simulation; participatory simulation
参与式仿真建模　participatory simulation modeling
参与式建模　participative modeling; participatory modeling
参与式设计　participatory design
参与式试验　participatory experiment
参与式智能体仿真　participatory agent simulation
参与者　actor; participant; stakeholder
参与者变量　participant variable
参与者需求　stakeholder requirements
参照　benchmark
参照模型　referent model

残差 residual error
残差变量 residual variable
残差方程 residual equation
操动件 actuator
操纵 management
操纵变量 driving variable
操作 manipulation; operate; operation
操作的变量 manipulated variable
操作对策 operational game
操作仿真 hands-on simulation
操作风险 operational risk
操作环境 operational environment
操作连通性的角度 connectivity of operations perspective
操作模型 operational model
操作时间对齐 operations time alignment
操作视图 operational view
操作误差 operative error
操作想定 operational scenario
操作性管理知识 operative management knowledge
操作训练 hands-on training; train as you operate
操作验证 operational validation
操作有效性 operational validity
操作员 operator
操作值 operating value
槽 slot
侧面 profile
测定 determination
测度 measure
测距误差 ranging error
测量 measurement

测量单位 measurement unit
测量的 measured
测量的绝对误差 absolute error of measurement
测量的输出变量 measured outcome variable
测量技术 measurement technique
测量精度 measurement accuracy
测量偏差 bias-error of a measurement
测量误差 measurement error
测量系统 instrumentation system
测量仪器造成的误差 measuring instrument error
测量值 measured value
测试 test; testing
测试方法学 test methodology; testing methodology
测试工具 test tool
测试管理 test management
测试和评估仿真 test and evaluation simulation
测试和评估主计划 test and evaluation master plan
测试和评价 test and evaluation; testing and evaluation
测试和训练使能架构 TENA (Test and Training Enabling Architecture)
测试技术 testing technique
测试阶段 testing phase
测试模型 test model
测试数据 testing data
测试数据集合 testing data set
测试执行 tests execution
策略 strategy
策略比较 policy comparison

策略实验　policy experimentation
层　layer
层次　hierarchical
层次仿真引擎　hierarchical simulation engine
层次分解　hierarchical decomposition
层次分析　dimensional analysis
层次复杂性测量　hierarchical complexity measure
层次化建模　hierarchical modeling
层次化联邦　hierarchical federation
层次化模型结构　hierarchical model structure
层次化耦合　hierarchical coupling
层次结构　hierarchy; hierarchical structure
层次结构多模型　hierarchical multimodel
层次结构离散事件系统规范仿真器　hierarchical DEVS simulator
层次结构顺序仿真器　hierarchical sequential simulator
层次结构顺序映射　hierarchical sequential mapping
层次结构形式　hierarchical structuring formalism
层次模型　hierarchical model
层次模型组合　hierarchical model composition
层次耦合的　hierarchically coupled
层次式仿真　hierarchical simulation
层次着色佩特里网　hierarchical colored Petri net
层次组织　hierarchical organization
插件　plug-in
插件模型　plug-in model
插值　interpolation
插值变量　interpolated variable
插值误差　interpolation error
查询　search
差错防范　error prevention
差分　difference; differencing
差分方程　difference equation
差分方程仿真　difference equation simulation
差分方程模型　difference equation model
差异性　discrepancy
拆分　split (v.)
拆环离散事件系统规范模型　loop-breaking DEVS model
产品　product
产品测试　product testing
产品定义　product definition
产品规格标准　product specification standard
产品规格说明　product specification
产品规格需求　product specification requirements
产品开发　product development
产品模型　product model
产品需求　product requirements
产品质量　product quality
产生　generation; industry
产业断裂　industry fragmentation
长度　length
长距离传输网络　long-haul network
长期轨迹预测　long term trajectory forecast
长期可预测性　long-term

predictability
常规　convention；routine
常规错误　regular error
常规的知识处理　routine knowledge processing
常规范式　conventional paradigm
常规仿真　conventional simulation；regular simulation
常规经验　conventional experience
常规模型　conventional model
常规耦合　conventional coupling
常规实验　conventional experimentation
常规输入　conventional input
常规相关性　conventional dependency
常规游戏　conventional game
常和信任博弈　constant-sum trust game
常量输入　constant input
常识性变量　common sense variable
常识性模型　common sense model
常识性推理　common sense reasoning
常数　constant；invariable
常微分、偏微分混合方程　mixed partial and ordinary differential equations
常微分方程　ordinary differential equation; ODE (Ordinary Differential Equation)
常微分方程仿真　ordinary differential equation simulation
常微分方程建模与仿真　ordinary differential equation M&S
常微分方程求解器　ODE solver
常用用户　frequent user

常值　constant value
场　field
场标记语言　field markup language
场方法　field method
场仿真　field simulation
场行为　field behavior
场面　tableau
超变量　hypervariable；hypervariate
超变量数据　hypervariable data
超大规模仿真　ultrascale simulation
超大涡仿真　very large eddy simulation
超大型仿真　very large simulation
超定微分代数方程　overdetermined differential algebraic equation
超定线性系统　overdetermined linear system
超定线性系统求解器　overdetermined linear-system solver
超对策　hypergame
超对策理论　hypergame theory
超高级的语言　very high-level language
超基变量　superbasic variable
超级计算机　supercomputer
超级联邦　hyper-federation
超级系统　supersystem
超级智能　superintelligence
超几何分布　hypergeometric distribution
超几何随机变量　hypergeometric random variable
超可加信任博弈　superadditive trust game
超类　superclass

超类化　superclassing
超启发式　hyper-heuristics
超实时系统　faster than real-time system
超速仿真　exascale simulation
超速计算　exascale computing
超速计算机　exascale computer
超文本支持文档　hypertext-supported documentation
超现实　hyperreality
超隐式的数值积分法　overimplicit numerical integration scheme
超隐式数值微分公式　over-implicit numerical differentiation formula
朝前　advance
撤回　retraction
沉浸　immersion
沉浸的　immersive
沉浸式仿真　immersive simulation
沉浸式建模　immersive modeling
沉浸式可视化　immersive visualization
沉浸式体验　immersive experience
沉浸式学习体验　immersive learning experience
沉浸式训练　immersive training
陈列　exposure
陈述性　declarative
陈述性方法　declarative approach
陈述性技术　declarative technology
陈述性知识　declarative knowledge
陈述性智能体技术　declarative agent technology
成本　cost
成本模型　cost model

成本需求　cost requirements
成比例的　scaled
成分的　componential
成分分析　componential analysis
成果　product
成果评估　product evaluation
成就　effort
成熟的理解系统　mature understanding system
成熟模型　mature model
成员变量　member variable
成员函数　membership function
成员身份　membership
呈现　presence
承载假设　load bearing assumption
承载值　bearing value
程度　degree
程式化模型　stylized model
程序　procedure; program
程序变量　program variable
程序错误　program error
程序分析　program analysis
程序管理　program management
程序管理器　program manager
程序化模型　programmed model
程序检查　program check
程序检验　program certification
程序可接受性　program acceptability
程序可靠性　program reliability
程序可靠性评估　assessment of program reliability
程序可理解性　program comprehensibility
程序鲁棒性　program robustness
程序鲁棒性的评估　assessment of

program robustness
程序敏感性误差　program-sensitive error
程序模块化　program modularity
程序模型　procedural model
程序认证　program certification
程序设计范式　programming paradigm
程序生成器　program generator
程序生成支持　program generation support
程序输入/输出　programmed input/output
程序文件　program documentation
程序误差　procedure error
程序相关的知识　procedure-related knowledge
程序校核　program verification
程序效率　program efficiency
程序效率评估　assessment of program efficiency
程序正确性　program correctness
程序正确性评估　assessment of program correctness
程序转换　program transformation
程序组　program group
程序组合　program composition
迟滞量化函数　hysteretic quantization function
持久对象　persistent object
持久节点　persistent node
持久图灵机　persistent Turing machine
持久性　persistence
持久性环境　persistent environment
持久状态　persistent state

持续的　persistent
持续规划　continual planning
持续时间　continuous time
持续时间统计　time persistent statistic
持续数据　persistent data
持续误差　persistent error
尺寸　size
尺度　scale
尺度不变的行为　scale-invariant behavior
尺度不变性　scale invariance
尺度连接　scale linking
冲突　warfare
冲突管理　conflict management
冲突规则　conflicting rule
充分　adequacy
充分的模型　adequate model
充足的　adequate
充足的数据　adequate data
重复博弈　repeated game
重复的行为　repetitive behavior
重复输出　duplicate output
重建　reconstruction
重命名错误　renaming error
重配置　reconfiguration
重配置设备　reconfiguration facility
重启动　restart (v.)
重启动点　restart point
重启动例程　restart routine
重启动请求　restart request
重校核　reverification
重写错误　overwrite error
重新出现的状态　re-emergent state
重新初始化不连续性　reinitialization discontinuity

重新开始　restart (*v.*)
重验证　revalidation
重用　reuse；reuse (*v.*)
重载　overloading
重载属性　overloading attribute
抽象　abstract；abstraction
抽象（注重过程）　abstracting
抽象本原　abstract primitive
抽象表示（法）　abstract representation
抽象层　abstract layer
抽象层次　level of abstraction
抽象对象类　abstract object class
抽象仿真　abstract simulation
抽象仿真器　abstract simulator
抽象仿真协议　abstract simulation protocol
抽象概念　abstraction
抽象规范　abstract specification
抽象化技术　abstraction technique
抽象级别　abstraction level
抽象结构　abstract architecture
抽象精化　abstraction refinement
抽象离散事件系统规范仿真器　abstract DEVS simulator
抽象论域　abstract domain
抽象模型　abstract model
抽象模型结构　abstract model structure
抽象实体　abstract entity
抽象顺序仿真器　abstract sequential simulator
抽象顺序映射　abstract sequential mapping
抽象系统模型　abstract-system model
抽象线程仿真器　abstract threaded simulator
抽象映射　abstraction mapping
抽象状态机　abstract state machine
抽象准则　abstraction criterion
出故障的　mulfunctioning
出故障的平均时间　mean time to failure
出现　emerge (*v.*)；emerged
出现的特征　emerged feature
出站　outbound
初步阶段　preliminary phase
初级语言　low-level language
初始标记　initial marking
初始的　initial
初始化　initialization；initialize (*v.*)
初始化部分　initialization section
初始化的变量　initialized variable
初始化机制　initialization mechanism
初始化阶段　initialization phase
初始化条件　initialization condition
初始化误差　initialization error
初始化值　initialization value
初始活动　primitive activity
初始输入　primary input
初始瞬值　initial transient
初始条件　initial condition
初始值　initial value
初始状态　initial state
初值问题　initial-value problem
处理　handling；process；processing
处理程序　handler
处理节点　processing node
处理器　handler；processor
处理误差　processing error
触发　trigger

触发变迁　firing of transition
触发的　triggered
触发规则　fired rule
触发行为　trigger behavior
触发阶段　firing phase; triggering phase
触发器　trigger
触发输入　trigger input
触发序列　firing sequence
触发状态　triggering state
触觉　haptic perception
触觉的　haptic; tactile
触觉反馈　haptic feedback; tactile feedback
触觉反馈硬件　haptic feedback hardware; tactile feedback hardware
触觉仿真器　haptic simulator
触觉感知　haptic sensation
触觉技术　haptic technology
触觉交互　tactile interfacing
触觉接口　haptic interfacing
触觉界面　tactile interface
触觉控制器　haptic controller
触觉论　haptics
触觉设备　haptic device
触觉输入　haptic input; tactile input
触觉输入/输出　tactile I/O
触觉数据　haptic data
触觉特性　tactile characteristic
触觉显示　tactile display
触觉信号　tactile signal
触觉约束　hard constraint
触摸式界面　touch sensory interface
触摸输入　touch input
传播误差　propagated error
传递　delivery; pass (*v.*)
传递函数仿真　transfer function simulation
传递性　transitivity
传动　transmission
传动装置　transmission
传感器　sensor; transducer
传感器融合　sensor fusion
传感器输入　sensor input
传感器数据　sensor data
传感器误差　sensor error
传感器系统　sensor system
传输　transmission; transmit (*v.*)
传输误差　transmission error
传统设计方法　traditional design approach
传统现实　traditional reality
串行仿真　serial simulation
串行连接　serial connection
串行模型　serial model
串联仿真　tandem simulation
创造的　created
创造的模型　created model
创造的实体　created reality
创造误差　creation error
创造性的方面　creative aspect
创造性系统　creative system
创制　poiesis
纯反馈耦合　pure feedback coupling
纯反馈耦合配方　pure feedback coupling recipe
纯软件仿真　pure software simulation
纯协调博弈　pure coordination game
词典　dictionary
词典模型　lexicographic model

词汇　vocabulary
词汇错误　lexical error
词汇的　lexical
词汇评价　lexical evaluation
词汇约束变量　lexically-bound variable
词汇作用域变量　lexically-scoped variable
磁盘查找错误　disk-seek error
磁盘错误　disk error
磁盘写错误　disk write error
次级、限值一下的刺激　subtreshold stimulus
次序　order
次要成分　subcomponent
刺激　stimulation; stimulus
刺激物　stimulator; stimulant
从底向上涌现　bottom-up emergence
从句　clause
从开始起　ab initio
从上至下设计　top-down design
从属模型　dependent model
从头仿真　ab initio simulation
从业者　practitioner
粗糙　coarseness
粗糙数据　coarse data
粗粒度　coarse-grain granularity
粗粒度的组件　coarse-grained component; coarser-grained component
粗粒度时间变量　coarse time-scale variable
促进　stimulate (v.)
猝发　burst
脆弱系统　brittle system

存储的能量　stored energy
存储器　memory; storage
存伪错误模型　error of accepting wrong model
存在的假设　existential assumption
存在的现实　existing reality
存在的依赖　existential dependence
存在的依赖关系　existential dependency
存在条件　existing condition
错过的假设　missed assumption
错觉　erroneous perception; illusion
错乱性　confusion
错误　false; mistake
错误标志　error flag
错误插入测试　fault insertion testing
错误处理　error handling; error processing
错误处理程序　error handler
错误猝发　error burst
错误代码　error code
错误的　incorrect; wrong
错误的概念　erroneous concept
错误的行为　erroneous behavior
错误的假设　incorrect assumption
错误的建模假设　faulty modeling assumption
错误的模型　wrong model
错误的预测　incorrect prediction
错误的自模型　wrong self-model
错误的组件　faulty component
错误分析　error analysis
错误俘获　error capture
错误概率　error probability
错误感知的现实　misperceived reality

错误行为 misbehavior
错误恢复 error recovery
错误检测代码 error detecting code
错误检测系统 error detecting system
错误检查 error checking
错误检查代码 error checking code
错误纠正 error correction
错误类型 error type; type of error
错误理解的现实 misunderstood reality
错误率 error ratio
错误耦合 miscouple (*v.*); miscoupled; miscoupling
错误评估 error assessment
错误设定误差 misspecification error
错误实现 misimplementation
错误输出 incorrect output
错误数据 error data
错误探测 error detection; mistake detection
错误条件 error condition
错误推理 erroneous reasoning
错误消息 error message
错误隐藏 error concealment
错误预防 mistake prevention
错误运行日志 error log
错误证明 mistake proofing
错误字符 error character

Dd

DNA 机 DNA machine
DNA 计算 DNA computing
Duhem-Quine 命题 Duhem-Quine thesis
Duhem-Quine 问题 Duhem-Quine problem
Duhem 问题 Duhem problem
打包 wrap (*v.*)
打破和局规则 tie-breaking rule
打印机错误 printer error
打印间隔 print interval
打字错误 typing error
大尺度 large scale
大尺度仿真模型 large-scale simulation model
大尺度模型 large scale model
大规模 massive scale
大规模并行处理 massive parallel processing
大规模多用户 massively multiuser
大规模多用户仿真 massively multiplayer simulation
大规模多用户游戏 massively multiplayer game; massively multiuser game
大规模多用户在线游戏 massively multiplayer online game
大规模多智能体系统 massively multiagent system
大规模仿真 massive-scale simulation; large-scale simulation
大规模仿真环境 large-scale simulation environment
大规模仿真软件开发 large-scale simulation software development
大规模仿真试验 large-scale simulation experiment
大规模分析 large-scale analysis
大规模模型 massive-scale model

大规模试验　large scale experiment
大规模数据分析　large-scale data analysis; massive-data analysis
大规模数据集　large-scale dataset
大规模系统　large-scale system
大粒度的组件　large-grained component
大脑控制　brain-controlled
大脑控制输入　brain-controlled input
大脑启发　brain-inspired
大脑启发式计算　brain-inspired computing
大脑启发式建模　brain-inspired modeling
大数据　big data
大尾数法　big endian
大尾数法数据格式　big endian data format
大涡模拟　large eddy simulation
大型多人　massively multiplayer
大型仿真　big simulation; large simulation
大型仿真数据集　large simulation data set
大型刚性系统　large-scale stiff system
大约的　approximate
大致估计　approximate (v.)
代表性需求　representational requirements
代价　cost
代价函数　cost function
代理　agent; brokering; proxy
代理触发式仿真　agent-triggered simulation
代理导向的　agent-directed
代理导向的仿真　agent-directed simulation
代理导向的框架　agent-directed framework
代理对代理　agent-to-agent
代理仿真　agent simulation; proxy simulation
代理仿真器　agent simulator
代理服务器　proxy
代理行为　agent behavior
代理行为图　agent-behavior diagram
代理合理性　agent rationality
代理机制　agent-based mechanism
代理技术　agent technology
代理间　inter-agent
代理间的交互　agent-to-agent interaction
代理间的通信　inter-agent communication
代理监控的　agent-monitored
代理监控的多仿真　agent-monitored multisimulation
代理监控的仿真　agent-monitored simulation
代理监控的环境　agent-monitored environment
代理监控的耦合　agent-monitored coupling
代理监控的预期多仿真　agent-monitored anticipatory multisimulation
代理建模　agent modeling
代理交互　agent interaction
代理交互协议　agent-interaction protocol

代理控制的仿真　agent-controlled simulation
代理控制的耦合　agent-controlled coupling
代理框架　agent framework
代理类　agent class
代理模型　agent model; brokered model; surrogate model
代理耦合　agent coupling
代理平台　agent platform
代理评估准则　agent evaluation criterion
代理启动的仿真　agent-initiated simulation
代理人　broker; proxy
代理设计方法学　agent design methodology
代理设计工具　agent design tool
代理特征　agent characterization
代理调和的仿真　agent-mediated simulation
代理调和的仿真游戏　agent-mediated simulation game
代理调和的环境　agent-mediated environment
代理系统　agent system
代理系统建模　agent system modeling
代理协调的　agent-coordinated
代理支持的　agent supported
代理支持的多种仿真　agent-supported multisimulation
代理支持的仿真　agent-supported simulation
代理支持的仿真框架　agent-supported simulation framework
代理支持的框架　agent-supported framework
代理仲裁的　agent-mediated
代理仲裁的解决方案　agent-mediated solution
代码　code
代码核查　code examination
代码生成　code generation
代码生成工具　code-generation tool
代码维护　code maintenance
代码验证　code verification
代数变量　algebraic variable
代数的　algebraic
代数方程模型　algebraic equation model
代数方程组　algebraic system of equations
代数环　algebraic loop
代数微分　algebraic differentiation
带环　band
带结构化　band-structured
带宽　bandwidth
带状结构矩阵　band-structured matrix
担保　assurance
单边博弈　one-sided game
单变量的　univariate
单步公式　single-step formula
单步积分法　single-step integration method
单步算法　single-step algorithm
单部件仿真　single-component simulation
单处理器仿真　single-processor simulation
单纯型算法　simplex algorithm

单次仿真时间步长　single-simulation time step
单次运行仿真研究　single-run simulation study
单次运行实验　single-run experiment
单点目标　single-point object
单独任务　separate task
单范式仿真系统　single-paradigm simulation system
单方面多模型　single-aspect multimodel
单方面仿真　single-aspect simulation
单方面模型　single-aspect model
单方面系统　single-aspect system
单个变量　singleton variable
单个全局变量　singleton global variable
单个事件　single-event
单个事件影响　single-event effect
单个消息　single-message
单个子整体　individual holon
单机仿真　stand-alone simulation
单极的　single-pole
单空间变量　single-space variable
单零交叉　single-zero-crossing
单目视觉理解　single-vision understanding
单人游戏　single-player game
单人游戏（博弈）技术　single-player gaming technique
单输出　single-output
单输入　single-input
单输入单输出模型　single-input single-output model
单输入单输出系统　single-input single-output system
单输入多输出模型　single-input multi-output model
单输入多输出系统　single-input multi-output system
单输入系统　single-input system
单体模型　monolithic model
单调进化仿真算法　monotonic evolutionary simulation algorithm
单调克里格　monotonic Kriging
单位换算　unit conversion
单向耦合　one-way coupling
单项误差　single-error
单演系统　monogenic system
单一变换　single-transition
单一传播　unicast
单一范式　single paradigm
单一输入　monotonous input
单一随机变量　unique random variable
单元　cell; unit
单元标记语言　cell markup language
单值函数　single-valued function
单转换子网　single-transition subnet
弹道仿真　trajectory simulation
弹道规划器　trajectory planner
弹性设计　resilient design
弹性系统　resilient system
当前的　current
当前模型　current model
当前时间　current time
当前事件　current event
当前数据　current data
当前值　current value
当前状态　current state
导出的行为　derived behavior

导数　derivative
导数的不连续性　derivative discontinuity
导向的　directed
到场　presence
到达过程　arrival process
到达事件　arrival event
盗版软件　warez
盗版软件游戏　warez game
道德错误　moral error
道德推理　moral reasoning
道德原则　moral principle
德尔菲技术　Delphi technique
登录 ID　login ID
登录权限　logon right
等待模型　waiting model
等高线图　contour plot
等价　equivalence; equivalencing
等价关系　equivalence relation
等价划分测试　equivalence partitioning testing
等价同型　equivalence morphism
等价物　equivalent
等效模型　equivalent model
等效输入　equivalent input
等效系统　equivalent system
等值线图　contour plot
低保真度仿真　low-fidelity simulation
低保真度仿真器　low-fidelity simulator
低层级　low level
低层级的组件　low-level component
低层级仿真　low-level simulation
低层级激励　low-level stimulus
低层级建模环境　low-level modeling environment
低层级数据　low level data
低分辨率模型　lower-resolution model; low-resolution model
低级语言　low-level language
低交互系统　low interactive system
低阶显式方法　low-order explicit method
低粒度　low granularity
低粒度模型　low-granularity model
低耦合　low coupling
低情境的复杂性　low situational complexity
低认知的复杂性　low cognitive complexity
低认知复杂性个体　low cognitive-complexity individual
低水平　low level
低位效应　low-order effect
低位优先数据格式　little endian data format
笛卡尔坐标　Cartesian coordinate
底节　basis
地方时　local time
地基　base
地理位置输入　geopositional input
地面实况　ground truth
地面数据处理中心　central station
地面真值　ground truth
地形学模型　topographic model
递归　recursion
递归的　recursive
递归定义　recursive definition
递归仿真　recursive simulation
递归模型　recursive model

递归值　redemption value
递阶　hierarchic
递阶的　hierarchical
递阶复杂系统　hierarchical complex system
递阶概念　hierarchical concept
递阶结构　hierarchical structure
递阶离散事件系统规范　hierarchical DEVS
递阶模块化离散事件系统规范　hierarchical modular DEVS
递阶模型校正　hierarchical model calibration
第二类模糊系统模型　type 2 fuzzy system model
第三变量　third variable
第三人称射击游戏　third-person shooter game
第一类模糊系统模型　type 1 fuzzy system model
典型错误　typical error
典型地　archetypally
典型模型　canonical model
典型输出　representative output
点　point
点对点仿真　peer-to-peer simulation
点对点模型　peer-to-peer model
点分布模型　point distribution model
点特征　point behavior
电脑适用性检查　computerized applicability check
电脑完整性检查　computerized completeness check
电脑一致性检查　computerized consistency check
电脑游戏　computer gaming
电子表格仿真　spreadsheet simulation
电子玻璃可视化　e-glass visualization
电子玻璃显示　e-glass display
电子经验　in silico experience
电子数据　in silico data
电子透镜显示　e-lens display
电子隐形眼镜显示　electronic contact lens display
电子游戏　electronic game
电子游戏产业　electronic game industry
电子游戏与仿真　electronic gaming and simulation
电子转录错误　electronic transcription error
调查　investigate (v.); investigation
调动　swap (v.)
调度　schedule (v.); scheduling
调度保持的离散事件系统规范　SP (Schedule-Preserving) DEVS
调度程序　scheduler;
调度机制　scheduling mechanism
调度器　schedular; scheduler
调换　transpose
迭代　iteration
迭代变量　iteration variable
迭代法　iterative method
迭代方案　iteration scheme
迭代建模　iterative modeling
迭代建模方法学　iterative modeling methodology
迭代连接　iteration connection
迭代循环　iteration loop
叠加的状态　superposed state

叠加原理　superposition principle
叠加状态　superposition state
顶点模型　vertex model
定标　calibration
定常变量　constant variable
定点迭代　fixed-point iteration
定理　theorem
定量变量　quantitative variable
定量测量　quantitative measure; quantitative measurement
定量的　quantitative
定量的概念　quantitative concept
定量的假设　quantitative assumption
定量的评估　quantitative assessment
定量度量　quantitative metric
定量仿真　quantitative simulation
定量仿真模型　quantitative simulation model
定量分析　quantitative analysis
定量规则　quantitative rule
定量混合仿真　quantitative mixed simulation
定量建模　quantitative modeling
定量决策变量　quantitative decision variable
定量模型　quantitative model
定量模型分析　quantitative model analysis
定量模型验证　quantitative model validation
定量评价　quantitative evaluation
定量术语　quantitative term
定量数据　quantitative data
定量算法评估　quantitative algorithm assessment
定量因果模型　quantitative causal model
定量有向图模型　quantitative digraph model
定量约束　quantitative constraint
定量值　quantitative value
定时的输入/输出自动机　timed input/output automaton
定时事件系统　timed-event system
定时系统　timed system
定时形式　timed formalism
定时转换　timed transition
定位　localize (v.); location
定位游戏　locative game
定向耦合　targeted coupling
定性变量　qualitative variable
定性测量　qualitative measure
定性的　qualitative
定性度量　qualitative metric
定性仿真　qualitative simulation
定性仿真模型　qualitative simulation model
定性仿真系统　qualitative simulation system
定性分析　qualitative analysis
定性概念　qualitative concept
定性规则　qualitative rule
定性混合模拟　qualitative mixed simulation
定性假设　qualitative assumption
定性建模　qualitative modeling
定性决策变量　qualitative decision-variable
定性模型　qualitative model
定性模型分析　qualitative model

analysis
定性模型验证　qualitative model validation
定性评估　qualitative assessment
定性评价　qualitative evaluation
定性术语　qualitative term
定性数据　qualitative data
定性算法评估　qualitative algorithm assessment
定性推理　qualitative reasoning
定性心理模型　qualitative mental model
定性因果模型　qualitative causal model
定性有向图模型　qualitative digraph model
定性约束　qualitative constraint
定性值　qualitative value
定义　definition
定义的变量　defined variable
定义的默认值　defined allowable value
定义误差　definition error
定义域　domain
定域化　localization
定制仿真　customized simulation
动步长调整策略　hyperactive step-size adjustment strategy
动画　animation
动画变量　animation variable
动画标识语言　animation markup language
动画模型　animation model
动画片绘制者　animator
动画数据　data animation

动画制作技术　animation technology
动机　motive
动力　momentum
动力学　dynamics
动力学变量　kinetic variable
动力学误差　dynamic error
动量　momentum
动量守恒　conservation of momentum
动态　dynamic
动态保真度建模　dynamic-fidelity modeling
动态本体　dynamic ontology
动态本体共享　dynamic ontology sharing
动态超对策　dynamic hypergame
动态程序检查　dynamic program check
动态错误分析　dynamic error-analysis
动态的行为　dynamic behavior
动态的假设　dynamic assumption
动态的检查　dynamic check
动态的组合　dynamic composition
动态的组合性　dynamic composability
动态地改变　dynamically changing
动态对策　dynamic game
动态仿真　dynamic simulation
动态仿真更新　dynamic simulation-update; dynamic simulation-updating
动态仿真连接　dynamic simulation-linking
动态仿真组合　dynamic simulation-composition
动态复杂性　dynamic complexity
动态个性化过滤器　dynamic personality filter

动态更新　dynamic update; dynamic updating
动态关系　dynamic relationship
动态光照　dynamic lighting
动态互操作水平　dynamic-interoperability level
动态互操作性　dynamic interoperability
动态划界　dynamically bound
动态划界的变量　dynamically-bound variable
动态环境　dynamic environment
动态技术　dynamic technique
动态建模　dynamic modeling
动态交互　dynamic interaction
动态脚本　dynamic scenario
动态结构　dynamic structure
动态结构多模型　dynamic-structure multimodel
动态结构化模型　dynamic structural-model
动态结构-离散事件系统规范　DS-DEVS
动态结构离散事件系统规范　dynamic structure DEVS
动态结构模型　dynamic-structure model
动态结构网络　dynamic-structure network
动态结构自动机　dynamic-structure automaton
动态解耦　dynamic decoupling
动态可组合的　dynamically composable
动态可组合仿真　dynamically-composable simulation
动态可组合联邦　dynamically-composable federation
动态离散事件系统规范　dynamic DEVS; dyn-DEVS
动态联邦耦合　dynamic federate-coupling
动态联邦制　dynamic federate-composition
动态模拟服务　dynamic simulation-service
动态模式更新机制　dynamic mode-update mechanism
动态模型　dynamic model; kinematic model
动态模型更新　dynamic model update; dynamic model updating
动态模型结构　dynamic model structure
动态模型结构的适当性　adequacy of dynamic model structure
动态模型耦合　dynamic model-coupling
动态模型区位　dynamic model locateon
动态模型文件　dynamic model-documentation
动态模型形式　dynamic modeling formalism
动态模型转换　dynamic model switching
动态模型组合　dynamic model-composition
动态模型组合性　dynamic model composability

动态耦合　dynamic coupling
动态耦合的　dynamically coupled
动态耦合多模型　dynamically-coupled multimodel
动态耦合模型　dynamically-coupled model
动态软件结构　dynamic software-architecture
动态实体　dynamic reality
动态事件序列　dynamic event sequence
动态数据　dynamic data
动态数据更新　dynamic data-update
动态数据结构系统　dynamic data-structure system
动态数据驱动的应用系统　dynamic data-driven application system
动态数据驱动仿真　dynamic data-driven simulation
动态网络　dynamic network
动态微仿真模型　dynamic microsimulation-model
动态文件　dynamic documentation
动态系统　dynamic system
动态系统仿真　dynamic system simulation
动态相似性　dynamic similarity
动态校核、验证与测试技术　dynamic VV&T technique
动态有效性　dynamic validity
动态知识　dynamic knowledge
动态转换　dynamic switching
动态子模型替换　dynamic submodel replacement
动态自然环境　dynamic natural environment
动态自适应　dynamic adaptation
动态作用域变量　dynamically-scoped variable
动作　action
动作游戏　action game
都柏林核心元数据模板　Dublin core metadata template
独立的　independent
独立的错误　separate error
独立的假设　independent assumption
独立的时间推进　independent time advancement
独立的校核与验证　independent V&V; independent verification and validation
独立性　independence
独立于观察者的复杂性　observer-independent complexity
独立域　domain independent
独立执行的模型　implementation-independent model
独立状态变量　independent state variable
独占执行　exclusive execution
读错误　read error
读者　reader
度量　measure; metric
度量意识　metric-aware; metric-awareness
端到端模型　end-to-end model
端口　port
端口互连　port interconnection
短期可预测性　short-term predictability

短时本体　short-lived ontology
短时间尺度变量　short time-scale variable
短时状态　short-lived state
段落　section
断定误差　ascertainment error
断裂　break (*v.*)
断言　assertion; predicate
断言检查　assertion checking
断言模型　assertional model
队列　queue
队列仿真　queue simulation
队列误差　queue error
对　pair
对策　game
对策关联　gaming association
对策管理　management of game
对策论　game theory
对策树　game tree
对称博弈　symmetric game
对称的军事演习　symmetrical wargame
对称仿真　symmetric simulation
对称分布　symmetrical distribution
对称函数　symmetrical function
对称性模型　symmetry model
对等的模型　peer-to-peer model
对换　swapping
对角矩阵　diagonal matrix
对流扩散方程　convection-diffusion equation
对偶　antithetic
对偶变量　antithetic variable; antithetic variate; dual variable
对偶仿真运行　antithetic simulation run
对偶式仿真研究　antithetic simulation study
对偶研究　antithetic study
对数单位模型　logit model
对数函数　logit function
对数量化积分　logarithmically quantized integrator
对数正太随机变量　lognormal random variable
对象　object
对象定义　object definition
对象仿真程序　object simulation program
对象仿真语言　object SL
对象管理架构　object-management architecture
对象建模　object modeling
对象交互　object interaction
对象交互控制　object interaction control
对象类　object class
对象类型　object type
对象流测试　object-flow testing
对象模型　object model
对象模型框架　object model framework
对象模型模版　object model template
对象模型转换　object model transformation
对象属性　object attribute
对象数据库　object database
对象所有权　object ownership
对象图　object diagram
对照测试　comparison testing

对照运行   antithetic run
对准   alignment
对准机理   pointing mechanism
钝化   deactivation
多   multi
多保真度仿真   multiple fidelity simulation
多边形   polygon
多变量   multivariable
多步法   multistep method
多步积分法   multistep integration method
多步积分算法   multistep integration algorithm
多采样系统   multisampling system
多参与者模型   multiplayer model
多参与者游戏   multiplayer game
多参与者游戏（博弈）技术   multiplayer gaming technique
多层安全   multilevel security
多层次   multilayer
多层的   multilevel
多层面的   multifaceted
多层面的多仿真   multiaspect multisimulation
多层面建模   multifaceted modeling
多层面建模形式   multifaceted modeling formalism
多层面模型   multifaceted model
多层面性   multifacet
多层网络   multilayer network
多层系统   multilayer system
多尺度   multiscale
多尺度层次模型   multiscale hierarchical model
多尺度法   multiscale method
多尺度仿真   multiscale simulation
多尺度分析   multiscale analysis
多尺度建模   multiscale modeling
多尺度建模工具   multiscale modeling tool
多尺度建模技术   multiscale modeling technique
多尺度建模与仿真   multiscale M&S
多尺度模型   multiscale model
多尺度问题   multiscale problem
多尺度现象   multiscale phenomenon
多处理器仿真   multiprocessor simulation
多传感器交互   multisensory interaction
多传感器输入   multisensory input
多传感器输入/输出   multisensory I/O
多次运行   multiple run
多次运行实验   multirun experiment
多点触控仿真   multitouch emulation
多点传输   multicast
多点建模   multipoint modeling
多队列   multiple queues
多对多关系   many-to-many relation
多对一关系   many-to-one relation
多范式   multiparadigm
多范式仿真   multiparadigm simulation
多范式仿真系统   multiparadigm simulation system
多范式仿真语言   mixed-formalism SL
多范式建模   multiparadigm modeling
多范式模型   multiparadigm model
多方法仿真   multimethod simulation
多方面   multiaspect

多方面多模型　multiaspect multimodel
多方面仿真　multiaspect simulation
多方面仿真模型　multiaspect simulation model
多方面建模　multiaspect modeling
多方面理解　multiaspect understanding
多方面模型　multiaspect model
多方面系统　multiaspect system
多仿真　multisimulation
多仿真博弈　multisimulation gaming
多仿真博弈策略　multisimulation gaming strategy
多仿真方法学　multisimulation methodology
多仿真管理对策　multisimulation management game
多仿真框架　multisimulation framework
多仿真器　multisimulator
多仿真系统层次结构　hierarchy of simulations
多仿真形式　multisimulation formalism
多仿真引擎　multisimulation engine
多分辨率　multiresolution
多分辨率多模型　multiresolution multimodel
多分辨率多视角建模　multiresolution multiperspective modeling
多分辨率多重仿真　multiresolution multisimulation
多分辨率方法　multiresolution approach
多分辨率仿真　multiresolution simulation
多分辨率仿真模型　multiresolution simulation model
多分辨率建模　multiresolution modeling
多分辨率建模与仿真　multiresolution M&S
多分辨率联邦成员　multiresolution federate
多分辨率模型　multiresolution model
多分量离散事件系统规范　multicomponent DEVS
多个工程　multiengineering
多个事件　multievent
多核缓存仿真器　multicore cache simulator
多级　multiechelon
多级的　multilevel
多级动态耦合　multilevel dynamic coupling
多级多模型　multistage multimodel
多级多模型框架　multilevel multimodel framework
多级方法　multilevel approach
多级仿真　multilevel simulation
多级集成　multistage integration
多级建模　multilevel modeling
多级交互　multilevel interaction
多级结构　multilevel structure
多级聚类　multilevel clustering
多级控制　multilevel control
多级模型　multiechelon model; multilevel model; multistage model
多级验证　multistage validation

多级有效性　multistage validity
多角度建模　multiperspective modeling
多阶段多仿真　multistage multisimulation
多阶段仿真　multistage simulation
多阶段仿真模型　multistage simulation model
多阶段建模　multistage modeling
多阶段建模形式　multistage modeling formalism
多阶段模型的仿真　simulation with multistage model
多进程　multiprocess
多零交点　multiple zero-crossing
多领域　multidomain
多路存取　multiple access
多路仿真器　multiple simulator
多媒体仿真　multimedia simulation
多媒体建模　multimedia modeling
多媒体建模语言　multimedia modeling language
多媒体模型　multimedia model
多媒体强化仿真　multimedia-enriched simulation
多媒体文件　multimedia documentation
多面　multi
多面的　multidimensional; multifacet
多模态的　multimodal
多模态方法　multimode method
多模态分布　multimodal distribution
多模态接口技术　multimodal interfacing
多模态模型　multimodal model

多模态输入　multimodal input
多模型　multimodel
多模型代理耦合　multimodel agent coupling
多模型仿真器　multimodel simulator
多模型框架　multimodel framework
多模型耦合　multimodel coupling
多模型配置　multimodel configuration
多模型智能体耦合　multimodel agent coupling
多目标模型　multiobjective model
多目标优化问题　multiobjective optimization problem
多耦合　multicoupling
多平台　multiplatform
多平台建模　multiplatform modeling
多平台模型　multiplatform model
多人模式　multiplayer
多人在线游戏　multiplayer online game
多任务处理　multitasking
多任务处理行为　multitasking behaveior
多任务处理模型　multitasking model
多任务的　multitasked
多时间尺度　multiple temporal scales
多时间尺度变量　multiple time-scale variable
多实例事件　multi-instance event
多视角　multiperspective
多视角仿真　multiperspective simulation
多视角模型　multiperspective model
多视觉理解　multivision understanding

多输出　multi-output；multiple output
多输入　multi-input；multiple input
多输入单输出　multi-input single-output；multiple input single output
多输入单输出模型　multi-input single-output model；multiple input single output model
多输入单输出系统　multi-input single-output system
多输入多输出　multi-input multi-output；multiple input multiple-output
多输入多输出模型　multi-input multi-output model；multiple input multiple output model
多输入多输出系统　multi-input multi-output system
多属性　multiattribute
多数据模型　multiple data model
多速率　multirate
多速率采样　multirate sampling
多速率仿真　multirate simulation
多速率集成　multirate integration
多态对象　multistate object
多态性　polymorphism
多体仿真　multibody simulation
多体系统　multibody system
多体系统动力学　multibody system dynamics
多通道用户界面　multimodal user interface
多同步点　multiple synchronization point
多维的　multidimensional
多维度建模　multidimensional modeling
多维空间　multidimensional space
多维数据集合　multidimensional data set
多维数据可视化　multidimensional data visualization
多维数据模型　multidimensional data model
多稳态　multistable state
多物理场仿真　multiphysics simulation
多物理场建模　multiphysics modeling
多物理场模型　multiphysics model
多相过程　multiphase process
多响应变量　multiple response variable
多项的　multinomial；multinomial；polynomial
多项式逼近　polynomial approximateon
多项式分布　multinomial distribution
多项式分对数　multinomial logit
多项式轨迹　polynomial trajectory
多项式模型　polynomial model
多形式的　multiformalism
多形式构成　polyformalism composition
多形式建模　multiformalism modeling；polyformalism modeling
多形式模型　multiformalism model；polyformalism model
多形式模型构成　polyformalism model composition
多形式耦合的　coupled multiformalism

多形态模型　polyformic model
多学科系统　multidisciplinary system
多样性　multiplicity
多义数据　ambiguous data
多用户　multiuser
多用户仿真　multiplayer simulation
多用户支持　multiplayer support
多语言　multilanguage
多语言编程　multilingual programming
多语言仿真　multilingual simulation
多域仿真系统　multidomain simulation system
多域建模　multidomain modeling
多域建模语言　multidomain modeling language
多元　multivariate
多元的　multicomponent
多元分析　multivariate analysis
多元高斯随机变量　multivariate Gaussian random variable
多元建模　plural modeling
多元模型　multicomportment model
多元随机变量　multivariate random variable
多源数据　multisourse data
多运行仿真研究　multiple run simulation study; multirun simulation study
多值函数　multivalued function
多智能体　multiagent
多智能体参与式仿真　multiagent participatory-simulation
多智能体仿真　multiagent simulation
多智能体模型　multiagent model

多智能体系统　multiagent system; MAS (MultiAgent System)
多智能体系统安全性　security in MAS
多智能体正交性　multiagent orthogonality
多智能体支持的仿真　multiagent-supported simulation
多种的　multiple
多重　multi
多重测量　multiple measure
多重尺度空间　multiple spatial scales
多重的　multiple
多重分解　multiple decomposition
多重感知　multiple perception
多重建模　multimodeling
多重建模范式　multiple modeling paradigm; multimodeling paradigm
多重建模方法学　multimodeling methodology
多重建模形式　multimodeling formalism
多重控制　multiple control
多重理解　multiunderstanding
多重实例化　multiple instantiation
多重试验框架　multiple experimental frame
多重突现行为　multiple emergent behavior
多重稳定性　multistability
多重想定实例　multiple scenario instantiation
多重性　multiplism
多重涌现　multiple emergence
多重预期　multiple anticipation

多重转换　multiple transition
多重转换子网　multiple transition subnet
多状态　multiple states; multistate
多准则决策　multicriteria decision; multicriteria decision making
多组件的　multicomponent
夺获　capture
惰性计算　lazy evaluation

Ee

II 型错误　type II error
II 型错误概率　type II error probability
厄兰随机变量　Erlang random variable
扼制时间翘曲的仿真　throttled time-warp simulation
恶劣问题　wicked problem
二级仿真　second degree simulation
二阶倒数模型　second derivative model
二阶精度　second-order accurate
二阶控制论　second-order cybernetics
二阶效应　second-order effect
二阶协同　second order synergy
二阶涌现　second-order emergence
二进制变量　binary variable
二进制数值　binary value
二类分对数　binary logit
二人超对策　two-person hypergame
二态的　dimorphic
二态的行为　dimorphic behavior
二维拓扑　bi-dimensional topology
二维网络　two-dimensional network
二项式分布　binomial distribution
二项式随机变量　binomial random variable
二元关系　binary relation
二元关系模型　binary relationship model
二元数据　bivariate data

Ff

发布　publication; publish ($v.$)
发出　emitting
发出的　emitted
发起人　sponsor
发射　emit ($v.$); launch
发射极　emitter
发射末端　end-firing phase
发射前的准备阶段　warm up period
发射延时　firing delay
发生　occurrence
发生的情境　emerging context
发生的条件　emerging condition
发生器　generator; producer
发送　delivery; transmit ($v.$)
发现　discovery
发现式学习　discovery learning
发现事件　detected event
发展　development
发展需求　development requirements
法律　law
法则　law; prescription
翻译　translate ($v.$)
翻译程序　translator
翻译错误　translation error
翻译方法学　translation methodology
反驳　disprove ($v.$)

反传 backward
反对称的 antisymmetric
反复试验 trial-and-error
反赫米特插值 inverse Hermite interpolation
反解释主义 anti-interpretivism
反馈 feedback
反馈环（回路） feedback loop
反馈控制连接 feedback control connection
反馈耦合 feedback coupling
反馈线性化 feedback linearization
反射的 reflective
反射的属性 reflected attribute
反射的物体 reflected object
反射式仿真 reflective simulation
反射式知识处理 reflective knowledge processing
反射涌现 reflexive emergence
反事实的仿真 counterfactual simulation
反相器 inverter
反相输入 inverted input; inverting input
反向 backward
反向插值技术 back-interpolation technique
反向仿真 backward simulation
反向推理模型 backward-reasoning model
反应的行为 reactive behavior
反应方法 reactive approach
反应规划 reactive planning
反应机制 reactive mechanism
反应技术 reactive technique
反应建模 reactive modeling
反应式决策 reactive decision
反应式模型维护 reactive model maintenance
反应式维护 reactive maintenance
反应式知识处理 reactive knowledge processing
反应式智能体 reactive agent
反应系统 reactive system
反应型关系 reactive relationship
反应型交互 reactive interaction
反转 invert (v.)
反自然博弈 game against nature
返回值 back value
泛北极区建模 holarctic modeling
泛化 generic
泛化抽象 generic abstraction
泛化误差 generalization error
泛在博弈 ubiquitous game
泛在仿真 ubiquitous simulation
泛在仿真器 ubiquitous simulator
泛在计算仿真 ubiquitous computing simulation
泛在计算仿真器 ubiquitous computing simulator
泛在系统 ubiquitous system
范畴 category
范例 paradigm
范式 canonical form; paradigm
范围 range; scope; span
范围蠕变 scope creep
方案 plan; scheme; variance
方差分析 analysis of variance
方差估计器 variance estimator
方差减缩技术 variance reduction

technique
方差缩减　variance reduction
方程　equation
方程模型　equation model
方程系统的因果化处理　causalization of an equation system
方法　approach; method; technique
方法裁剪　methodology tailoring
方法论　methodology
方法论本体　methodological ontology
方法论体系　methodological framework
方法论约束　methodological constraint
方法论知识　methodological knowledge
方法误差　method error
方框图　block diagram
方面　aspect
方式　paradigm
方位　aspect
方位误差　azimuth error
方向变换　direction transformation
方向耦合　directly coupled
仿生方法　bio-inspired method
仿生仿真　biomimetic simulation
仿生工具　bio-inspired tool
仿生建模　bio-inspired modeling
仿生模拟　bio-inspired simulation
仿生模型　bio-inspired model
仿生系统　bio-inspired system
仿形　profiling
仿形误差　copying error
仿真　emulation; simulation; simulate (v.)
仿真、测试与评估过程　simulation, test, and evaluation process
仿真包　simulation package
仿真保真度　simulation fidelity
仿真本体　simulation ontology
仿真编程语言　simulation programming language
仿真标记语言　simulation mark up language
仿真标准　simulation standard
仿真博弈　simulation game; simulation gaming
仿真博弈工程　simulation gaming project
仿真博弈工具　simulation gaming tool
仿真博弈过程　simulation gaming process
仿真博弈软件　simulation gaming software
仿真博弈应用　simulation gaming application
仿真部署　simulation deployment
仿真参考标记语言　simulation reference markup language
仿真参数　simulation parameter
仿真参与者　simulation participant; simulation stakeholder
仿真操作　simulation operation
仿真操作者　simulation operator
仿真策略　simulation strategy
仿真策略组　simulation policy group
仿真层　simulation layer
仿真插件　simulation plug-in
仿真尝试　simulation effort
仿真成本　cost of simulation
仿真程序　simulating program;

simulation program
仿真程序管理　simulation program management
仿真程序设计　simulation programming
仿真程序生成器　simulation program generator
仿真处理器　simulation processor
仿真代理　simulation brokering; simulation proxy
仿真代码　simulation code
仿真道德规范　simulation ethics
仿真的　emulated; simulated; simulative
仿真的表示（法）　simulated representation
仿真的并行化　parallelization of simulation
仿真的回溯因果关系　simulated retrocausality
仿真的经济影响　economic impact of simulation
仿真的设计　simulative design
仿真的执行同步　simulation execution synchronization
仿真动画　simulation animation
仿真对抗　simulated warfare
仿真对象　simulated object
仿真对象模型　simulation object model
仿真范式　simulation paradigm
仿真方法　simulation method
仿真方法学　simulation methodology
仿真分辨率　simulation resolution
仿真分析　simulation analysis;
simulation analytics
仿真分析工具　simulation analytics tool
仿真分支　simulation branching
仿真服务　simulation service
仿真服务器　simulation server
仿真服务提供方　simulation service provider
仿真辅助的　simulation-aided
仿真复杂性　simulation complexity
仿真概念　simulation concept
仿真更新　simulation update; simulation updating
仿真更新时间　simulation update time
仿真工程　simulation engineering; simulation project
仿真工程管理　simulation project management
仿真工具　simulation tool
仿真攻击　simulated attack
仿真关联　simulation association; simulation relation
仿真管理　simulation management
仿真管理能力　simulation management capability
仿真管理器　simulation manager
仿真规范　simulation specification
仿真规范环境　simulation specification environment
仿真规范库　simulation specification repository
仿真规范语言　simulation specification language
仿真规则　simulation rule
仿真过程　simulation process

仿真行为　emulated behavior; simulated behavior
仿真恒量　simulation consultant
仿真后报告　postsimulation report
仿真后分析　postsimulation analysis
仿真后阶段　postsimulation phase
仿真互操作标准　simulation interoperability standard
仿真互操作性　simulation interoperability
仿真环境　simulated environment; simulation environment; simulative environment
仿真混合　simulation mash-up
仿真机　simulation machine
仿真机制　simulation mechanism
仿真基础设施　simulation infrastructure
仿真基准　simulation benchmark
仿真即服务　simulation as a service
仿真计划　simulation plan
仿真技能　simulation skill
仿真技术　simulation technique; simulation technology
仿真加速　acceleration of simulation
仿真假设　simulation hypothesis
仿真架构　simulation architecture
仿真监控　simulation monitoring
仿真建模　simulation modeling
仿真建模的基础设施　simulation-modeling infrastructure
仿真交付　simulation delivery
仿真交互　simulation interaction
仿真阶段　simulation phase
仿真接口　simulation interface
仿真接口问题　simulation interface issue
仿真接口系统　simulation interface system
仿真节点　simulation node
仿真结构　simulation structure
仿真结果　simulation result
仿真结果分析　analysis of simulation result
仿真结果解释　simulation results comprehensibility
仿真结果质量　simulation results quality
仿真进化　simulated evolution; simulation evolution
仿真经济　economics of simulation
仿真精度　simulation accuracy
仿真开发　simulation development
仿真开发程序　simulation development program
仿真开发环境　simulation development environment
仿真开发生命周期　simulation development life cycle
仿真开发者　simulation developer
仿真可靠性　simulation reliability
仿真可理解性　simulation comprehensibility
仿真可视化联邦成员　simulation visualization federate
仿真可信度　simulation credibility
仿真可用性　simulation availability
仿真可重用性　simulation reusability
仿真可组合性　simulation composability

仿真客户　simulation customer
仿真课程　simulation curriculum
仿真控制　simulation control
仿真控制程序　simulation control program
仿真库　simulation library; simulation repository
仿真框架　simulation framework
仿真理论　simulation theory
仿真粒度　simulation granularity
仿真连接　simulation linkage
仿真联合　simulation federation
仿真联盟　simulation confederation
仿真领域需求　simulation domain requirements
仿真鲁棒性　simulation robustness
仿真论证　simulation argument
仿真谬论　simulation fallacy
仿真模块化　simulation modularity
仿真模式　simulation mode
仿真模式控制　simulation mode control
仿真模型　simulated model; simulation model; emulation model
仿真模型的互操性　interoperability of simulation model
仿真模型互操作　simulation model interoperability
仿真模型评估　simulation model assessment
仿真模型设计质量　simulation model design quality
仿真模型响应　simulation-model response
仿真目的　purpose of simulation

仿真能力　simulation capability
仿真平台　simulation platform
仿真平台自由度　simulation platform degree of freedom
仿真企业　simulation enterprise
仿真启发式　simulation heuristic
仿真器　simulator
仿真器程序　simulator program
仿真器服务器　simulator server
仿真器频谱　spectrum of simulators
仿真器设计　simulator design
仿真器协调　simulator coordination
仿真器在训练中的应用　use of simulator in training
仿真器中间件　simulator middleware
仿真情境　simulational context
仿真求解技术　simulative solution technique
仿真驱动　simulation-driven
仿真驱动的教育　simulation-driven education
仿真驱动的实验　simulation-driven experimentation
仿真驱动的训练　simulation-driven training
仿真驱动的优化　simulation-driven optimization
仿真趋势　trends in simulation
仿真认识论　simulation epistemology
仿真认证　simulation certification
仿真任务　simulation task
仿真任务规范　specification of simulation task
仿真入口　simulation portal
仿真软件　simulation software

仿真软件开发 simulation software development
仿真软件评估 simulation software assessment
仿真商业实践 simulation business practice
仿真设计 simulation design
仿真设计的系统 simulation for system design
仿真设计方法 simulative design approach
仿真设计方法学 simulation design methodology
仿真设计环境 simulative design environment
仿真生成 simulation-generated
仿真生命周期 life cycle of simulation; simulation life cycle
仿真时间 simulated time; simulation time
仿真时间步长 simulation time step
仿真时延 simulation latency
仿真时钟 simulation clock
仿真实例 simulation instance
仿真实时 simulated real-time
仿真实体 simulated entity; simulation entity
仿真实现 simulation implementation
仿真实验 simulation experiment
仿真实验可复现性 simulation experiment reproducibility
仿真实验描述 simulation experiment description
仿真实验描述标记语言 simulation experiment description markup language
仿真实验设计 design of simulation experiment
仿真使用 simulation use
仿真世界 simulated world
仿真市场 simulation market
仿真事件视界 simulated event horizon
仿真试验的统计设计 statistical design of simulation experiments
仿真视界 simulation horizon
仿真视界需求 simulation horizon requirement
仿真视图 simulated view
仿真适配 simulation adaptation
仿真输出 simulation output
仿真输出数据分析联邦 simulation output-data-analysis federate
仿真输入 simulated input
仿真术语 simulation term; simulation terminology
仿真数据 simulated data; simulation data
仿真数据的可接受性 simulated data acceptability
仿真数据管理 simulation data management
仿真数据可接受性 acceptability of simulated data
仿真数据评估 assessment of simulated data
仿真数据正态性 normality of simulation data
仿真素养 simulation literacy
仿真算法 simulation algorithm

仿真随机采样　simulated random sampling
仿真套件　simware
仿真挑战　simulation challenge
仿真投资　simulation investment
仿真图形用户界面　simulation GUI
仿真团队　simulation team
仿真团队评估　assessment of simulation team
仿真推荐的可实现性　implementability of simulation recommendation
仿真拓扑　simulation topology
仿真网格　simulation grid
仿真网格架构　simulation grid architecture
仿真网络服务　simulation web service
仿真维护程序　simulation maintenance program
仿真文档　simulation documentation
仿真文献　simulation literature
仿真稳定性　simulation stability
仿真问题　simulation problem
仿真问题求解环境　simulative problem solving environment
仿真误差　simulation error
仿真误差评估　simulation error assessment
仿真系统　simulated system; simulateon system
仿真系统工程　simulation systems engineering
仿真系统工程的知识体系　body of knowledge of simulation systems engineering
仿真系统工程师　simulation systems engineer
仿真系统工程知识体系　simulation systems engineering body of knowledge
仿真现代化　simulation modernization
仿真现实　simulated reality
仿真相关的　simulation-related
仿真响应　simulation response
仿真响应函数　simulation response function
仿真响应面　simulation response surface
仿真想定　simulation scenario
仿真项目报告　simulation project report
仿真效益　benefit of simulation
仿真效用　simulation utility
仿真协会　simulation society
仿真协调　simulation coordination
仿真协议　simulation protocol
仿真胁迫　simulation coercion
仿真需求　simulation requirement
仿真序列　simulation sequence
仿真学习可接受性　simulation study acceptability
仿真学习平台　simulated learning platform
仿真训练器开发　simulation trainer development
仿真研究　simulation study
仿真研究合格性　acceptability of simulation study
仿真研究监控　simulation study monitoring

仿真研究可接受性　acceptability of simulation study
仿真演练　simulation exercise
仿真验证　simulation validation
仿真业务　simulation business
仿真仪器　emulated instrument
仿真引擎　simulation engine
仿真影响　impact of simulation
仿真应用　simulation application
仿真应用类型　simulation utility type
仿真应用联邦成员　simulation application federate
仿真用户　simulation user
仿真用户问题　simulation user issue
仿真优化　optimization within simulation; simulation optimization
仿真优化方法学　simulation optimization methodology
仿真优化问题　simulation optimization problem
仿真游戏管理　management of simulation game; simulation game management
仿真游戏开发　simulation game development
仿真游戏设计　simulation game design
仿真语言　simulation language; SL (Simulation Language)
仿真语言处理器　simulation language processor
仿真语言评估　assessment of SL
仿真语义学　simulation semantics
仿真预热阶段　simulation warm up period
仿真预热时间　simulation warm up time
仿真域　simulation domain
仿真元建模　simulation metamodeling
仿真元模型　simulation metamodel
仿真运行　simulation run
仿真运行次数　number of simulation runs
仿真运行监控　simulation runtime monitoring; simulation run monitoring
仿真运行控制　simulation run control
仿真运行库　simulation runtime library
仿真运行长度　length of the simulation run; simulation run length
仿真再现性　simulation reproducibility
仿真增强现实　simulation-augmented reality
仿真展示　simulation exposure
仿真展示效果　effects of simulation exposure
仿真正确性　simulation correctness
仿真证明　simulation proof
仿真支撑工具　simulation support tool
仿真支撑实体　simulation support entity
仿真支持　simulation support
仿真支持的　simulation supported
仿真支持的博弈　simulation-supported game
仿真支持的战争博弈　simulation-supported war game
仿真知识　simulation knowledge
仿真知识共享　simulation knowledge sharing
仿真知识获取　simulation knowledge

acquisition
仿真执行 simulation execution
仿真执行环境 simulation execution environment
仿真执行速率 simulation execution speed
仿真值 value of simulation
仿真质量保证 simulation quality assurance
仿真中的抽象化 abstracting in simulation
仿真中的克里格插值 Kriging interoplation in simulation
仿真中间件 simulation middleware
仿真重用 simulation reuse
仿真重置 simulation reconfiguration
仿真周期 simulation cycle
仿真主义 simulism
仿真专家 simulationist; simulation expert
仿真状态 simulation status
仿真资产 simulation asset
仿真资产管理 simulation asset management
仿真资产重用 reuse of simulation asset
仿真资源 simulation resource
仿真组合 simulation composition
仿真组件 simulation component
仿真组件接口 simulation component interface
仿真组件重用方法学 simulation component reuse methodology
仿真组态 simulation configuration
仿真组织 simulation organization

访问 access
访问错误 access error
放大的副本 scaled-up replica
放在一边 allocate ($v.$)
飞地模型 enclave model
飞地模型建立 enclave model set-up
飞行训练 flight training
非绑定 unbundling
非本地代码 nonnative code
非本质博弈 inessential game
非本质的状态事件 unessential state event
非必要事件条件 nonessential event condition
非遍历行为 nonergodic behavior
非标称涌现 nonnominal emergence
非标称有意涌现 nonnominal intentional emergence
非标准单元 nonstandard cell
非标准数据元素 nonstandard data element
非补偿模型 noncompensatory model
非采样误差 nonsampling error
非参数方法 nonparametric method
非参数化实验 nonparametrized experimentation
非参数随机变量 nonparametric random variate
非操作变量 nonmanipulated variable
非层次的模型 nonhierarchical model
非常规输入 unconventional input
非承重假设 non-load-bearing assumption
非传统数据 nontraditional data
非迭代协同仿真 noniterative co-

simulation
非定域性　nonlocality
非动力学模型　nonkinetic model
非动态行为　nondynamic behavior
非对称函数　asymmetrical function
非对称军事演习　asymmetrical wargame
非法查找错误　illegal seek error
非法错误　illegal error
非法的　illegal
非法进入错误　illegal access error
非反射性涌现　nonreflexive emergence
非仿真　nonsimulation
非仿真模型　nonsimulation model
非感知仿真　not perceptual simulation
非刚性仿真　nonstiff simulation
非功能性　nonfunctional property
非关系数据　nonrelational data
非合作博弈　noncooperative game
非合作的　noncooperative
非基本变量　nonbasic variable
非基础知识　nonfoundational knowledge
非结构化数据　unstructured data
非结构化网格　unstructured grid
非解析解的模型　analytically insolvable model
非经验知识　nonempirical knowledge
非均匀化事件　ununiformized event
非均匀化速率　ununiformized rate
非可行状态　unfeasible state
非控制状态　noncontrolled state
非隶属度　nonmembership
非连贯的模型　disjunctive model
非连贯的耦合　disjunctively coupled
非连接仿真　disconnected simulation
非连接配置　unconnected configuration
非连续变化模型　discontinuous-change model
非连续处理　discontinuity handling
非连续的仿真　discontinuous simulation
非连续的模型　discontinuous model
非连续系统　discontinuous system
非零和博弈　non-zero-sum game
非零和博弈模拟　non-zero-sum simulation
非零和仿真博弈　non-zero-sum simulation game
非命题知识　nonpropositional knowledge
非模糊值　nonfuzzy value
非逆事件　nonadverse event
非平稳的　nonstationary
非平稳数据　nonstationary data
非齐次边界条件　inhomogeneous boundary condition
非齐次系统　inhomogeneous system
非侵入性的　nonintrusive
非侵入性方法　nonintrusive method
非确定性变量　nondeterministic variable
非确定性仿真　nondeterministic simulation
非确定性函数　nondeterministic function
非确定性离散事件系统规范　nondeterministic DEVS

非确定性模型　nondeterministic model
非确定性试验　nondeterministic experiment
非确定性图灵机仿真　nondeterministic Turing machine simulation
非认知变量　noncognitive variable
非时变功能模型　untimed functional model
非时变模型　untimed model
非实验仿真　nontrial simulation
非视线仿真　non-line-of-sight simulation
非适宜仿真　inappropriate simulation
非收敛仿真　nonconvergent simulation
非数值变量　nonnumerical variable
非数值的　nonnumeric
非数值方法　nonnumerical method
非数值仿真　nonnumerical simulation
非数值计算　nonnumerical computation
非数字处理　nonnumeric processing
非调度事件　unscheduled event
非突变的多模型　nonmutational multimodel
非突发的　nonemergent
非完整模型　nonholonomic model
非完整系统　nonholonomic system
非完整约束　nonholonomic constraint
非唯一随机变量　nonunique random variable
非稳定解　unstable solution
非稳态　unstable state
非稳态变量　nonhomeostatic variable

非吸收状态　nonabsorbing state
非线性的　nonlinear
非线性动力系统　nonlinear dynamical system
非线性二阶微分模型　nonlinear-second derivative model
非线性仿射输入　nonlinear affine input
非线性仿射输入系统　nonlinear affine input system
非线性归纳变量　nonlinear induction variable
非线性函数　nonlinear function
非线性控制　nonlinear control
非线性连接　nonlinear connection
非线性模型　nonlinear model
非线性耦合　nonlinear coupling; nonlinearly coupling
非线性特性　nonlinearity
非线性统计耦合　nonlinear statistical coupling
非线性稳定性　nonlinear stability
非线性系统　nonlinear system
非线性系统仿真　nonlinear system simulation
非线性因果关系　nonlinear causality
非线性映射　nonlinear mapping
非线性优化　nonlinear optimization
非线性状态空间模型　nonlinear state-space model
非相关处理　irrelevant processing
非相似的模型　dissimilar model
非相似仿真　dissimilar simulation
非严格竞争博弈　non-strictly-competitive game

非因果方程　acausal equation
非因果关系的　acausal
非应用程序　nongame application
非娱乐游戏　nonentertainment game
非语音音频输入　nonspeech audio input
非预期模型　nonanticipatory model
非预期系统　nonanticipatory system
非正规定义　informal definition
非正式化方法　informal technique
非正式化概念模型　informal conceptual model
非正式化描述　informal specification
非正式化校核、验证与测试方法　informal VV&T technique
非正式模型　informal model
非正式模型描述　informal model description
非中心卡方随机变量　noncentral Chi-Squared random variable
非终止仿真　nonterminating simulation
非周期的　antipositivism; aperiodic
非周期性的　nonperiodic
非周期性输出　nonperiodic output
非周期性系统　aperiodic system
非主导解决方案　nondominated solution
非自反关系　irreflexive relationship
非自反约束　irreflexive constraint
非自治模型　nonautonomous model
费用　cost
分包管理　subcontract management
分辨率　resolution
分辨率级别　level of resolution
分辨率水平　resolution level
分辨率误差　resolution error
分布　distribute (v.)
分布参数系统　distributed-parameter system
分布参数系统仿真　distributed-parameter system simulation
分布函数　distribution function
分布交互仿真　distributed interactive simulation
分布交互仿真兼容的　DIS compatible
分布交互仿真协议数据　DIS protocol data
分布式　distributed
分布式参数模型　distributed-parameter model
分布式处理器　distributed processor
分布式代理仿真　distributed agent simulation
分布式的并行计算　distributed and parallel computing
分布式仿真　distributed simulation
分布式仿真环境　distributed simulation environment
分布式仿真技术　distributed simulation technology
分布式仿真架构　distributed simulation framework
分布式仿真器　distributed simulator
分布式仿真体系结构　distributed simulation architecture
分布式仿真协议　distributed simulation protocol
分布式非对称仿真　distributed asymmetric simulation

分布式高性能计算　distributed high-performance computing
分布式计算　distributed computing
分布式计算模型　distributed computational model
分布式建模与仿真　distributed M&S
分布式交互仿真　DIS (Distributed Interactive Simulation)
分布式懒惰仿真　distributed lazy simulation
分布式离散事件系统规范仿真　distributed DEVS simulation
分布式联邦　distributed federation
分布式联邦成员　distributed federate
分布式模型　decentralized model; distributed model
分布式模型仓库　distributed model-repository
分布式模型检查　distributed model checking
分布式模型开发　distributed model development
分布式任务训练　distributed mission training
分布式实时仿真　distributed real-time simulation
分布式实验框架　distributed experimental frame
分布式事件驱动仿真　distributed event-driven simulation
分布式算法　distributed algorithm
分布式调度机制　distributed scheduling mechanism
分布式问题求解　distributed problem solving
分布式系统　decentralized system; distributed system
分布式系统仿真语言　distributed-system SL
分布式协调　decentralized coordination
分布式协同　distributed collaboration
分布式虚拟环境　distributed virtual environment
分布式训练　distributed training
分布式演练　distributed exercise
分布式异构仿真　distributed heterogeneous simulation
分布式应用程序　distributed application
分布式执行　distributed execution
分布式自组织系统　decentralized selforganizing system
分布系统设计　distributed system design
分布系统体系结构规范　distributed systems architecture specification
分布状态　distribution
分层采样　stratified sampling
分层的　hierarchic; hierarchical
分层的方法　layered approach
分层的模型　layered model
分层聚类　hierarchical clustering
分层决策　hierarchical decision
分层形式　hierarchical form
分叉　bifurcation
分簇仿真　cluster simulation
分段的　partitioned
分段多项式轨迹　piecewise polynomial trajectory

分段连续段　piecewise continuous segment
分段模型　compartmental model
分段时间序列　segmenting time series
分段线性　piecewise linear
分断　breaking
分对数　logit
分对数分解　exploded logit
分发　distribute (v.); distribution
分割　partition (v.); sectioning; segment
分解　decompose (v.); decomposition
分解的模型　decompositional model; disaggregated model
分解方法　decomposition method
分解轨迹　decomposed trajectory
分解活动　decomposed activity
分解模型　decomposition model
分解树　decomposition tree
分类　classification; taxonomy
分类变量　categorical variable
分类代码　classification code
分类量表　ordinal scale
分类模型　classification model
分类器错误　sorter error
分类属性　class attribute
分类树　taxonomy tree
分类误差　classification error
分类学　taxonomy
分离　disjoint; disjointness; split (v.)
分离的模型　disjunctive model
分离的事件　split event
分离涡仿真　detached eddy simulation
分离误差　separate error
分派　allocate (v.)

分配　allocation
分配误差　allocation error
分配值　assigned value
分歧参数　bifurcation parameter
分区度量　partitioning metric
分区分析　partition analysis
分散的输入　distracting input
分散注意　divided attention
分时输入／输出系统　time-shared input/output system
分析　analysis
分析的　analytic
分析的（解析的）轨迹　analytic trajectory
分析方法　analytical method
分析方法学　analysis methodology
分析仿真　analytic simulation
分析仿真技术　analytic simulation technique
分析复杂性　analytic complexity
分析建模　analytical modeling
分析模型　analytical model
分析器　analyzer
分析认知模型　analytical cognitive model
分析数据压缩　analytical data compression
分析误差　analysis error
分析者　analyst
分支　bifurcated; branched; branching
分支测试　branch testing
分支方法学　bifurcated methodology
分支化的仿真　branched simulation
分子通信　molecular communication

丰富现实　enriched reality
风格　style
风格指南　style guide
风险　risk
风险变量　risk variable
风险仿真　risk simulation
风险管理　risk management
风险级别　risk level
风险计算　risk calculation
风险降低　risk reduction
风险模型　risk model
风险评估　risk assessment
风险影响　risk impact
风险预测　risk prediction
封闭解　closed-form solution
封闭模型　closed-form model
封闭式仿真　closed-form simulation
封闭系统　closed system
封闭型博弈　closed game
封闭性原则　closure principle
封装　encapsulation; packaging; wrapping
封装的　encapsulated
封装的复杂性　encapsulated complexity
封装器　wrapper
蜂窝状的　cellular
冯-诺依曼模型　Von Neuman model
否　false
否定后件律规则　modus tollens rule
服务　service
服务封装　service encapsulation
服务封装的　service encapsulated
服务规范　service specification
服务规范需求　service specification requirements
服务规则　service discipline
服务化　servitization
服务建模　service modeling
服务可靠性　service reliability
服务可用性　service availability
服务器　server
服务器错误　server error
服务器仿真器　server simulator
服务器需求　server requirements
服务视图　service view
服务水平　level of service
服务提供方　service provider
服务性能　service performance
服务需求　service requirements
服务质量　service quality
服务总线　service bus
俘获　capture
符号 DEVS　symbolic DEVS
符号处理　symbolic processing
符号处理器　symbolic processor
符号方法　symbolic approach
符号仿真　semiotic simulation; symbolic simulation
符号分析　symbolic analysis
符号化知识　symbolic knowledge
符号化执行　symbolic execution
符号回归　symbolic regression
符号计算　symbolic computing
符号技术　symbolic technique
符号模型　symbolic model
符号模型处理　symbolic model processing
符号评价　symbolic evaluation
符号实体　symbolic entity

符号算法　symbolic algorithm
符号索引约简算法　symbolic index reduction algorithm
符号调试　symbolic debugging
符号微分法　symbolic differentiation
符号微分方程求解器　symbolic differential equation solver
符号心理仿真　semiotic mental simulation
符号预处理　symbolic preprocessing
符合　compliance
符合标准　compliance with standardization; compliance with standards
符合的　compliant
符合分布交互仿真的　DIS compliant
符合分布交互仿真协议的仿真系统　DIS-compliant simulation system
符合聚合级仿真协议的仿真系统　ALSP-compliant simulation system
幅角传递　argument passing
辐角　argument
辅助　aided
辅助变量　auxiliary variable; instrumental variable; supplementary variable
辅助参数　auxiliary parameter
辅助的　auxiliary
辅助的假设　auxiliary assumption
辅助的模型维护　supplementary model maintenance
辅助的维护　supplementary maintenance
辅助服务　auxiliary service
辅助化的变量　instrumented variable

辅助值　auxiliary value
辅助作用　ancillary function
父模型　father model
负超几何随机变量　negative hypergeometric random variable
负二项随机变量　negative binomial random variable
负反馈回路　negative-feedback loop
负涌现　negative emergence
负涌现行为　negative emergent behavior
负约束　negative constraint
负载模型　load model
负值　negative value
附加　attached
附加的变量　additional variable; supplementary variable
附加价值　additional value
附加信任游戏　additive trust game
附加游戏　additive game
附加值　added value; plus value
附件的　appended
附属　attached
附属变量　attached variable
附属选单　attached menu
附着　attach (v.); attached
复变量　complex variable
复合　composite; composition
复合博弈　compound game
复合仿真　composite simulation
复合行为　composite behavior
复合活动　composite activity
复合建模　composite modeling
复合模型　composite model
复合模型运行　composite-model

execution
复合视频输入　composite video input
复合体封装　complexity encapsulation
复合误差　composite error
复合消息　composite message
复合语言　composition language
复杂的　complex; complicated
复杂的行为　complex behavior
复杂范式　complexity paradigm
复杂进化系统　complex evolutionary system
复杂模型　complicated model; complex model
复杂情景　complicated scenario; complex scenario
复杂事件　complex event
复杂数据　complex data
复杂数值变量　complex-valued variable
复杂系统　complex system; complicated system
复杂系统建模　complex system modeling
复杂现象　complex phenomenon
复杂性　complexity
复杂性测量　complexity measure
复杂性理论　complexity theory
复杂自适应系统　complex adaptive system
复杂自适应系统建模　complex adaptive system modeling
复制　duplication; replication; replicate (*v.*)
复制次数　number of replication
复制的模型　replicated model
复制的实验　replicated experiment
复制分析　replication analysis
复制品　replica
复制实验　replication experiment
复制性仿真　replicative simulation
复制性模型有效性　replicative model validity
复制性实验　replicative experiment
复制验证　replicative validation
复制有效的　replicatively valid; replicative validity
副本　replica
副框架　subframe
副帧　subframe
赋予……特色　characterize (*v.*)
赋值　assignment
傅里叶分析　Fourier analysis
富有洞察力的模型　insightful model
覆盖错误　override error

Gg

GEST 耦合　GEST coupling
Goore 游戏　Goore game
Gumbel 随机变量　Gumbel random variable
伽马分布　gamma distribution
伽马随机变量　gamma random variable
改编　adaptation
改进　improvement
改进的仿真算法　improved simulation algorithm
改进的科赫-林氏系统　modified Koch L-system

改进的模型　improved model
改进的欧拉积分法　modified Euler integration method
改进牛顿迭代法　modified Newton iteration
概率　probability
概率变量　probabilistic variable
概率的　probabilistic; probable
概率仿真过程　probabilistic simulation process
概率分布　probability distribution
概率分布函数　probability distribution function
概率分析　probabilistic analysis
概率风险评估　probabilistic risk assessment
概率过程　probabilistic process
概率函数　probabilistic function
概率互仿真　probabilistic bisimulation
概率假设　probabilistic assumption
概率建模　probabilistic modeling
概率灵敏度分析　probabilistic sensibility analysis
概率密度函数　probability density function
概率模型　probabilistic model
概率事件　probabilistic event
概率数据库　probabilistic database
概率双相似　probabilistic bisimilarity
概率误差　probability error; probable error
概率伊辛模型　probabilistic Ising model
概率值　probability value
概念　concept; notion
概念表示（法）　conceptual representation
概念抽象　conceptual abstraction
概念的　conceptual
概念的变化　conceptual-change
概念仿真　conceptual simulation
概念仿真模型　conceptual simulation model
概念仿真器　conceptual simulator
概念分析　conceptual analysis
概念互操作模型　conceptual interoperability model
概念互操作水平　conceptual interoperability level
概念互操作性　conceptual interoperability
概念化　conceptualization
概念基础　conceptual foundation
概念架构　conceptual framework
概念建模　conceptual modeling
概念建模技术　conceptual modeling technique
概念建模框架　conceptual modeling framework
概念建模形式　conceptual modeling formalism
概念建模语言　conceptual modeling language
概念可重用仿真模型　conceptual reusable simulation model
概念描述　conceptual description
概念模式　conceptual schema
概念模型　conceptual model
概念模型开发　conceptual model development

概念模型品质　conceptual model quality
概念模型属性　conceptual model attribute
概念模型校核　conceptual model verification; verification of conceptual model
概念模型验证　conceptual model validation; validation of conceptual model
概念上下文　conceptual context
概念设计　conceptual design
概念识别　concept identification
概念实验　conceptual experiment
概念实验框架　conceptual experimental frame
概念属性　conceptual attribute
概念数据　notional data
概念数据模型　conceptual data model
概念误差　conceptual error
概念系统　conceptual system
概念想定　conceptual scenario
概念想定元模型　conceptual scenario metamodel
概念需求　conceptual requirements
概念验证　concept validation
概念验证博弈　proof-of-concept game
概念验证仿真　proof-of-concept simulation
概念因果图　conceptual causal diagram
概念域　conceptual domain
概念元模型　conceptual metamodel
概念正确性　conceptual validity
概念证明　proof-of-concept

概念智能体模型　conceptual agent model
感官触摸　sensory touch
感官的　sensory
感官接口　sensory interface
感官经验　sensory experience
感官模型　sensory model
感官输入　perceptual input; sensory input
感官数据　sensory data
感官数据转换　sensory data conversion
感官知觉　sensory perception
感情响应　emotive response
感性认识　perceptual knowledge
感应　induction
感应变量　sensed variable
感应的　induced; inductive
感应输入　sensed input
感知　perceive (v.); perception
感知变量　perceptual variable
感知到的　perceived
感知到的内源输入　perceived endogenous input
感知到的情况　perceived situation
感知到的外源输入　perceived exogenous input
感知到的涌现　perceived emergence
感知到的真理　perceived truth
感知到的正确性　perceived correctness
感知的　perceptual
感知的对象　percept
感知的模型　perceived model
感知的目标　perceived goal

感知的内部事实　perceived internal fact
感知的内部输入　perceived internal input
感知的事实　perceived fact
感知的输入　perceived input
感知的外部事实　perceived external fact
感知的外部输入　perceived external input
感知的支付函数　perceived payoff function
感知仿真　perceptual simulation
感知面　aspect of perception
感知深度　depth of perception
感知事件　perceived event
感知数据　perceived data
感知水平　level of perception
感知误差　perception error
感知系统　aware system
感知现实　perceived reality
干扰　interference; interrupt (*v.*)
干扰因子　confounder
干涉变量　moderator variable
干燥实验室　dry laboratory
刚度　stiffness
刚度比　stiffness ratio
刚度矩阵　stiffness matrix
刚体仿真　rigid-body simulation
刚性不连续的模型　stiff discontinuous model
刚性的　stiff
刚性动力学系统　stiff dynamic system
刚性仿真　stiff simulation
刚性模型　stiff model
刚性微分方程　stiff differential equation
刚性稳定步长控制　stiffly stable step-size control
刚性稳定的　stiffly stable
刚性稳定算法　stiffly stable algorithm
刚性稳定隐式算法　stiffly stable implicit algorithm
刚性系统　stiff system
刚性系统仿真　stiff system simulation
刚性系统积分算法　stiff system integration algorithm
纲要　schemata
高保真的仿真　high-fidelity simulation
高保真仿真器　high-fidelity simulator
高层抽象　high-level abstraction
高层次　high level
高层次的组件　high-level component
高层建模环境　high-level modeling environment
高层体系架构规则　HLA rule
高层体系结构　high-level architecture; HLA (High-Level Architecture)
高层体系结构参数　HLA parameter
高层体系结构合格　HLA compliance
高层体系结构合格测试　HLA-compliance test
高层体系结构合格的认证设施　HLA-compliance certification facility
高层体系结构兼容的　HLA compatible
高层体系结构接口规范　HLA interface specification
高层体系结构联邦　HLA federation

高层体系结构相容的　HLA-compliant
高层体系结构相容的仿真　HLA-compliant simulation
高层体系结构相容的仿真系统　HLA-compliant simulation system
高层体系结构相容的仿真研究　HLA-compliant simulation study
高层体系结构相容的接口　HLA-compliant interface
高层体系结构协议数据　HLA protocol data
高层体系结构意识　HLA awareness
高分辨率仿真　high-resolution simulation
高分辨率建模　high-resolution modeling
高分辨率模型　high-resolution model
高分辨率战斗建模　high-resolution combat modeling
高估误差　overestimation error
高级（层）激励　high-level stimulus
高级的模型　high-level model
高级仿真　high-level simulation
高级仿真环境　high-level simulation environment
高级符号知识　high-level symbolic knowledge
高级概念模型　high-level conceptual model
高级工具　high-level tool
高级佩特里网　high-level Petri net
高级系统模型　high-level system model
高级语言　high-level language
高交互系统　highly interactive system
高阶　high order
高阶方法　high-order method
高阶非线性系统　high-order nonlinear system
高阶龙格-库塔算法　high-order Runge-Kutta algorithm
高阶模型　high-order model
高阶前向差分算子　high-order forward-difference operator
高阶算法　high-order algorithm
高阶系统　high-order system
高阶协同　high-order synergy
高阶影响　high-order effect
高粒度　high granularity
高粒度模型　high-granularity model
高情境的复杂性　high situational complexity
高认知的复杂性　high cognitive complexity
高认知复杂性个体　high-cognitive-complexity individual
高斯copula仿真　Gaussian copula simulation
高斯变量　Gaussian variable
高斯仿真　Gaussian simulation
高斯分布　Gaussian distribution
高斯分布仿真　Gaussian distribution simulation
高斯随机变量　Gaussian random variable
高斯随机函数模拟　Gaussian random function simulation
高斯误差　Gaussian error
高斯消元法　Gaussian elimination
高斯序列模拟　Gaussian sequential

simulation
高速仿真　high-speed simulation
高速缓存模拟器　cache memory simulator
高速计算机　high-speed computer
高速实验　high-speed experimentation
高速数据传送　high-speed data transfer
高完整性系统　high-integrity system
高性能仿真　high-performance simulation
高性能仿真系统　high-performance simulation system
高性能计算　high-performance computing
高性能计算仿真　HPC simulation
高性能计算机　high-performance computer
高性能计算机上仿真　simulation on high-performance computer
高指标问题　higher-index problem
高指数模型　higher-index model
格林尼治标准时间　Greenwich mean-time
格式　format; format (*v*.); style
格式化　formatting
格式化层　formatting layer
格式化错误　formatting error
格式描述层　format description layer
格子化模型　lattice model
隔开　partition (*v*.)
隔离物　partition
个人游戏　personal game
个人知识　personal knowledge
个人资料　personal data

个体　individual
个体变量　individual variable
个体的行为　individual behavior
个体仿真　individual simulation
个体控制　individual control
个体模型　individual model
个性过滤器　personality filter
个性化技术　personalization technology
个性化系统　personalized system
个性化用户界面（接口）　personalized user interface
个性化游戏　personalized game
给定误差　specified error
根本原因　root cause
根协调员　root coordinator
根源　seed
跟踪　trace; trace (*v*.)
跟踪能力　traceability
跟踪数据　trace data
跟踪误差　tracking error
更高层次　higher level
更高分辨率模型　higher-resolution model
更高阶变换　higher-order transformation
更高阶次模型　higher-order model
更高水平模型　higher-level model
更新　update; update (*v*.); updating
更新的模型　updated model
更新的数据　updated data
更新方法学　update methodology
更新区域　update region
工程　project
工程（学）　engineering

工程仿真　engineering simulation
工程仿真器　engineering simulator
工程设计　engineering design
工程突现　engineered emergence
工程需求　engineering requirements
工具　tool
工具互操作性　tool interoperability
工具集成　tool integration
工具集成框架　tool integration framework
工具误差　tool error
工具支持　tool support
工业规模　industrial scale
工业规模仿真　industrial-scale simulation
工业化分类代码　industrial classification code
工业领域　industry scope
工艺　technique; technology
工艺仿真　technical simulation
工艺误差　fabrication error
工作存储器　working memory
工作记忆　working memory
工作流建模　workflow modeling
公差　tolerance
公共联邦功能　common federation functionality
公共领域　public domain
公共领域博弈　public domain game
公共领域仿真　public domain simulation
公共耦合　common coupling
公共数据库　common database
公共形式　common formalism
公开的知识　open knowledge
公开可用的　publicly available
公开可用的仿真　publicly available simulation
公开可用的仿真器　publicly available simulator
公开可用的模型　publicly available model
公开可用的软件　publicly available software
公开提供的游戏技术　publicly available game technology
公理　axiom
公平仿真　fair simulation
公认的体系　acknowledged system-of-systems
公认假设　recognized assumption
公认误差　acknowledged error
公式　formula
公式化　formulation
公文处理法仿真　in-basket simulation
公务　business
公用数据　public data
公允价值　sound value
公众行为　public behavior
功率分流　power split (*v.*)
功率流　power flow
功能　function
功能保真度　functional fidelity
功能测试　functional testing
功能的　functional
功能仿真　functional simulation
功能分解　functional decomposition
功能分解图　functional decomposition diagram
功能分析　functional analysis

功能封装　functionality encapsulation
功能封装的　functionality encapsulated
功能过程　functional process
功能过程的改进　functional process improvement
功能建模　function modeling; functional modeling
功能接口　functional interface
功能结构　functional structure
功能模型　functional model
功能模型属性　functional model attribute
功能区域　functional area
功能上地　functionally
功能上正确的　functionally correct
功能设计空间　functional design space
功能失常　malfunction
功能属性　functional attribute
功能数据管理员　functional data administrator
功能特性　functional property
功能系统　functional system
功能系统设计　functional system design
功能相似性　functional similarity
功能校核　functional verification
功能性　functionality
功能需求　functional requirement
功能演化　functional evolution
功能游戏　functional game
功能域区　functional domain area
攻击级别　aggression level
共点　concurrent
共轭变量　yoked variable
共轭的　yoked
共轭的关系　yoked relation
共轭梯度算法　conjugate gradient algorithm
共模输入　common mode input
共生变量　coenetic variable
共生仿真　symbiotic simulation
共生行为　symbiotic behavior
共识主动性的　stigmergic
共通启发式演算法　metaheuristics
共同支付函数　common payoff function
共享　sharing
共享变量　shared variable
共享沉浸体验　shared immersive experience
共享概念化　shared conceptualization
共享环境　shared environment
共享机密数据随机变量　shared secret data random variable
共享经验　shared experience
共享模型　shared model
共享软件　shareware
共享事件　shared event
共享数据　shared data
共享状态　shared state
共享状态事件　shared state event
共用　common use
供应链建模　supply chain modeling
构成　construct (v.); formation
构成主义　constructivism
构成主义范式　constructivist paradigm
构成主义方法论　constructivist methodology
构成主义理论　constructivist theory

构成主义者　constructivist
构架风格　architectural style
构建管理　architecture management
构思出的问题质量　formulated problem quality
构造　construction
构造仿真　structural simulation
构造仿真集成　constructive simulation integration
构造类模型　constructive model
构造模型保真度　structural model fidelity
构造评估　structural assessment; structure assessment
构造实例　composition instance
构造性仿真　constructive simulation
构造性实体　constructive entity
构造性学习仿真　constructive training simulation
构造性训练环境　constructive training environment
估定价值　appraisal value; appraised value
估计　estimation; estimated; estimate (v.)
估计理论　estimation theory
估计器　estimator
估计误差　estimation error
估量　measure (v.)
骨架驱动的仿真　skeleton-driven simulation
鼓舞人心的现实　inspirational reality
固定变量　fixed variable
固定变量模型　stabilized variable model

固定的元模型　fixed metamodel
固定结构目标　goal with fixed structure
固定结构自动机　fixed-structure automaton
固定时间步长　fixed time-step
固定输入　fixed input
固定拓扑仿真　fixed-topology simulation
固定误差　permanent error
固定值　fixed value
固件　firmware
固件错误　firmware error
固有的　built-in
固有误差　fixed error; intrinsic error
故意错误　willful error
故意歪曲事实　deliberately distorted reality
故障　failure; fault; mulfunction
故障避免　failure avoidance
故障插入测试　failure insertion testing
故障的行为　faulty behavior
故障仿真　fault simulation; faulty simulation
故障分析　failure analysis; fault analysis
故障管理　fault management
故障行为　fault behavior
故障检测　failure detection; fault detection
故障检测机制　failure detection mechanism
故障模型　faulty model
故障容限（容错）　fault tolerance
故障树　fault tree

故障组件　faulty component
挂钟时间　wall clock time
关键　key
关键变量　key variable
关键过程变量　key-process variable
关键事件　critical-event
关键事件对策　critical-event game
关键事件仿真　critical-event simulation
关键值　key value
关键组成　critical component
关联　association; relevance
关联的模型　interrelated model; linked model
关联的数据　linked data
关联经验　connected experience
关联控制　associative control
关联理解　associative understanding
关联列表　linked list
关联模型　conjunctive model
关联耦合　conjunctive coupling
关联耦合配方　conjunctive coupling recipe
关联佩特里网　associative Petri net
关联实体　associative entity
关联性假设　relational assumption
关联值　associated value
关系　relation; relationship
关系词典　relational dictionary
关系等价系统　relationally equivalent system
关系概念　relational concept
关系模型　relational model
关系数据库　relational database
关于模型的推理　reasoning about models
关注点分离　separation of concerns
观测　observation; observe (v.)
观测变量　observational variable; observed variable
观测到的　observed
观测到的行为　observed behavior
观测函数　observation function
观测机制　observation mechanism
观测模型　observation model
观测频率　observed frequency
观测事件　observation event
观测数据　observational data
观测误差　observational error; observation error
观测误差统计　observation error statistic
观测信息　observed information
观测知识　observational knowledge
观测值　observed value
观测状态　observation state
观察器　viewer
观察者　viewer
观察者模型　observer model
观察者效应　observer effect
观察者依赖的复杂性　observer-dependent complexity
管理　administration; management; managing
管理对策　management game
管理风险　management risk
管理服务　management service
管理游戏社区　managing gaming community
管理员　administrator

管理者　manager
管理知识　management knowledge
管线拓扑　pipeline topology
惯例　observance
光线追踪　ray tracing
光线追踪模型　ray-tracing model
光学计算　optical computing
广播　broadcast
广度优先　breadth first
广泛的　broad-based
广告游戏　advergame
广义变量　generalized variable
广义队列　generalized queue
广义多重建模方法学　generalized multimodeling methodology
广义混合模式仿真　generalized mixed-mode simulation
广义离散事件　generalized discrete-event
广义离散事件系统规范　GDEVS
广义模型　generalized model
广义水平集去模糊化　generalized level set defuzzification
广义随机佩特里网　generalized stochastic Petri net
广义同步　generalized synchronization
广义线性混合模型　generalized linear mixed model
广义线性模型　generalized linear model
广义坐标　generalized coordinate
归并　merge (v.)
归档错误　filing error
归纳　induction
归纳变量　induction variable
归纳的　inductive
归纳断言　inductive assertion
归纳法　inductive method
归纳法仿真　inductive simulation
归纳仿真证明　inductive simulation proof
归纳建模　inductive modeling
归纳推理　inductive reasoning
归属令牌　attributed token
归一化仿真　uniformization simulation
归一化事件　uniformized event
归一化数值　normalised value
归因　attributed
归因误差　attribution error
规避　avoidance
规定　prescribe (v.)
规定式仿真　prescriptive simulation
规范　specification
规范的　normative
规范的行为　normative behavior
规范的描述格式　canonical description format
规范方法　specification method
规范仿真　normative simulation
规范环境　specification environment
规范库　specification repository
规范描述　canonical description
规范条件　specification condition
规范维护说明　specification maintenance
规范误差　specification error
规范系统　normative system
规范形式　specification formalism
规范性决策　normative decision
规范语言　specification language

规格说明阶段　specification phase
规格校验　specification verification
规格验证　specification validation
规划　planning; program; programming
规划器　planner
规划者　planner
规律　discipline
规模　scale
规一化稳定性区域　normalized stability region
规约　reduction
规则　rule
规则标记语言　rule markup language
规则的结论部分　conclusion part of a rule
规则的前件部分　antecedent part of a rule
规则库　rule base
规则模糊化　rule fuzzification
规则模型　rule model
规则曲面细分　regular tessellation
规则去模糊化　rule defuzzification
规则引擎　rule engine
硅片上　in silico
轨道稳定性　trajectory stability
轨迹　trajectory
轨迹比较　trajectory comparison
轨迹行为　trajectory behavior
轨迹驱动的　trace-driven
轨迹驱动的仿真器　trace-driven simulator
轨迹驱动的模型　trace-driven model
轨迹驱动的输入测试　trace-driven input testing
诡辩（法）　sophism
诡异谷　uncanny valley
国际的联盟　international confederation
国家联邦　national federation
国家模型　national model
过程　procedure; process
过程变量　process variable
过程成熟度模型　process maturity model
过程抽象化　process abstraction
过程仿真　process simulation
过程改进　process improvement
过程改进建模　process improvement modeling
过程集成　process integration
过程建模　process modeling
过程交互　process interaction
过程交互仿真　process interaction simulation
过程可靠性　process reliabilism
过程控制　process control
过程模型　process model
过程误差　process error
过程性知识　procedural knowledge
过程性知识处理　procedural knowledge processing
过程质量　process quality
过度简化　oversimplification
过渡　transition
过渡变量　transition variable
过渡过程　transient
过渡阶段　transient phase
过渡条件　transition condition
过零　zero-crossing

过零函数　zero-crossing function
过滤的数据　filtered data
过滤的值　filtered value
过去的行为　past behavior
过去的经验　past experience
过去的趋势　past trend
过时的模型　obsolete model
过时的软件游戏　abandonware game
过时数据　obsolete data
过于简单化的　oversimplified
过载　overloading

Hh

哈密顿场　Hamiltonian field
哈密顿函数　Hamiltonian
海量存储器　mass storage
含糊　ambiguity; vagueness
含糊的　ambiguous; vague
函数　function
函数逼近　function approximator
函数值　function value
航位推算法　dead reckoning
行业认同　industry identity
好的建模实践　good modeling practice
耗时的活动　time-consuming activity
合成　composite
合成的　synthetic
合成环境　synthetic environment
合成环境内容　synthetic environment content
合成环境数据　synthetic environment data
合成建模　compositional modeling
合成空间　synthetic space
合成模拟　synthetic analog
合成器　composer
合成输入　synthetic input
合成系统　constituent system
合成现实　synthetic reality
合成演练　combined exercise
合成战场　synthetic battlefield
合成战场空间　synthetic battlespace
合成组件　synthetic component
合法值　legal value
合格　compliance
合格标准　standard for acceptability
合格测定　compliance determination
合格测试　compliance test; compliance testing
合格率　acceptability
合格证书　certificate of compliancy; compliance certificate
合规的　ethical
合乎规格的评估　ethical assessment
合理的　admissible; reasonable
合理的行为　rational behavior
合理仿真　reasonable simulation
合理模型　reasonable model
合理性　rationality
合弄结构　holarchy
合弄结构模型　holarchic model
合适的仿真　appropriate simulation; suitable simulation
合同管理　contract management
合一错误　unification error
合作博弈　cooperative game; coalitional game; collaboration game
合作博弈理论　cooperative game theory

合作的 coopetitive
合作仿真 cooperative simulation
合作关系 cooperative relationship
合作竞争 coopetition
合作竞争对策 coopetition game
合作竞争仿真 coopetition simulation
合作开发 cooperative development
合作系统 cooperative system
合作性仿真 coopetitive simulation
合作性关系 coopetitive relationship
合作性行为 coopetitive behavior
合作学习仿真 cooperative learning simulation
和平博弈 peace game
和平仿真 peace simulation
和平行动 operations peace
和平支持 peace support
核查 examination
核定测试 authorization testing
核函数 kernel function
核实 verified; verify (v.)
核实者 verifier
核心本体论 core ontology
核心竞争力对策 core competency game
核心主题 core topic
黑板模型 blackboard model
黑盒测试 black box testing
黑盒校核 black box verification
黑盒验证 black box validation
黑塞矩阵 Hessian matrix
黑塞形式 Hessian
黑箱 black box
黑箱方法 black box approach
黑箱建模方法 black box approach in modeling
黑箱模型 black box model
痕迹 trace
恒星时 sidereal time
宏观步长 macro-step size
宏观层行为 macro-level behavior
宏观层模式 macro-level pattern
宏观层现象 macro-level phenomenon
宏观行为 macroscopic behavior; macro behavior
宏观紧急行为 macroscopic emergent behavior
宏观模型 macroscopic model
宏观误差 macro error
宏模型 macro model
后处理 postprocessing
后处理步骤 postprocessing step
后端 back-end
后端分析 back-end analysis
后端接口（界面） back-end interface
后继 successor
后继仿真 successor simulation
后继仿真研究 successor simulation study
后继关系 successor relation
后继模型 descendant model; successor model
后继站 successor
后面的 following
后事件 postevent
后台 background
后台任务 background task
后台执行 background execution
后现代主义 postmodernism
后相关因素分析 prognostic analysis

后向插值方法　back-interpolation method
后向插值算法　back-interpolation algorithm
后向估计　backcasting
后向估计方法　backcasting methodology
后向积分适用性　applicability of backward integration
后向龙格-库塔算法　backward Runge-Kutta algorithm
后向模型　backcasting model
后向时间因果关系　backwards time causality
后向验证　backward validation
后项　consequent
后续事件　following event
后验　a posteriori
后验模型　a posteriori model
后验实验　a posteriori experiment
后验误差估计　a posteriori error estimate; a posteriori error estimation
后验知识　a posteriori knowledge
后执行状态　postexecution state
候选键　candidate key
候选码　candidate key
候选模型　candidate model
候选值　candidate value
候选子模型　candidate submodel
忽略　pass (*v.*)
弧多样性　arc multiplicity
互补变量　complementary variable
互操作　interoperation; interoperate (*v.*)
互操作标准　interoperability standard
互操作层　interoperability layer
互操作服务　interoperability service
互操作适配器　interoperation adaptor
互操作性　interoperability
互操作性模型　interoperability model
互操作性水平　interoperability level
互仿真　bisimulation
互仿真等价的　bisimulation equivalent
互仿真等价性　bisimulation equivalence
互换性　substitutability
互连　interconnection
互连技术　interconnection technology
互联的模型　interconnected model
互熵方法　cross-entropy method
互相似的　bisimilar
互相似系统　bisimilar system
互相似性　bisimilarity
划分　partition; partitioning
划分测试　partition testing
化学感知　chemical sensation
化学通信　chemical communication
画面　tableau
话题　topic
话题模型　topic model
环保方法　environmental method
环境　environment
环境变量　environment variable; environmental variable
环境表示（法）　environment representation; environmental representation
环境仿真　environmental simulation
环境建模　environmental modeling
环境交互　environmental interaction

环境模型 environment model; environmental model
环境实体 environmental entity
环境事件 environmental event
环境数据 environmental data
环境特征 environmental feature
环境调和的 environment-mediated
环境调和的自适应系统 environment-mediated self-adaptive system
环境调和的自组织系统 environment-mediated self-organizing system
环境误差 environment error
环境效应 environmental effect
环境效应模型 environmental effect model
还原 reduction; restore (v.)
还原论 reductionism
还原论范式 reductionism paradigm
还原论方法学 reductionist methodology
还原论视点 reductionist point of view
缓冲 buffer
缓冲输入 buffered input
缓冲值 buffer value
缓存模拟器 cache simulator
缓和不稳定的突现行为 moderated unstable emergent behavior
缓和稳定的突现行为 moderated stable emergent behavior
幻想 fantasy
幻象模型 ghost model
幻影 simulacra; simulacre; simulacrum
换位误差 transposition error
黄金分割法 golden section method

灰盒测试 gray box testing
灰箱 gray box
灰箱方法 gray box approach
灰箱建模法 gray box approach in modeling
恢复 recovery
回归 regression
回归变量 regressed variable
回归测试 regression testing
回归方程中的从属变量 regressand
回归分析 regression analysis
回归技术 regression technique
回归量 regressor
回归模型 regression model
回归元建模技术 regression metamodeling technique
回归元模型 regression metamodel
回合制博弈 turn-based game
回路 loop
回溯存档 retrodocumentation
回溯仿真 retrosimulating; retrosimulation; retrosimulate (v.)
回溯仿真的 retrosimulated
回溯因果关系 retrocausality; retrocausation
回溯原因的 retrocausal
回缩 retraction
绘图 plot
绘图间隔 plot interval
混叠 aliasing
混沌的 chaotic
混沌动力系统 chaotic dynamic system
混沌动力学 chaotic dynamics
混沌仿真 chaotic simulation

混沌理论　chaos theory
混沌模型　chaotic model
混沌误差　chaotic error
混沌系统　chaotic system
混沌状态　chaotic state
混合　hybrid
混合博弈仿真　hybrid gaming simulation
混合方法　mixed method
混合方法学　mixed methodology
混合方法学模型　mixed methodology model
混合仿真　blended simulation; hybrid simulation; mixed simulation
混合仿真模型　hybrid simulation model
混合仿真器　hybrid simulator
混合仿真软件　hybrid simulation software
混合仿真系统　hybrid simulation system
混合仿真研究　hybrid simulation study
混合仿真语言　hybrid simulation language; hybrid SL
混合分辨率模型　mixed-resolution model
混合分对数　mixed logit
混合符号和数值方法　mixed symbolic and numerical approach
混合计算机　hybrid computer
混合计算机仿真　hybrid computer simulation
混合粒度　mixed granularity
混合粒度模型　mixed-granularity model
混合连续系统仿真语言　hybrid continuous-system SL
混合模式仿真　mixed-mode simulation
混合模式集成　mixed-mode integration
混合模型　hybrid model; mixed model
混合时间模型　mixed-time model
混合式耦合　mixed coupling
混合式耦合方法　mixed coupling recipe
混合算法　blended algorithm
混合同步　hybrid synchronization
混合析因设计　hybrid factorial design
混合系统　hybrid system; mixed system
混合现实　hybrid reality; mixed reality
混合现实系统　mixed-reality system
混合现实训练仿真器　mixed-reality training simulator
混合信号仿真　mixed-signal simulation
混合形式　mixed formalism
混合形式模型　mixed-formalism model
混合型主体　hybrid agent
混合学习仿真　blended learning simulation
混合优化　hybrid optimization
混合源仿真语言　hybrid source SL
混合状态　mixed state
混合状态模型　mixed-state model
混合自动机　hybrid automaton
混聚　mash-up

混淆 confound
混杂变量 confounding variable
混杂系统解 hybrid systemic solution
混杂因素 confounding factor
活动 action; activity
活动仿真语言 activity SL
活动规则 activity rule
活动力 activity
活动模型 activity model
活动扫描 activity scan; activity scanning
活动扫描法 activity-scanning approach
活动实体 active entity
活动图 activity diagram
活动形状模型 active shape model
活动性语言 activity language
活动周期图 activity cycle diagram
活动状态 active state
活化 activation
活化误差 activation error
活化作用 activation function
活力 activity
活锁 livelock
活跃的 animated
活跃的多模型 active multimodel
活跃值 active value
获得 sourcing
"或"输入 OR input
霍因积分法 Heun's integration method

Ji

机电系统 mechatronic system
机构 organization
机会 chance; opportunity
机会游戏 game of chance
机理 mechanism
机器仿真 machine simulation
机器可理解的模型 machine intelligible model
机器误差 machine error
机器学习 machine learning
机器智能 machine intelligence
机器中心模型 machine-centric model
机械化推理 mechanized reasoning
机械设计建模 mechanical design modeling
机械误差 mechanical error
机械系统 mechanical system
机械学模型 mechanistic model
机械装置 mechanism
机制 mechanism; scheme
积分 integration
积分变量 integration variable
积分步长 integration step size
积分法 integration method
积分方程 integral equation; integrator equation
积分精度 integration accuracy
积分器 integrator
积分算法 integration algorithm
积分问题 integral problem
积分误差 integration error
积极的 active
积极的涌现 positive emergence
积极的涌现行为 positive emergent behavior
积极行为 active behavior

基本案例仿真　base case simulation
基本变量　base variable; basic variable; essential variable
基本对象模型　base object model
基本仿真器　basic simulator
基本级联式配方　essentially cascade coupling recipe
基本结构　infrastructure
基本模型　base model
基本耦合　basic coupling
基本佩特里网　basic Petri net
基本输入/输出系统　basic input/output system
基本误差　intrinsic error
基本原理　fundamental principle
基础　foundation
基础的　basic
基础的行为　basic behavior
基础环境　base context
基础框架　base frame
基础离散事件系统规范模型　base DEVS model
基础配置　basic configuration
基础设施　infrastructure
基础设施的视角　infrastructure perspective
基础设施系统　infrastructure system
基础知识　foundational knowledge
基底　basis
奇函数　odd function
基极　base
奇偶检验误差　parity error
基维亚特图表　Kiviat chart
基线仿真　baseline simulation
基线仿真想定　baseline simulation scenario
基线想定　baseline scenario
基线想定仿真　baseline scenario simulation
基线预测　baseline forecast; baseline forecast
基因型　genotype
基因型表达　genotypic representation
基因型建模　genotypic modeling
基于　based
基于DNA的　DNA-based
基于DNA的仿真　DNA-based simulation
基于DNA的建模　DNA-based modeling
基于DNA的模型　DNA-based model
基于DNA的耦合　DNA-based coupling
基于案例的仿真　case-based simulation
基于本体的　ontology-based
基于本体的词典　ontology-based dictionary
基于本体的多智能体仿真　ontology-based multiagent simulation
基于本体的方法学　ontology-based methodology
基于本体的仿真　ontology-based simulation
基于本体的仿真方法学　ontology-based simulation methodology
基于本体的工具　ontology-based tool
基于本体的框架　ontology-based framework
基于本体的描述　ontology-based

基于本体的模型规范　ontology-based model specification
基于本体的智能体仿真　ontology-based agent simulation
基于变异的　mutation-based
基于标准的　standard-based
基于标准的语言　standard-based language
基于参与的　participation-based
基于草图的仿真　sketch-based simulation
基于测试的　test-based
基于测试建立的信任　test-based confidence
基于层的　layer-based
基于层的分割　layer-based segmentation
基于插件的　plug-in-based
基于程序的　program-based
基于程序的仿真　program-based simulation
基于传感器的系统　sensor-based system
基于代理的　agent-based; surrogate-based
基于代理的博弈　agent-based game; agent-based gamification; agent-based gaming
基于代理的参与性仿真　agent-based participatory simulation
基于代理的程序设计范式　agent-based programming paradigm
基于代理的多仿真　agent-based multisimulation
基于代理的方法　agent-based approach
基于代理的方法学　agent-based methodology
基于代理的仿真　agent-based simulation
基于代理的仿真游戏　agent-based simulation game
基于代理的分布式数据挖掘　agent-based distributed data mining
基于代理的建模　agent-based modeling
基于代理的解决方案　agent-based solution
基于代理的模型　agent-based model; surrogate-based model
基于代理的佩特里网　agent-based Petri net
基于代理的数据挖掘　agent-based data mining
基于代理的网格计算　agent-based grid computing
基于代理的系统　agent-based system
基于代理的应用程序　agent-based application
基于定位的　location-based
基于定位的体验　location-based experience
基于定位的游戏　location-based game
基于动作的错误探测　action-based error detection
基于端口的　port-based
基于端口的建模　port-based modeling
基于端口的建模范式　port-based modeling paradigm

基于对象的　object-based
基于对象的模型　object-based model
基于多智能体的仿真　multiagent-based simulation
基于方程的模型　equation-based model
基于方法论的分类　methodology-based classification
基于方法学的　methodology-based
基于方法学的软件工具　methodology-based software tool
基于仿真采办的成本分析　cost analysis for simulation-based acquisition
基于仿真的　simulation-based
基于仿真的安全　simulation-based security
基于仿真的标定　simulation-based calibration
基于仿真的博弈　simulation-based game
基于仿真的采办　simulation-based acquisition
基于仿真的发现　simulation-based discovery
基于仿真的方法　simulation-based method
基于仿真的方法学　simulation-based methodology
基于仿真的分布式训练　simulation-based distributed training
基于仿真的分析　simulation-based analysis
基于仿真的概念证明　simulation-based proof-of-concept
基于仿真的个体训练　simulation-based individual training
基于仿真的工程　simulation-based engineering
基于仿真的工具　simulation-based tool
基于仿真的故障分析　simulation-based failure analysis
基于仿真的管理　simulation-based management
基于仿真的规划　simulation-based planning
基于仿真的环境　simulation-based environment
基于仿真的计算发现　simulation-based computational discovery
基于仿真的建模　simulation-based modeling
基于仿真的教学　simulation-based teaching
基于仿真的教育　simulation-based education
基于仿真的教育技术　simulation-based education technique
基于仿真的教育系统　simulation-based education system
基于仿真的经验　simulation-based experience
基于仿真的决策　simulation-based decision making
基于仿真的决策支持　simulation-based decision support
基于仿真的科学　simulation-based science
基于仿真的可靠性评估　simulation-

based reliability assessment
基于仿真的控制　simulation-based control
基于仿真的控制图　simulation-based control graph
基于仿真的理解　simulation-based understanding
基于仿真的鲁棒设计　simulation-based robust design
基于仿真的评估　simulation-based assessment; simulation-based estimation
基于仿真的评估工具　simulation-based assessment tool
基于仿真的评价　simulation-based evaluation
基于仿真的企业　simulation-based enterprise
基于仿真的起始　simulation-based initiative
基于仿真的区域　simulation-based field
基于仿真的设计　simulation-based design
基于仿真的设计方法　simulation-based design approach
基于仿真的设计方法学　simulation-based design methodology
基于仿真的设计优化　simulation-based design optimization
基于仿真的实时预测　simulation-based real-time prediction
基于仿真的实验　simulation-based experimentation; simulation-based experiment

基于仿真的数据挖掘　simulation-based data mining
基于仿真的算法　simulation-based algorithm
基于仿真的调度　simulation-based scheduling
基于仿真的通信方法　simulation-based communication approach
基于仿真的团队训练　simulation-based teamwork training
基于仿真的推理　simulation-based inference
基于仿真的问题求解　simulation-based problem solving
基于仿真的问题求解环境　simulation-based problem solving environment
基于仿真的系统　simulation-based system
基于仿真的系统采办　simulation-based system acquisition
基于仿真的系统分析　simulation-based systems analysis
基于仿真的系统工程　simulation-based systems engineering
基于仿真的协同工程　simulation-based collaborative engineering
基于仿真的性能分析　simulation-based performance analysis
基于仿真的性能评估　simulation-based performance evaluation
基于仿真的虚拟环境　simulation-based virtual environment
基于仿真的序优化　simulation-based ordinal optimization

基于仿真的学科　simulation-based discipline
基于仿真的学习　simulation-based learning
基于仿真的学习系统　simulation-based learning system
基于仿真的训练　simulation-based training
基于仿真的训练系统　simulation-based training system
基于仿真的严肃游戏　simulation-based serious game
基于仿真的研究　simulation-based research
基于仿真的验证　simulation-based validation
基于仿真的遗传算法　simulation-based genetic algorithm
基于仿真的优化　simulation-based optimization
基于仿真的优化设计　simulation-based optimal design
基于仿真的娱乐　simulation-based entertainment
基于仿真的预测　simulation-based forecasting
基于仿真的预测显示　simulation-based predictive display
基于仿真的预估　simulation-based prediction
基于仿真的原型　simulation-based prototype
基于仿真的原型设计　simulation-based prototyping
基于仿真的运行支持　simulation-based operational support
基于仿真的增强现实　simulation-based augmented reality
基于仿真的战略博弈　simulation-based strategy game
基于仿真的诊断　simulation-based diagnosis
基于仿真的置信度　simulation-based confidence
基于仿真的最大化　simulation-based maximization
基于仿真的最小化　simulation-based minimization
基于仿真器的训练　simulator-based training
基于非经验的知识　non-experience-based knowledge
基于服务的　service-based
基于服务的仿真　service-based simulation
基于服务的仿真系统方法　service-based simulation system approach
基于感知的建模　perception-based modeling
基于感知的耦合　awareness-based coupling; perception-based coupling
基于高层体系结构的仿真　HLA-based simulation
基于格子的模型　lattice-based model
基于个体的仿真　individual-based simulation
基于个体的建模　individual-based modeling
基于个体的模型　individual-based model

基于共生仿真的训练　symbiotic simulation-based training
基于规范的　specification-based
基于规则的　rule-based
基于规则的仿真　rule-based simulation
基于规则的模型　rule-based model
基于规则的系统　rule-based system
基于规则的系统嵌入仿真　rule-based system embedded simulation
基于规则的信息交换　rule-based information exchange
基于轨迹的　trace-based
基于轨迹的建模　trace-based modeling
基于过程的　process-based
基于过程的离散事件仿真　process-based discrete-event simulation
基于合成环境的　synthetic-environment based
基于合成环境的采办　synthetic-environment-based acquisition
基于互联网的　Internet-based
基于回滚的并行离散事件仿真　rollback-based parallel discrete-event simulation
基于回滚的仿真　rollback-based simulation
基于回滚的离散仿真　rollback-based discrete simulation
基于活动的　activity-based
基于活动的仿真　activity-based simulation
基于活动的建模　activity-based modeling

基于活动的模型　activity-based model
基于机器的形式化方法　machine-based formal method
基于计算机的　computer-based
基于计算机的博弈　computer-based game
基于计算机的仿真　computer-based simulation
基于计算机的交互式训练　interactive computer-based training
基于计算机的培训　computer-based training
基于计算机的实验　computer-based experiment
基于交互式电影的学习　interactive-movie-based learning
基于角色的仿真　actor-based simulation
基于角色的仿真引擎　actor-based simulation engine
基于角色的体验　role-based experience
基于脚本的　script-based
基于结果的误差检测　outcome-based error detection
基于经验的　experience-based
基于经验的和非经验的知识　experience- and nonexperience-based knowledge
基于经验的知识　experience-based knowledge
基于经验推导的解析模型　empirically-derived analytical model
基于经验推导的模型　empirically-derived model

基于可穿戴式计算机的仿真 wearable computer-based simulation
基于框架的 frame-based
基于离散事件的多重建模方法学 discrete-event-based multimodeling methodology
基于离散事件系统规范的 DEVS-based
基于离散事件系统规范的工具 DEVS-based tool
基于离散算法的 discrete-arithmetic-based
基于离散算法的仿真 discrete-arithmetic-based simulation
基于理论的 theory-based
基于量化的 quantization-based
基于量化的逼近 quantization-based approximation
基于量化的方法 quantization-based method
基于量化的积分 quantization-based integration
基于模板的 template-based
基于模板的建模 template-based modeling
基于模板的文档 template-based documentation
基于模仿的 imitation-based
基于模仿的博弈 imitation-based gaming
基于模糊规则的模型 fuzzy rule-based model
基于模式的 pattern-based
基于模式的转换 pattern-based transformation

基于模型的 model-based
基于模型的测试 model-based testing
基于模型的测试技术 model-based testing technique
基于模型的方法 model-based method
基于模型的方法学 model-based methodology
基于模型的仿真 model-based simulation
基于模型的仿真监视器 model-based simulation monitor
基于模型的分析 model-based analysis
基于模型的复杂系统 model-based complex system
基于模型的工程 model-based engineering
基于模型的工具 model-based tool
基于模型的互操作性 model-based interoperability
基于模型的活动 model-based activity
基于模型的技术 model-based technique
基于模型的解决方案 model-based solution
基于模型的开发 model-based development
基于模型的离散事件系统规范方法学 model-based DEVS methodology
基于模型的评估 model-based review
基于模型的评估过程 model-based review process
基于模型的评估诊断 model-based review diagnosis
基于模型的软件 model-based

基于模型的试验　model-based experiment
基于模型的推理　model-based reasoning
基于模型的系统　model-based system
基于模型的系统工程　model-based systems engineering
基于模型的现实　model-based reality
基于模型的校核　model-based verification
基于模型的协同设计　model-based codesign
基于模型的验证　model-based validation
基于模型的预估　model-based prediction
基于模型的主动性　model-based initiative
基于模型驱动工程的过程模型　MDE-based process model
基于内省的耦合　introspection-based coupling
基于能力的　capability-based
基于能力的评估　capability-based assessment
基于能力工程的　capability engineering-based experimentation
基于耦合的　coupling-based
基于耦合的方法　coupling-based technique
基于情景的　scenario-based
基于情景的对策管理　scenario-based game management
基于情景的模型综合　scenario-based model synthesis
基于情景的设计　scenario-based design
基于情景的学习　scenario-based learning
基于区域的　region-based
基于人工智能的程序设计范式　AI-based programming paradigm
基于人工智能的建模　AI-based modeling
基于认识论的分类　epistemology-based classification
基于软件的　software-based
基于软件的方法　software-based approach
基于软件的仿真　software-based simulation
基于软件的离散事件仿真　software-based discrete-event simulation
基于软件的连续系统仿真　software-based continuous system simulation
基于神经网络的模型　neural network-based model
基于生物的　bio-based
基于时间的　time-based
基于时间的扩展　time-base expansion
基于时间的信号　time-based signal
基于实际系统的训练　real-system-based training
基于实验的　experiment-based
基于市场的　market-based
基于事件的　event-based
基于事件的程序　event-based program
基于事件的程序设计　event-based

基于事件的程序设计范式　event-based programming paradigm
基于事件的仿真　event-based simulation
基于事件的分布式系统　distributed event-based system
基于事件的离散仿真　event-based discrete simulation
基于事件的任务　event-based task
基于事件的语言　event-based language
基于事件的智能体仿真　event-based agent simulation
基于事实的　fact-based
基于事实的决策　fact-based decision making
基于事务的佩特里网　transaction-based Petri net
基于事物的　transaction-based
基于输入/输出的　I/O-based
基于输入/输出的模型　I/O-based model
基于输入/输出的系统模型　I/O-based system model
基于树的建模　tree-based modeling
基于数据的　data-based
基于数据的建模　data-based modeling
基于数据的模型　data-based model
基于速率的　rate-based
基于速率的仿真　rate-based simulation
基于算法的　algorithm-based
基于随机代理的模型　stochastic agent-based model
基于随机仿真的　stochastic-simulation-based
基于图的　graph-based
基于图的变换　graph-based transformation
基于图的表征　graph-based representation
基于图的多模型规范形式　graph-based multimodel specification formalism
基于图的规范形式　graph-based specification formalism
基于图的模型　graph-based model
基于图的模型转换　graph-based model transformation
基于图像的　image-based
基于图像的建模　image-based modeling
基于图像的模型　image-based model
基于团队的　team-based
基于团队的训练　team-based training
基于推理的智能系统　reasoning-based intelligent system
基于万维网的　web-based
基于万维网的标准　web-based standard
基于万维网的多用户仿真　web-based multi-user simulation
基于万维网的仿真　web-based simulation
基于万维网的仿真博弈　web-based simulation game
基于万维网的分布式仿真　distributed web-based simulation
基于万维网的基础设施　web-based

基于万维网的计算　web-based computing
基于万维网的建模　web-based modeling
基于万维网的建模系统　web-based modeling system
基于万维网的商业博弈　web-based business game
基于万维网的学习　web-based learning
基于万维网的训练　web-based training
基于万维网服务的仿真　web-service-based simulation
基于万维网使能的建模与仿真应用程序　web-enabled M&S application
基于网格的　grid-based; mesh-based
基于网格的仿真　grid-based simulation; mesh-based simulation
基于网格的计算　grid-based computing
基于网格环境　grid-based environment
基于网格协同仿真环境　grid-based collaborative simulation environment
基于网络的信任博弈　network-based trust game
基于文本的　text-based
基于文本的屏幕设计　text-based screen design
基于物理的　physically-based; physics-based
基于物理的仿真　physics-based simulation
基于物理的建模　physically-based modeling; physics-based modeling
基于物理的模型　physics-based model
基于物理的设计　physics-based design
基于系统动力学的建模　system dynamics-based modeling
基于系统理论的　systems-theory-based
基于系统理论的方法　systems-theory-based approach
基于系统理论的仿真　systems-theory-based simulation; system-theory-based simulation
基于系统理论的建模与仿真　systems-theory-based M&S
基于系统理论的离散事件仿真　systems-theory-based discrete-event simulation; system-theory-based discrete-event simulation
基于系统理论的连续系统仿真　systems-theory-based continuous system simulation; system-theory-based continuous system simulation
基于显示的仿真　display-based simulation
基于现实的　reality-based
基于现实的约束　reality-based constraint
基于相位的　phase-based
基于相位的模型　phase-based model
基于想定的虚拟环境　scenario-based virtual environment
基于向量的元胞模型　vector-based cellular model

基于像素的模型　pixel-based model
基于效果的建模　effects-based modeling
基于效用的　utility-based
基于效用的系统　utility-based system
基于协同组件的仿真　collaborative component-based simulation
基于选单的　menu-based
基于因特网的数据收集　Internet-based data collection
基于游戏的　game-based
基于游戏的工具　game-based tool
基于游戏的教育技术　game-based education technique
基于游戏的训练　game-based training
基于语法的　grammar-based
基于语法的程序生成器　grammar-based program generator
基于语义的　semantic-based
基于语义的工具　semantic-based tool
基于语义的组合　semantics-based composition
基于预测耦合　anticipation-based coupling
基于预期的建模　anticipation-based modeling
基于元模型的方法　metamodel-based approach
基于元模型的仿真　metamodel-based simulation
基于元模型的模型　metamodel-based model
基于元模型的智能体模型　metamodel-based agent model
基于原型的模型　prototype-based model
基于云的　cloud-based
基于云的仿真　cloud-based simulation
基于云的仿真框架　cloud-based simulation framework
基于云的建模框架　cloud-based modeling framework
基于云的框架　cloud-based framework
基于知识的　knowledge-based
基于知识的程序生成支持　knowledge-based program generation support
基于知识的动态仿真组合　knowledge-based dynamic simulation composition
基于知识的多代理系统　knowledge-based MAS
基于知识的方法　knowledge-based approach
基于知识的仿真　knowledge-based simulation
基于知识的仿真器　knowledge-based simulator
基于知识的建模支持　knowledge-based modeling support
基于知识的决策　knowledge-based decision
基于知识的可视化　knowledge-based visualization
基于知识的模型处理支持　knowledge-based model-processing support
基于知识的评估　knowledge-based assessment

基于知识的实验支持　knowledge-based experimentation support
基于知识的系统　knowledge-based system
基于知识的支持　knowledge-based support
基于智能体的　agent-based
基于智能体的博弈　agent-based game; agent-based gamification; agent-based gaming
基于智能体的参与性仿真　agent-based participatory simulation
基于智能体的程序设计范式　agent-based programming paradigm
基于智能体的多仿真　agent-based multisimulation
基于智能体的方法　agent-based approach
基于智能体的方法学　agent-based methodology
基于智能体的仿真　agent-based simulation
基于智能体的仿真游戏　agent-based simulation game
基于智能体的分布式数据挖掘　agent-based distributed data mining
基于智能体的机制　agent-based mechanism
基于智能体的建模　agent-based modeling
基于智能体的解决方案　agent-based solution
基于智能体的模型　agent-based model
基于智能体的佩特里网　agent-based Petri net
基于智能体的数据挖掘　agent-based data mining
基于智能体的网格计算　agent-based grid computing
基于智能体的应用程序　agent-based application
基于重要性采样的　importance-sampling-based
基于重要性采样的仿真　importance-sampling-based simulation
基于重要性采样的仿真方法　importance sampling-based simulation method; importance-sampling-based simulation technique
基于周期的　cycle-based
基于周期的仿真　cycle-based simulation
基于状态的　state-based
基于状态的描述　state-based description
基于状态的模型　state-based model
基于状态的稳定性　state-based stability
基于状态的系统描述　state-based system description
基于状态的系统模型　state-based system model
基于状态的形式化方法　state-based formal method
基于子整体的模型　holon-based model
基于自动机的　automata-based
基于自动机的技术　automata-based technique

基于自动机的描述　automata-based description
基于组件的　component-based
基于组件的范式　component-based paradigm
基于组件的方法　component-based approach
基于组件的仿真　component-based simulation
基于组件的仿真范式　component-based simulation paradigm
基于组件的仿真系统　component-based simulation system
基于组件的分布式仿真　component-based distributed simulation
基于组件的构架　component-based architecture
基于组件的建模　component-based modeling
基于组件的建模范式　component-based modeling paradigm
基于组件的开发　component-based development
基于组件的协同仿真　component-based collaborated simulation
基于组元的测试　component-based testing
基元　basis; primitive
基站随机变量　base station random variable
基准　benchmark; reference; benchmark (v.)
基准点定位误差　fiducial localization error
基准分析　benchmarking analysis
基准解决方案　baseline solution
基准模型　benchmark model
基准问题　benchmark problem
基准误差　fiducial error
基准值　base value
基准注册错误　fiducial registration error
基座　base
激发　fire; firing
激发状态　motivational state
激活　activate (v.)
激活变量　activation variable
激活的规则　enabled rule
激活的模型　activated model
激活模型　activation model
激活能量　activation energy
激活损失误差　loss-of-activation error
激活值　activation value
激活准则　activation criterion
激励　excitation; stimulation; stimulate (v.)
激励的　stimulant
激励器　stimulator
激励源　stimulus
激情触发　hot emotional trigger
级　stage; level
级间分离　staging
级联耦合　cascade coupling
级联耦合方法　cascade coupling recipe
级联失效　cascaded failure
极点　pole
极端事件　extreme event
极端输入测试　extreme input testing
极化网络　polarized network

极化误差　polarization error
极限　limit value
极限编程　extreme programming
极限规模仿真　extreme scale simulation
极限规模计算　extreme scale computing
极限规模计算机　extreme scale computer
极限环　limit cycle
极坐标　polar coordinate
即插即用的可组合性　plug-and-play composability
即插即用功能　plug-and-play functionality
即时过渡优先规则　immediate transition priority rule
集成　integrate (v.); integration
集成方法　integration method
集成费用　integration cost
集成环境　integrated environment
集成架构　integrated architecture
集成建模　integrated modeling
集成建模和仿真框架　integrated modeling and simulation framework
集成精度　integration accuracy
集成框架　integration framework
集成模型　integrated model
集成数据　integration data
集成算法　integration algorithm
集成误差　integration error
集合论　set theory
集合论模型　set-theoretic model
集群　cluster
集群编队　cluster formation
集群模型　cluster model
集群切片　cluster sectioning
集体行为　collective behavior
集体控制　collective control
集中　concentration
集中的　lumped
集中式仿真　center-based simulation
集中式模型　convergence model
集中式系统　centralized system
集中式协作　centralized coordination
集中式自组织系统　centralized self-organizing system
集中调度机制　centralized scheduling mechanism
集中同步的方法　centralized synchronous approach
集中注意　focused attention
集总　lumped
集总参数模型　lumped-parameter model
集总离散事件系统规范模型　lumped DEVS model
集总连续时间马尔可夫链　lumped continuous-time Markov chain
集总模型　lumped model
几何参数　geometric parameter
几何分布　geometric distribution
几何建模　geometric modeling
几何模型　geometric model
几何随机变量　geometric random variable
几何形状建模　geometric shape modeling
几何学　geometry
几何学与运动学离散事件系统规范

GK-DEVS
计划　plan; project; scheduling; scheme; schema; schedule (v.)
计划行为　planned behavior
计划事件　scheduled event; scheduling an event
计划外情境　unplanned context
计量经济模型　econometric model
计算　computing; calculation; computation; reckoning; calculate (v.)
计算闭包　computational closure
计算不稳定性　computational instability
计算出来的　calculated
计算的　computational
计算的复杂性　computational complexity
计算的因果关系　computational causality
计算发现　computational discovery
计算方法　computation method; computational method
计算仿真　computational simulation
计算过程　computational process
计算互操作性　computational interoperability
计算活动　computational activity
计算机　computer
计算机处理的模型校核　computerized model-verification
计算机处理速率　computer processing speed
计算机仿真　computer simulation; in silico simulation
计算机仿真博弈　computer simulation game
计算机辅助博弈　computer-assisted game
计算机辅助的　computer-aided; computer-assisted
计算机辅助的仿真　computer-assisted simulation
计算机辅助的建模　computer-aided modeling
计算机辅助的模型　computer-assisted model
计算机辅助的模型校核　computer-aided model verification
计算机辅助的校核、验证与确认　computer-aided VV&A
计算机辅助方法学　computer-assisted methodology
计算机辅助仿真　computer-aided simulation
计算机辅助仿真证明　computer-assisted simulation proof
计算机辅助理解　computer-assisted understanding
计算机辅助设计　computer-aided design
计算机辅助实验　computer-aided experimentation; computer-aided experiment
计算机辅助训练　computer-aided training
计算机辅助用途　computer-assisted use
计算机观测　in silico observation
计算机管理对策　computerized

management game
计算机化　computerization
计算机化的仿真　computerized simulation
计算机化的模型　computerized model
计算机化文件　computerized documentation
计算机化误差　computerization error
计算机介导仿真　computer-mediated simulation
计算机可处理的　computer processable
计算机可处理的模型　computer-processable model
计算机可读的模型　computer-readable model
计算机可解释的计算模型　computer-interpretable computational model
计算机可解释的模型　computer-interpretable model
计算机模拟　in silico analog
计算机模型　computer model; in silico model
计算机驱动的仿真　computer-driven simulacre
计算机软件　computer software
计算机设置　in silico set-up
计算机生成　computer-generated
计算机生成的影像　computer-generated imagery
计算机实验　in silico experiment; in silico experimentation
计算机实验设置　in silico experimental set-up
计算机图形学　computer graphics

计算机网络　computer network
计算机网络仿真　computer-network simulation
计算机误差　computer error
计算机硬件　computer hardware
计算机游戏　computerized game; computer game
计算机游戏开发　computer game development
计算机游戏控制　computer game control
计算机游戏设计　computer game design
计算机诱导性仿真　captologic simulation
计算机诱导性研究　captology
计算机战争游戏　computer war game
计算机资源　computer resource
计算几何学　computational geometry
计算建模　computational modeling
计算结果　computational result
计算可复制性　computational replicability
计算可行模型　computationally feasible model
计算联邦　computational federation
计算逻辑　computational logic
计算密集的　computation-intensive
计算模型　computational model
计算器　evaluator
计算强加　computationally imposed
计算强加的简化　computationally imposed simplification
计算时间　computation time
计算时间间隔　calculation interval

计算实验 computational experiment; computational experimentation
计算数值 computed value
计算稳定性 computational stability
计算误差 computational error; calculation error
计算系统 computational system
计算校核 calculation verification
计算涌现 computational emergence
计算值 calculated value
计算智能 computational intelligence
计算智能体 computational agent
计算资源 computational resource
记号 notation
记录误差 clerical error
记忆变量 memory variable
记忆模型 memory model
记忆状态模型 memory state model
技能 skill
技能增强 skill enhancing
技术 technique; technology
技术的基础设施 technical infrastructure
技术独立的模型 technology-independent model
技术仿真 technical simulation
技术分析 technology analysis
技术风险 technical risk
技术规范 specification
技术互操作级别 technical interoperability level
技术互操作性 technical interoperability
技术模型属性 technological model attribute

技术奇点 technological singularity
技术上过时的模型 technologically obsolete model
技术属性 technological attribute
技术数据 technical data
技术需求 technical requirements; technology requirements
技术有效性 technical validity
技术增强的 technology-enhanced
技术增强型仿真 technology-enhanced simulation
继承 inherit (v.); inheritance
继承的误差 inherited error
寄存器 register
寄存器错误 register error
寄生行为 parasitic behavior
加恩-萨尔松数据流建模 Gane-Sarson data flow modeling
加工 processing
加工过的 processed
加密 encryption; encrypt (v.)
加密错误 encryption error
加密的 encrypted
加密数据 encryption data
加权平均 weighted average
加权值 weighted value
加热 warm up (v.)
加热的 warm up
加速 speedup
加速度误差 acceleration error
加速仿真 accelerated simulation
加载 load (v.)
加载误差 loading error
家族 family
假的 false

假定的行为　presumed behavior
假定目标　assumed goal
假定值　assumed value
假警报错误　false alarm error
假设　assumption; hypothesis; postulate; hypothesize (v.)
假设的　assumed
假设的参照物　referent of an assumption
假设的范围　scope of an assumption
假设的命题　proposition of an assumption
假设分析　what-if analysis
假设函数　assumption function
假设检验　hypothesis test; hypothesis testing
假设空间　hypotheses space
假设命题　hypothetical proposition
假设事件　hypothetical event
假设误差　assumption error; hypothesis error
假设系统　hypothetical system
假设研究　what-if study
假输入　fake input
假说有效性　hypothesis validity
假现实　fake reality
假消息　fake news
假装　pretend (v.); pretending
价值　merit; value
价值度量　measure of merit
价值无关的决策　value-free decision
价值系统　value system
价值系统的合规评估　ethical assessment of value system
驾驶仿真器　driving simulator
驾驶员训练　driver training
架构　architecture; schema
架构开发　architectural development
架构效率　architectural efficiency
间断变化的变量　discontinuous-change variable
间断生成　intermittent generation
间断性处理算法　discontinuity-processing algorithm
间断性定位　discontinuity localization
间隔　interval
间接测量技术　indirect measurement technique
间接的理解　mediated understanding
间接的通信　indirect communication
间接仿真　indirect simulation
间接建模　mediated modeling
间接交互　indirect interaction
间接理解　indirect understanding
间接耦合　indirect coupling
间接耦合的　indirectly coupled
间接输入　indirect input
间接输入设备　indirect input device
间接协调　indirect coordination
间接有向图模型　mediated digraph model
间隙　gap
间歇的　intermittent
间歇式仿真　intermittent simulation
间歇式仿真语言　intermittent SL
监测　monitoring
监测数据　monitoring data
监控变量　monitored variable
监控程序　monitoring program
监控的　monitored

监控软件　monitor software
监视器　monitor
监听　auditory monitoring
兼容的　compatible
兼容的接口　compatible interface
兼容形式　compatible formalism
兼容性　compatibility
兼容性标准　compatibility standard
兼容性测定　compatibility determination
兼容性规范　compatible specification
兼容性模型　compatibility model
检测效率　detection efficiency
检查　inspection; check; review; audit (*v.*); check (*v.*); review (*v.*)
检查点　checkpoint
检查过的　checked
检查算法　check algorithm
检定　verification
检偏器　analyzer
检索　retrieval
检索错误　retrieving error
检验　checking; verified; test; inspect (*v.*); verify (*v.*)
检验器　checker
检验者　verifier
减缓时间　slow time
剪裁　tailor (*v.*); tailoring
剪枝　pruning
简并测试　degeneracy test
简单　simplicity
简单变量　simple variable
简单博弈　simple game
简单故障　simple failure
简单行为　simple behavior

简单克里格　simple Kriging
简单输出　simple output
简单应急行为　simple emergent behavior
简单涌现　simple emergence
简化　reduction; simplification; simplify (*v.*)
简化（还原）论者　reductionist
简化抽象　simplifying abstraction
简化的节点集　reduced nodeset
简化的模型　reduced model; simplified model
简化方法学　simplification methodology
简化复杂性的模型　reduced-complexity model
简化过程　simplification procedure
简化假设　simplifying assumption
简化论　reductionism
简化论范式　reductionism paradigm
简化误差　simplification error
简化系统　reduced system
简化原则　principle of simplicity
简明的时间模拟　condensed-time simulation
见解　point of view
建成　activate (*v.*)
建档　documenting
建构性训练　constructive training
建立　construct (*v.*); set-up
建模　model (*v.*); modeling
建模、仿真和可视化　modeling, simulation and visualization
建模本体　modeling ontology
建模标准　modeling standard

建模层　modeling layer
建模场景　modeling scenario
建模代理　modeling agent
建模的创造性方面　creative aspects of modeling
建模的重复方面　repetitive aspects of modeling
建模动力学　modeled dynamics
建模范式　modeling paradigm
建模范式的适用性　suitability of a paradigm for modeling
建模方法　modeling approach; modeling method
建模方法基础　basis for modeling method
建模方法学　modeling methodology
建模方法学的适用性　suitability of modeling methodology
建模方法学规范　norms of modeling methodology
建模方法学评估　assessment of modeling methodology
建模仿真和可视化　MS&V (Modeling Simulation and Visualization)
建模概念　modeling concept
建模工具　modeling tool
建模工具可用性　modeling tool usability
建模工具评价　modeling tool evaluation
建模工具设计　modeling tool design
建模工具实现　modeling tool implementation
建模工具有效性　modeling tool validity
建模顾问　modeling consultant
建模关系　modeling relation; modeling relationship
建模惯例　modeling convention
建模过程　modeling process
建模函数　modeling function
建模环境　modeling environment
建模环境支持　modeling environment support
建模活动　modeling activity
建模基础设施　modeling infrastructure
建模技术　modeling technique
建模假设　modeling assumption
建模阶段　modeling stage
建模结构　modeling structure
建模界面　modeling interface
建模框架　modeling framework
建模理论　modeling theory
建模能力　modeling capability
建模器　modeler
建模器风险　modeler's risk
建模器评估　assessment of modeler
建模软件　modeling software
建模实践　modeling practice
建模视角　modeling perspective
建模术语　modeling term
建模素养　modeling literacy
建模算法　modeling algorithm
建模挑战　modeling challenge
建模完备性　modeling maturity
建模问题　modeling issue
建模误差　modeling error
建模系统　modeling system
建模系统结构　modeling system structure

建模形式化　modeling formalism
建模应用　modeling application
建模与仿真　modeling and simulation; M&S (Modeling and Simulation)
建模与仿真本体　M&S ontology
建模与仿真程序管理器　M&S program manager
建模与仿真从业者　M&S practitioner
建模与仿真从业者职业生涯　M&S practitioner career
建模与仿真的逼真度　M&S fidelity
建模与仿真的互操作性　M&S interoperability
建模与仿真的可重复性　M&S reproducibility
建模与仿真的历史回顾　historical overview of M&S
建模与仿真的美学计算　aesthetic computing for M&S
建模与仿真的认识论　epistemology of M&S
建模与仿真的使用　use of M&S
建模与仿真的延展性　M&S scalability
建模与仿真的知识体系　M&S body of knowledge
建模与仿真范畴　M&S category
建模与仿真范式　M&S paradigm
建模与仿真分辨率　M&S resolution
建模与仿真服务　M&S service
建模与仿真辅助工具　M&S adjunct tool
建模与仿真工程师　M&S engineer
建模与仿真工具　M&S tool
建模与仿真工作组　M&S working group
建模与仿真共享软件　M&S shareware
建模与仿真规划　M&S planning
建模与仿真过程　M&S process
建模与仿真基础设施　M&S infrastructure
建模与仿真基础训练课程　M&S basic training course
建模与仿真级别的可验证性　level of M&S validatability
建模与仿真结构　M&S architecture
建模与仿真开发工具　M&S development tool
建模与仿真开发者　M&S developer
建模与仿真课程　M&S course；M&S curriculum
建模与仿真库　M&S repository
建模与仿真框架　M&S framework
建模与仿真理论　M&S theory
建模与仿真历史记录　M&S history
建模与仿真联邦　M&S federation
建模与仿真能力　M&S capability
建模与仿真群　M&S group
建模与仿真认证　M&S accreditation
建模与仿真软件设计　M&S software design
建模与仿真设备　M&S facility
建模与仿真生命周期　M&S life cycle
建模与仿真生命周期管理　M&S life cycle management
建模与仿真使用的数据　data consumed by M&S
建模与仿真市场　M&S market
建模与仿真提议者　M&S proponent

建模与仿真信息源　M&S information source
建模与仿真需求　M&S requirements
建模与仿真应用　M&S application
建模与仿真应用出资者　M&S application sponsor
建模与仿真应用质量　M&S application quality
建模与仿真用户　M&S user
建模与仿真原理　M&S principle
建模与仿真赞助商　M&S sponsor
建模与仿真知识体系　body of knowledge of modeling and simulation; M&S BOK (M&S Body of Knowledge)
建模与仿真执行代理　M&S executive agent
建模与仿真执行委员会　executive council for M&S
建模与仿真中的做与不做　dos & don'ts in M&S
建模与仿真主计划　M&S master plan
建模与仿真专家　M&S professional
建模与仿真总体规划　MSMP (Modeling and Simulation Master Plan)
建模与仿真组织　M&S organization
建模语言　modeling language
建模语言的适用性　suitability of a language for modeling
建模语言的语义　semantics of modeling language
建模语言离散事件系统规范　ml-DEVS
建模语言语法　syntax of modeling language

建模语义　modeling semantics
建模域　modeling domain
建模元语　modeling primitive
建模原理　modeling principle
建模约束　modeling constraint
建模支持　modeling support
建模中的白箱法　white box approach in modeling
建模中的不确定性　uncertainty in modeling
建模专家　modeling expert
建模自然环境　modeling natural environment
建设性的情绪行为　constructive emotional behavior
建议　proposition
建筑物　building
建筑学　architecture
渐进模型拟合　progressive model fitting
渐进式建模　approximation in modeling
渐近标准误差　asymptotic standard error
渐近稳定的系统　asymptotically stable system
渐近稳定性　asymptotic stability
渐近有效性　gradual validity
鉴别　authenticate (v.); authentication
鉴定　appraisal
鉴定价值　appraised value
鉴定器　assessor
键合图　bond graph
键合图仿真　bond graph simulation
键合图建模　bond graph modeling

键合图模型　bond graph model
键合图因果关系　bond graph causalization
键盘仿真器　keyboard emulator; keyboard simulator
键盘输入　keyboard input
将来的行为　future behavior
将来值　future value
降阶　order reduction; reduced order
降阶的模型　reduced-order model
交叉　crossing
交叉变量　across variable
交叉核实　cross validation
交叉模型比较　cross model comparison
交叉模型验证　cross model validation
交叉模型有效性　cross model validity
交叉实体　intersection entity
交叉网络通信　cross-network communication
交叉验证过程　cross-validation procedure
交叉有效性　cross validity
交付　delivery
交付模型　delivered model
交互　interaction
交互变量　interaction variable
交互参数　interaction parameter
交互代理　interacting agent
交互的模型　interactive model
交互点　interaction point
交互方式　interaction style
交互仿真　mutual simulation
交互仿真数据　intersimulation data
交互复杂性　interactional complexity
交互复杂性测量　interactive complexity measure
交互环境　interactive environment
交互绘图　interactive plot
交互简化　interactional simplicity
交互建模　interaction modeling
交互模型　interaction model
交互能力　interaction capability; interactive capability
交互能力一体化　cross functional integration
交互实体　interacting entity
交互示范　interactive demonstration
交互式帮助　interactive help
交互式的　interactive
交互式仿真　interaction-based simulation; interactive simulation
交互式仿真博弈　interactive simulation game
交互式仿真环境　interactive simulation environment
交互式仿真模型　interactive simulation model
交互式仿真语言　interactive SL
交互式计算　interactive computation
交互式建模　interactive modeling
交互式可视化　interactive visualization
交互式体验　interactive experience
交互式图形化仿真　interactive graphical simulation
交互式系统　interaction-based system; interactive system
交互式虚拟现实　interactive virtual reality

交互式训练　interactive training
交互式意图建模　interactive intent modeling
交互式游戏　interactive game
交互式转换系统　interactive transition system
交互图　interaction graph
交互现实　interactive reality
交互性的行为　interactive behavior
交互性结构　interaction structure
交互一致性　interaction coherence
交互影响分析　cross impact analysis
交互作用　inter-action
交换　exchange; switching; swap (*v.*); exchange (*v.*); interchange
交换标准　interchange standard
交换方法　swapping method
交换模型　interchange model
交换数据　exchange data
交替　alternate (*v.*)
交替的　alternate
交替状态　alternating state
交战　warfare
交战训练　engagement training
角度测量误差　angular error
角色　actor
角色扮演　role-playing
角色扮演仿真　role playing simulation; role-play simulation
角色扮演游戏　role-playing game
角色程序设计模型　actor-programming model
角色模型　actor model
矫顽磁力仿真　coercivity simulation
脚本　script
脚本编写　scripting
脚本测试　scripted testing
脚本的　scripted
脚本化想定　scripted scenario
脚本角色扮演游戏　scripted role-playing game
脚本游戏　scripted game
脚本语言　scripting language
校核　verification; verified; verify (v.)
校核、验证和测试　VV&T (Verification, Validation and Testing)
校核、验证与测试技术　VV&T technique
校核、验证与测试证书文档　documentation of VV&T certification
校核、验证与确认　verification, validation and accreditation（VV&A）
校核、验证与确认过程　verification, validation and accreditation process
校核、验证与确认数据成本　VV&A data cost
校核、验证与确认文档　VV&A documentation
校核、验证与确认相关的　VV&A-related
校核、验证与确认相关的问题　VV&A-related issue
校核、验证与认证　verification, validation and certification（VV&C）
校核操作者　verification operator
校核代理　verification agent
校核和验证范式　verification and validation paradigm
校核和验证计划　verification and validation plan

校核计划　verification plan
校核技术　verification technique
校核评估　verification assessment
校核算法　verification algorithm
校核误差　verification error
校核系统　verification system
校核性度量　verification metric
校核需求　verification requirement
校核与验证　V&V（Verification and Validation）
校核与验证提议者　verification and validation proponent
校核准则　verification criterion
教练　training
教练机　trainer
教学　teaching
教学仿真　pedagogical simulation
教学模拟　instructional simulation
教学系统　instructional system
教学系统设计　instructional system design
校验　checking
教育　educate (*v.*); education
教育的　educational; educative; edutainment
教育方法　education technique
教育仿真　simulation for education
教育游戏　educational game
校正　adjustment; calibrated; correction
校正错误问题的误差　error of solving wrong problem
校正后数据　adjusted data
校正误差　rectification error
校正值　adjusted value
校正子　corrector
校准　calibration; calibrating; calibrate (*v.*)
校准的仿真模型　calibrated simulation model
校准模型　calibration model; calibrated model
校准数据　calibrated data; calibration data
校准误差　calibration error
阶　step
阶段　stage; phase
阶段的组合　staged composition
阶跃变化　step-change
阶跃函数　step function
阶跃输入　step input
接触镜显示　contact lens display
接合　linkage
接近　access; approach; approximate (*v.*)
接口　interface
接口测试　interface testing
接口层　interface layer
接口代理　interface agent
接口定义　interface definition
接口功能　interface functionality
接口规范　interface specification
接口规范语言　interface specification language
接口技术　interface technique; interfacing; interface technology
接口模块　interface module
接口属性　interface attribute
接口问题　interface issue
接口组件　interface component
接入可靠性　access reliability

接收操作　receive operation
接收到的操作　received operation
接收器　acceptor; receptor
接　受　accept (*v.*); acceptance; accepted; accepting
接受的数据　accepted data
接受的状态　accepted state
接受概率　acceptance probability; probability of acceptance
接受者　acceptor
接受值　accepted value
接受状态　accepting state
街机游戏　arcade game
节点　node
结构　structure
结构比较　structural comparison
结构变化　structural-change
结构辨识　structure identification
结构辨识方法学　structure identification methodology
结构不连续性　structural discontinuity
结构不确定性　structural uncertainty
结构的对比　comparison of structure
结构方程建模　structural equation modeling
结构方程模型　structural equation model
结构仿真　structure simulation
结构仿真语言　structural SL
结构分析　structural analysis; structure analysis
结构复杂性　structural complexity
结构关联矩阵　structure incidence matrix
结构化测试　structural testing
结构化的　structured
结构化方法学　structured methodology
结构化仿真知识　structured simulation knowledge
结构化集合　structured set
结构化技术　structural technique
结构化建模　structured modeling
结构化可还原性　structured reducibility
结构化理解　structural understanding
结构化实体　structured entity
结构化数据　structured data
结构化文档　structured documentation
结构化系统　structured system
结构化信息　structured information
结构化预演　structured walkthrough
结构建模　structure modeling
结构可信度　structure credibility
结构论域　structural domain
结构模型　structural model
结构模型比较　structural model comparison
结构模型有效性　structural model validity
结构耦合　structural coupling
结构耦合的　structurally coupled
结构奇异模型　structurally singular model
结构奇异系统　structurally singular system
结构奇异性　structural singularity
结构奇异性消除　structural singularity elimination
结构属性　structural property

结构条件　structural condition
结构推论　structural inference
结构完全的模型　richly structured model
结构稳定性　structural stability
结构误差　architecture error
结构相似性　structural similarity
结构性能　structural behavior
结构验证　structural validation
结构因素　structural factor
结构涌现　structural emergence
结构有向图　structure digraph
结构有效的模型　structurally valid model
结构有效性　structural validity
结构转换函数　structure-transition function
结果　consequent; resultant; outcome; result; effect
结果变量　outcome variable
结果度量　outcome metric
结果缓存　result caching
结果模型　resultant model
结果驱动的　outcome-driven
结果驱动的仿真　outcome-driven simulation
结果上的耦合　resultant coupling
结果调查　result investigation
结果文档　documentation of result
结果系统　resulting system
结果演示　presentation of result
结果验证　result validation
结论　consequence
结束状态　ending state
截断　truncation

截断变量　truncated variable
截断分布　truncated distribution
截断随机量　truncated random variable
截断误差　truncation error
截断因变量　truncated dependent variable
截断正态分布　truncated normal distribution
解的唯一性　uniqueness of solution
解的稳定性　solution stability
解集合　solution set
解聚　disaggregate; disaggregation; disaggregated; disaggregate (*v.*)
解决方案　solution
解决方案的延展性　solution scalability
解决方案文档　solution documentation
解决与验证管理　resolution and validation management
解决与验证管理工具　resolution and validation management tool
解空间　solution space
解耦　decoupling; uncoupling; uncouple (*v.*); decouple (*v.*)
解耦的　decoupled
解耦仿真　uncoupled simulation
解耦合的　uncoupled
解耦模型　decoupled model
解剖学标记语言　anatomical markup language
解释　interpreted; explanation; interpret (*v.*); explain (*v.*)
解释变量　explained variable; explanatory variable
解释的　explanatory; hermeneutic

解释仿真　interpretational simulation
解释紧急行为　explained emergent behavior
解释模型　interpretation model; interpreted model
解释误差　interpretation error
解释型仿真　interpretive simulation
解释型仿真语言　interpretive SL
解释性仿真　explanatory simulation
解释性模型　explanatory model
解释学　hermeneutics
解释主义　interpretivism
解释主义范式　interpretivist paradigm
解释主义方法论　interpretivist methodology
解释主义理论　interpretivist theory
解释主义者　interpretivist
解析　analysis; parse
解析的　analytic
解析的解决方案　analytical solution
解析的稳定性　analytical stability
解析建模　analytical modeling
解析解的模型　analytically solvable model
解析模型　analytic model
解析器　parser
解析算法　analytic algorithm
解析稳定解　analytically stable solution
解析域稳定性　domain of analytical stability
介观模型　mesoscopic model
介质　media
界面　interface
界面测验　interface testing
界面分析　interface analysis
界面模块　interface module
界限　bound
紧急出口仿真　emergency egress simulation
紧急功能　emergent function
紧急功能性　emergent functionality
紧急输出　emergent output
紧急输出操作　emergent output function
紧急输入　emergent input
紧急物资　emergent substance
紧急性宏观行为　emergent macro-behavior
紧急转换　emergent transition
紧急转换函数　emergent transition function
紧急状况　emergent condition
紧急状态　emergent state
紧耦合　tight coupling
紧耦合的　tightly coupled
紧耦合的模型　tightly coupled model
紧耦合的系统　tightly coupled system
进场　approach
进程　process
进程交互模型　process interaction model
进度　advance
进行实验　experimenting
进行试验　experiment (v.)
进化　evolution; evolve (v.)
进化博弈　evolutionary game
进化的现实　evolutionary reality
进化多模型　evolutionary multimodel
进化方法　evolutionary method

进化仿真　evolutionary simulation
进化仿真算法　evolutionary-simulation algorithm
进化计算　evolutionary computation
进化解　evolutionary solution
进化模型　evolutionary model
进化数据　evolutionary data
进化算法　evolutionary algorithm
进化维护　evolutionary maintenance
进化系统　evolutionary system
进化系统仿真　evolutionary-system simulation
进化验证　evolutionary validation
进化涌现　evolutionary emergence
近实时　near real-time
近似　approximate (v.)
近似变量　approximate variable
近似的　approximate
近似仿真　approximate simulation
近似离散事件系统规范同型　approximate DEVS morphism
近似零方差仿真　approximate zero-variance simulation
近似同型　approximate morphism
近似推理模型　approximate reasoning model
近似误差　approximation error
近似系统　approximate system; proximate system
近似正确　approximately correct
近似值　approximate value; approximated value
近似值正确性　approximation correctness
近似最优　near-optimal

近因　proximate cause
晋升　advancement
浸入式环境　immersive environment
禁忌搜索　tabu search
禁忌搜索方法　tabu search method
禁止跃迁　forbidden transition
经典仿真　classical simulation
经典一般平衡模型　classical general equilibrium model
经过　passing
经过的　passing
经过审核的模型　vetted model
经过校核的模型　verified model
经过校核的系统　verified system
经过训练学到的行为　learned behavior
经过验证的模型　validated model
经过验证的系统　validated system
经历过的　experienced
经验　experience
经验变量　experiential variable
经验参数　experience parameter
经验测量　empirical measure
经验成果　empirical consequence
经验导向的　experience-directed
经验的　empirical; experiential
经验的概念　experiential concept
经验的认证　empirical confirmation
经验分布　empirical distribution
经验分析　empirical analysis
经验工具　experiential tool
经验管理　experience management
经验建模　empirical modeling
经验建模工具　empirical modeling tool

经验建模算法　empirical modeling algorithm
经验建模应用　empirical modeling application
经验建模语言　empirical modeling language
经验建模原理　empirical modeling principle
经验开放性　openness to experience
经验库　experience base
经验论　empiricism
经验论者　experientialist
经验模型　empirical model; experience model; experiential model
经验模型开发　empirical model development
经验模型设计　empirical model design
经验模型验证　empirical model validation
经验模型有效性　empirical model validity
经验评价　empirical evaluation
经验上地　experientially
经验上下文　experiential context
经验试验　empirical experiment
经验数据　empirical data
经验推导的　empirically derived
经验相关的参数　experience-related parameter
经验性研究　empirical study
经验学习　experiential learning
经验研究　experiential study
经验依赖　empirical dependability
经验源　empirical source
经验约束　experience constraint

经验知识　empirical knowledge; experiential knowledge
经验主义　experientialism
经验主义的　empiric
经营　management
经证明的假设　justified assumption
晶片　postsilicon
精度　accuracy; precision
精度代价　accuracy cost
精度级别　level of accuracy
精化　refinement
精确　exactness
精确参数　accurate parameter
精确测量　accurate measurement
精确的　accurate; exact; precise
精确的模型　exact model
精确的耦合　precise coupling
精确仿真　accurate simulation
精确建模　accurate modeling
精确解　accurate solution
精确解法　accurate solution
精确论域　accuracy domain
精确模型　accurate model
精确性　veracity
精确值　precise value
精神仿真　mental simulation
精细性　tractability
精心制作　elaboration
精益沙盘模拟　lean simulation
精准离散化方案　accurate discretization scheme
景观模型　landscape model
竞技行为　competitive behavior
竞技游戏　competitive game; contest game

竞争　competition
竞争博弈　competition game
竞争仿真　competition simulation
竞争关系　competitive relationship
竞争模拟　competitive simulation
竞争模拟博弈　competitive simulation game
竞争模型　competitive model
竞争系统　competitive system
竞争性仿真　competitive simulation
竞争性关系　competitive relationship
竞争性行为　competitive behavior
竞争学习模型　competitive learning model
静态　quiescent state
静态变量　static variable
静态博弈　static game
静态程序检查　static program check
静态的　static
静态的行为　static behavior
静态的检查　static check
静态仿真　static simulation
静态分析　static analysis
静态关系　static relationship
静态技术　static technique
静态假设　static assumption
静态建模形式化　static modeling formalism
静态结构　static structure
静态结构多模型　static-structure multimodel
静态结构模型　static-structure model
静态解耦　static decoupling
静态模型　static model
静态模型结构　static model structure

静态模型结构的适当性　adequacy of static model structure
静态模型文档　static model documentation
静态耦合　static coupling
静态耦合的多模型　statically coupled multimodel
静态实境　static reality
静态输出　static output
静态数据　static data
静态网络　static network
静态微观仿真模型　static microsimulation model
静态误差　static error
静态校核、验证与测试技术　static VV&T technique
静态值　static value
静态作用域变量　statically scoped variable
静止的　stationary
镜像仿真　mirror simulation
镜像世界仿真　mirror world simulation
纠缠　entanglement
纠缠态　entangled state
纠错代码　error correcting code
纠错分析　error-correcting analysis
纠正　correcting
旧的　used
就地观测　in situ observation
居中的　centered
局部包容性　local containment
局部变量　local variable
局部的　local
局部的复杂性　local complexity

局部仿真　partial simulation
局部化　localization
局部积分误差　local integration error
局部精度　local accuracy
局部均衡仿真　partial equilibrium simulation
局部均衡模型　partial equilibrium model
局部连接　subcoupling
局部量子变量　local quantum variable
局部模型　local model; partial model
局部时　local time
局部时间表　local schedular
局部特性　local property
局部稳定性　local stability
局部误差　local error
局部误差容限　local error tolerance
局部现实主义　local realism
局部相对精度　local relative accuracy
局部信息　local information
局部隐含变量　local hidden variable
局部知识库　local knowledge base
局部值　local value
局部转移函数　local transition function
局限性　limitation
局域网　local area network
矩封闭　moment closure
矩封闭技术　moment-closure technique
矩封闭近似　moment-closure approximation
矩阵　matrix
矩阵的逆　matrix inversion
矩阵特征值　eigenvalue matrix

矩阵值变量　matrix-valued variable
句法　syntax
句法互操作性　syntactic interoperability
句法互操作性水平　syntactic interoperability level
句法可组合性　syntactic composability
句法评价　syntactic evaluation
句法突现行为　syntactic emergent behavior
句法一致性　syntactic conformance
句子变量　sentential variable
拒绝　rejection
拒绝错误　rejection error
拒绝概率　rejection probability
具体的　visualized
具体的模型　concrete model
具体的原语　concrete primitive
具体数据　specific data
具体值　concrete value
具有知识属性的佩特里网　knowledge-attributed Petri net
距离变换　distance transformation
聚合的模型　aggregated model
聚合的数据　aggregated data
聚合级仿真　aggregate-level simulation
聚合级仿真协议　aggregate-level simulation protocol
聚合级仿真协议兼容的　ALSP compatible
聚合级仿真协议数据　ALSP data
聚合级仿真协议相容的　ALSP compliant

聚合模型　aggregate model
聚合形式建模　aggregated modeling
聚集　aggregation
聚集层次　level of aggregation
聚类　clustering
聚类算法　clustering algorithm
聚类状态　clustering state
决策　decision; decision making
决策变量　decision variable
决策变量空间　decision variable space
决策仿真　decision simulation; simulation for decision making
决策技能增强　decision making skill enhancing
决策技术　decision skill
决策模型　decision model
决策树模型　decision-tree model
决策误差　decision error
决策者　decision maker
决策者选择评估　assessment of decision maker's alternatives
决策支持　decision support
决策支持仿真　simulation for decision support
决策准则　decision criterion
决定　decision
决定了的　determined
决定论　determinism
决定性的变量　deterministic variable
决胜函数　tie-breaking function
决心　determination
决议　resolution
决议管理　resolution management
决议细化　resolution refinement
绝对测量误差　absolute measurement-error
绝对事实　absolute truth
绝对稳定性（A 稳定性）　absolute stability
绝对误差　absolute error
绝对误差控制　absolute error control
绝对现实　absolute reality
绝对有效性　absolute validity
绝对值　absolute value
绝热系统仿真　adiabatic-system simulation
军事博弈　military gaming
军事仿真博弈　military simulation gaming
军事行动　operation
军事演习　wargame
军事游戏　military game
均方根误差　root mean square error
均匀分布　uniform distribution
均匀分布的随机数　uniform random number
均匀随机变量　uniform random variable
均匀系统　homogeneous system

Kk

Keller-Segel 模型　Keller-Segel model
卡方随机变量　Chi-Squared random variable
卡莱曼线性化　Carleman linearization
开断　breaking
开发　development; develop (v.)
开发步骤　development step
开发的模型　developmental model

开发范式　development paradigm
开发方法学　development methodology
开发风险　development risk
开发过程　development process
开发模型　development model
开发生命周期　development life cycle
开发性建模　exploitative modeling
开发者　developer
开放的分级系统　open hierarchical system
开放联邦　open federation
开放式仿真　open-form simulation
开放式模型　open-form model
开放系统　open system
开关　switch
开合　switching
开环　open loop
开环博弈　open loop game
开环仿真　open loop simulation
开环控制系统　open loop control system
开环连接　open loop connection
开火　fire; firing
开始启动阶段　start firing phase
开型博弈　open game
开源　open source
开源编程　open source programming
开源多核功能高速缓存仿真器　open source multicore functional cache memory simulator
开源仿真　open source simulation
开源仿真器　open source simulator
开源模型　open source model
开源软件　open source software
开源游戏技术　open source game technology
考证　research
柯西变量　Cauchy variable
科技视角　technical view
科学　science
科学错误　scientific error
科学性仿真　scientific simulation
科学知识　scientific knowledge
可安装性　installability
可编程分析　programmable analysis
可编译代码　compilable code
可变变量　mutable variable
可变参数系统　variable parameter system
可变的　variable
可变的时间步长　variable time-step
可变的输入　variable input
可变更的目标　modifiable goal
可变联接的耦合　coupling with variable connection
可变模型　changeable model
可变误差　variable error
可变形的模型　deformable model
可变性　variability
可变组件模型的耦合　variable-component-model coupling
可变作用力　effort variable
可辨别的　discernible; identifiable
可辨识系统　identifiable system
可表达性　expressiveness
可采纳的片段　admissible segment
可参考性　referability
可操作的　actionable
可操作的知识　actionable knowledge

可操作数据　actionable data
可操作信息　actionable information
可操作性　operability
可测变量　measurable variable
可测量变量　instrumentable variable
可测量的　measurable
可测量的行为　measured behavior
可测试的　testable
可测试的结果　testable consequence
可测试的应急行为　testable emergent behavior
可测试性　testability
可持续的　sustainable
可持续的仿真　sustainable simulation
可持续的群体　sustainable group
可持续的应急行为　sustainable emergent behavior
可重复的　repeatable; reproducible
可重复仿真　reproducible simulation
可重复环境　repeatable environment
可重复计算实验　reproducible computational experiment
可重复实验　reproducible experiment
可重复性　repeatability
可重复研究　reproducible research
可重构的　reconfigurable
可重构的软件架构　reconfigurable software architecture
可重构仿真　reconfigurable simulation
可重构仿真器　reconfigurable simulator
可重构合成环境　reconfigurable synthetic environment
可重构系统　reconfigurable system
可重构训练　reconfigurable training

可重现的　reproducible
可重现涌现　reproducible emergence
可重用的　reusable
可重用的仿真组件　reusable simulation component
可重用的联邦　reusable federation
可重用的模型　reusable model
可重用的组件　reusable component
可重用仿真库　reusable simulation library
可重用仿真模型　reusable simulation model
可重用性　reusability
可重用性框架　reusability framework
可重用组件方法论　component-reuse methodology
可处理的　processable
可触知的互动　tangible interaction
可穿戴仿真　wearable simulation
可穿戴式计算机　wearable computer
可达树　reachability tree
可达图　reachability graph
可达性分析　reachability analysis
可导出、可微分的性质或状态　derivability
可定制的系统　customizable system
可定制化仿真　customizable simulation
可定制性　customizability
可读屏幕　readable screen
可读性　readability
可仿真的　simulatable
可仿真模型　simulatable model
可访问的　accessible
可访问的万维网　accessible web

可访问模型　accessible model
可访问数据　accessible data
可访问性　accessibility
可访问性错误　accessibility error
可分解的　decomposable
可分解的活动　decomposable activity
可分解对策　decomposable game
可分解性　decomposability
可复制的实验结果　replicable experimental result
可复制结果　replicable result
可复制模型　replicable model
可复制性　replicability
可复制性感知　replicability-aware
可复制性感知建模与仿真　replicability-aware M&S
可复制性建模　replicability-aware modeling
可覆盖树　coverability tree
可感知的　perceivable
可跟踪的仿真　tractable simulation
可更改的　modifiable
可更新的分段连续模型　updatable piece-wise continuous model
可更新的过程模型　updatable process model
可更新的离散变化模型　updatable discrete-change model
可更新的离散模型　updatable discrete model
可更新的连续变化模型　updatable continuous-change model
可更新的连续模型　updatable continuous model
可更新的模型　updatable model

可更新的事件模型　updatable event model
可更新的无记忆模型　updatable memoryless model
可共享性　shareability
可供选择的模型　alternative model
可供选择的输入　alternative input
可观测变量　observable variable
可观测的　observable
可观测行为　observable behavior
可观测事件　observable event
可观测系统　observable system
可观测性　observability
可观测装置　observable plant
可还原性　reducibility
可行的　feasible
可行的数值　feasible value
可行的现实　feasible reality
可行性　feasibility
可行性研究　feasibility study
可行状态　feasible state
可合成想定　synthesizable scenario
可核实的　verifiable
可核实的子类　verifiable subclass
可忽略误差　negligible error
可互操作的　interoperable
可互操作的博弈　interoperable game; interoperable gaming
可互操作的仿真　interoperable simulation
可互操作的仿真环境　interoperable simulation environment
可互操作的仿真模型　interoperable simulation model
可互操作的联邦　interoperable

federation
可互操作的模型　interoperable model
可互操作的战争游戏　interoperable war game；interoperable war gaming
可恢复的　recoverable
可恢复的误差　recoverable error
可恢复性　recoverability
可及　accessible
可计算变迁　computable transition
可计算的　computable
可计算模型　computable model
可计算性　computability
可检测到的错误　detectable failure
可检验的　verifiable
可检验性　testability
可简化的　reducible
可简化的多模型　reducible multimodel
可简化的模型　simplifiable model
可简化性　reducibility
可简约的不确定性　reducible uncertainty
可接近的　accessible
可接受的　acceptable
可接受的有效范围　acceptable validity range
可接受数据　acceptable data
可接受性　acceptability
可接受性参数　acceptability parameter
可接受性评估　acceptability assessment
可接受性阈值　acceptability threshold
可接受性准则　criterion for acceptability
可接受值　acceptable value

可接受准则　acceptability criterion
可解性　solvability
可解性问题　solvability issue
可进化的　evolvable
可进化的元模型　evolvable metamodel
可进化模型　evolvable model
可进化系统　evolvable system
可进化性　evolvability
可决定的　decidable
可靠的　dependable；reliable
可靠的逼近　reliable approximation
可靠的产品　reliable product
可靠的仿真　reliable simulation；trustworthy simulation
可靠的服务　reliable service
可靠的模型　reliable model
可靠的模型库　reliable model repository
可靠性　dependability；reliability
可靠性分析　reliability analysis
可靠性建模　reliability modeling
可靠性模型　reliability model
可靠性评价　dependability evaluation；reliability evaluation
可靠性认识论　reliabilist epistemology
可靠性问题　reliability issue
可靠性预测　reliability prediction
可靠者　reliabilist
可靠主义　reliabilism
可控的　controllable
可控的假设　controllable assumption
可控的组件　controllable component
可控系统　controllable system
可控性　controllability

可控因素　controllable factor
可控制的变量　controllable variable
可控制的输入变量　controllable input variable
可控装置　controllable plant
可扩展的　scalable
可扩展的解决方案　scalable solution
可扩展多模型　extensible multimodel
可扩展仿真　extensible simulation; scalable simulation
可扩展仿真的基本结构　extensible simulation-infrastructure
可扩展分析　scalable analysis
可扩展建模　scalable modeling
可扩展建模与仿真框架　extensible M&S framework
可扩展框架　extensible framework
可扩展联邦　extensible federation
可扩展模型　extendable model; extensible model; scalable model
可扩展数据　scalable data
可扩展性　extensibility; scalability
可扩展性约束　scalability constraint
可扩展有限状态机　extendable finite-state-machine; extensible finite-state-machine
可扩展有限状态模型　extendable finite-state-model
可理解的模型　comprehensible model
可理解的输入　comprehensible input
可理解性　comprehensibility; intelligibility; understandability
可连接的代码　linkable code
可量化测度　quantifiable measure
可量化关系　quantifiable relationship

可能的　probable; probability
可配置的模型　configurable model
可配置性　configurability
可评估的仿真　evaluative simulation
可强制的　coercible
可强制仿真　coercible simulation
可切换的理解　switchable understanding
可确认的　certifiable
可容许的　admissible
可实施的设计空间　implementable design space
可实施设计　implementable design
可实现性　implementability
可视化　visualization; visualize (v.)
可视化编程环境　visual programming environment
可视化参数　visualization parameter
可视化的　visualized
可视化仿真　visual simulation
可视化规范　visual specification
可视化环境　visual environment
可视化技术　visualization technique; visualization technology
可视化建模　visual modeling
可视化建模语言　visual modeling language
可视化交互仿真　visual interactive simulation
可视化交互仿真环境　visual interactive simulation environment
可视化交互环境　visual interactive environment
可视化交互建模　visual interactive modeling

可视化能力　visualization capability
可视化软件　visualization software
可适应的　adaptable
可适应界面　adaptable interface
可适用的　applicable
可塑设计　buildable design
可塑设计空间　buildable design space
可探测　detectable
可替代仿真　alternative simulation
可替代性　substitutability
可替换的　substitutable
可调参数　tunable parameter; adjustable parameter
可听化　auralization
可推断的出现　deducible emergence
可推论出的实验框架　derivable experimental frame
可推论的　derivable
可玩的　gameable
可微变量　derivability variable
可维护的　maintainable
可维护性　maintainability
可维修的　maintainable
可维修性　maintainability
可协变的　covariable
可信的　credible
可信的仿真模型　reliable simulation model
可信的模型　credible model
可信的输入　credible input
可信的游戏角色　believable game character
可信度　trustworthiness
可信仿真　credible simulation
可信解　credible solution

可信系统　dependable system
可信性　credibility
可信性测定　credibility determination
可信性度量　credibility metric
可修改属性　modifiable property
可修改性　modifiability
可修正的　correctable
可修正性　correctability
可选事件　alternative event
可选项　option
可延伸的　extensible
可演进的软件构架　evolvable software architecture
可验证性　validatability; verifiability
可移动性　transferability
可移植的　portable
可移植性　portability
可应用的　applicable
可用的　usable
可用性　availability; usability; utility
可用性测试　usability testing
可用性范围　scope of usability
可用性工具　usability tool
可用性需求　usability requirements
可诱导的　derivable
可预测的　predictable
可预测行为　predictable behavior
可预测数据　predictable data
可预测突现行为　predictable emergent behavior
可预测系统　predictable system
可预测性　predictability
可预防的　preventable
可预防性　preventability
可证伪假设　falsifiable hypothesis

可证伪性　falsifiability
可支持性　supportability
可执行的　executable
可执行的场景　executable scenario
可执行的错误　executable error
可执行的仿真模型　executable simulation model
可执行的假设　executable hypothesis
可执行的模型　executable model
可执行的认知模型　executable cognitive-model
可执行的实验　executable experiment
可执行构架　executable architecture
可执行选项　implementation option
可转移性　transferability
可追溯的　traceable
可追溯仿真　retrospective simulation
可追溯性　traceability
可追溯性矩阵　traceability matrix
可追溯性评估　traceability assessment
可追踪的　traceable
可组合的　composable
可组合的模型　composable model
可组合合成环境　composable synthetic environment
可组合环境　composable environment
可组合框架　composable framework
可组合联邦　composable federation
可组合式仿真　composable simulation
可组合水平　composability level
可组合想定　composable scenario
可组合性　composability
可组合性方法学　composability methodology
可组合性技术　composability technology
可组合性理论　composability theory
可组合约束　composability constraint
克里格　Kriging
克里格插值　Kriging interpolation
克里格方程　Kriging equation
克里格仿真　Kriging simulation
克里格假设　Kriging assumption
克里格建模　Kriging modeling
克里格模型　Kriging model
克里格模型算法　Kriging model algorithm
克里格模型选择　Kriging model selection
克里格模型验证　Kriging model validation
克里格模型优化　Kriging model optimization
克里格误差　Kriging error
克里格元建模　Kriging metamodeling
克里格元模型　Kriging metamodel
客观的　objective
客观复杂性　objective complexity
客观经验　objective experience
客观数据　objective data
客观现实　objective reality
客观性　objectivity
客观验证　objective validation
客户　customer
客户化　customization
客户机-服务器模型　client-server model
客户接受度　customer acceptance
客户数据　data customer
课程　course; curriculum

课堂学习　classroom learning
课题　project
空间　space
空间变量　spatial variable
空间导数　spatial derivative
空间的　spatial
空间动力学　spatial dynamics
空间仿真　spatial simulation
空间分辨率　spatial resolution
空间分布式系统　spatially distributed system
空间关系　spatial relation
空间假设　spatial assumption
空间建模　spatial modeling
空间结构　spatial structure
空间离散化　space discretization; spatial discretization
空间离散时间离散模型　discrete-space discrete-time model
空间离散时间连续模型　discrete-space continuous-time model
空间连续模型　continuous-space model
空间连续时间离散模型　continuous-space discrete-time model
空间连续时间连续模型　continuous-space continuous-time model
空间连续系统　continuous-space system
空间模型　spatial model
空间曲面细分　space tessellation
空间数据建模　dimensional data modeling; spatial data modeling
空间特性　spatial characteristic
空间条件　spatial condition
空间突现行为　spatially emergent behavior
空间微观仿真模型　spatial microsimulation model
空间误差　spatial error
空间相关误差　spatially correlated error
空间相似度　spatial similarity degree
空间相似关系　spatial similarity relation
空间相似性　spatial similarity
空间泄露　space leak
空间一致性　spatial consistency
空值　null value
控制　control; control (*v.*)
控制变量　control variable
控制分析　control analysis
控制精度　control accuracy
控制流分析　control flow analysis
控制论　cybernetics
控制论的方法　control-theoretic approach
控制模型　control model
控制耦合　control coupling
控制片段　control segment
控制器　controller
控制算法　control algorithm
控制台游戏　console game
控制误差　control error
控制系统　control system
控制站　control station
控制状态　control state
库　base; repository
库存变量　stock variable
库存的流量变量　stock and flow

variable
库驱动的仿真　library-driven simulation
跨变　across variable
跨度　span
跨领域建模　interdisciplinary modeling
跨学科　interdisciplinary
跨学科的主题　interdisciplinary topic
跨学科建模　transdisciplinary modeling
跨学科协同　transdisciplinary collaboration
跨整体合作　inter-holonic cooperation
块结构仿真语言　block-structured SL
块模型　block model
块设计　blocked design
块图　block diagram
块下三角形式　block-lower-triangular form
快速仿真　fast simulation
快速仿真建模　fast simulation modeling
快速时间　fast time
快速时间常数　fast time constant
快速原型法　rapid prototyping
宽度　width
框架　frame; framework
框架的适用性　frame applicability
框架库　frame base
框架模型的适用性　frame-model applicability
困境　dilemma
扩散参数　spread parameter
扩散的行为　diffusive behavior
扩展　expansion; extension
扩展的时间　expanded time
扩展概念　extension notion
扩展机制　extension mechanism
扩展时间仿真　expanded-time simulation
扩展随机佩特里网　extended stochastic Petri net
扩展知识属性佩特里网　extended knowledge-attributed Petri net
扩展状态转移图　extended state-transition diagram

Ll

拉丁方设计　Latin square design
拉格朗日建模　Lagrangian modeling
拉格朗日模型　Lagrangian model
拉力因素　pull factor
拉普拉斯变换　Laplace transformation
拉普拉斯方程　Laplace equation
兰彻斯特方程　Lanchester equation
懒惰仿真　lazy simulation
懒惰离散事件系统规范事件　lazy DEVS event
懒惰离散事件系统规范演变　lazy DEVS evolution
滥用的行为　abusive behavior
乐观并行仿真　optimistic parallel simulation
乐观仿真　optimistic simulation
乐观仿真架构　optimistic simulation architecture
乐观仿真算法　optimistic simulation algorithm

乐观时间管理 optimistic time management
乐观事件 optimistic event
乐观同步 optimistic synchronization
乐观同步协议 optimistic synchronization protocol
乐观优化 optimistic optimization
乐观执行 optimistic execution
雷达输入 radar input
雷斯科拉-瓦格纳模型 Rescorla-Wagner model
类 class
类比 analog; analogy
类比理解 analogical understanding
类比模型 analogical model; analogy model
类比特性 analog property
类变量 class variable
类层次结构 class hierarchy
类词 class word
类生成 class generation
类似 analogy
类似的 analogical
类同式仿真 quasi-identity simulation
类图 class diagram
类推 analogy
类推的 analogical
类型 stamp; style; type
类型变量 type variable
类型化变量 typed variable
类-子类关系 class-subclass relationship
累积的误差 accumulated error
累积分布函数 cumulative distribution function
累积模型 accumulation model
累积误差 accumulation error; cumulative error
累加值 accumulated value
离散 discretize (v.)
离散变化 discrete-change
离散变化的变量 discrete-change variable
离散变化仿真 discrete-change simulation
离散变化模型 discrete-change model
离散变量 discrete variable
离散变量仿真优化 discrete-variable simulation optimization
离散的 discrete
离散的网 discrete net
离散点 discretization point
离散多模型 discrete multimodel
离散仿真 discrete simulation
离散仿真环境 discrete-simulation environment
离散仿真语言 discrete SL; discrete-simulation language
离散分布 discrete distribution
离散傅里叶变换 discrete Fourier transform
离散轨迹 discrete trajectory
离散化变量 discretized variable
离散化的 discretization; discretized
离散化方法 discretization method
离散化机制 discretization scheme
离散化误差 discretization error
离散空间 discrete space
离散空间模型 discrete-space model
离散控制的变量 discrete-control

离散连续混合系统　mixed discrete and continuous system
离散马尔可夫过程　discrete Markov process
离散模型　discrete model
离散时间　discrete time
离散时间差分方程模型　discrete-time difference equation model
离散时间的变量　discrete-time variable
离散时间的连续仿真　discrete-time continuous simulation
离散时间的流机制　discrete-time flow mechanism
离散时间仿真　discrete-time simulation
离散时间仿真系统　discrete-time simulation system
离散时间控制器　discrete-time controller
离散时间模型　discrete-time model
离散时间系统　discrete-time system
离散时间系统理论　discrete-time systems theory
离散时间线性系统　discrete-time linear system
离散时间状态空间模型　discrete-time state-space model
离散事件　discrete-event
离散事件抽象　discrete-event abstraction
离散事件仿真　DES (Discrete Event Simulation)
离散事件仿真中的克里格元建模　Kriging metamodeling in discrete-event simulation
离散事件建模　discrete-event modeling
离散事件列表　discrete-event list
离散事件模型　discrete-event model
离散事件系统　discrete-event system
离散事件系统规范　DEVS (Discrete Event System Specification)
离散事件系统规范表示的通用性　universality of DEVS representation
离散事件系统规范表示的唯一性　uniqueness of DEVS representation
离散事件系统规范参数同型　DEVS parameter morphism
离散事件系统规范的分层形式　DEVS hierarchical form
离散事件系统规范的内部转换函数　DEVS internal transition function
离散事件系统规范的全局状态转换函数　DEVS global state transition function
离散事件系统规范的时间推进函数　DEVS time advance function
离散事件系统规范的外部转换函数　DEVS external transition function
离散事件系统规范动态结构　DEVS dynamic structure
离散事件系统规范发生器　DEVS generator
离散事件系统规范仿真　DEVS simulation
离散事件系统规范仿真器　DEVS simulator
离散事件系统规范仿真协议　DEVS

simulation protocol

离散事件系统规范跟协调器　DEVS root-coordinator

离散事件系统规范路径　DEVS pathways

离散事件系统规范模型　DEVS model; discrete-event system specification model

离散事件系统规范耦合　DEVS coupling

离散事件系统规范实验框架的实现　DEVS experimental frame realization

离散事件系统规范系统实体结构　DEVS system entity structure

离散事件系统规范系统同型　DEVS system morphism

离散事件系统规范协调器　DEVS coordinator

离散事件系统规范形式　DEVS formalism

离散事件系统规范转换器　DEVS transducer

离散事件线仿真　discrete-event line simulation

离散事件形式　discrete-event formalism

离散属性　discrete attribute

离散数据　discrete data

离散随机变量　discrete random variable

离散系统　discrete system

离散系统仿真　discrete-system simulation

离散系统建模器　discrete-system modeler

离散值的变量　discrete-valued variable

离散状态变量　discrete-state variable

离散状态方程　discrete-state equation

离散状态模型　discrete-state model

离散状态系统　discrete-state system

离散阻尼　discrete damping

离线存储设备　off-line storage device

离线的　off-line

离线仿真　off-line simulation

李克特量表数据　Likert-scale data

李克特量表数据分析　Likert-scale data analysis

李克特量表数据解释　Likert-scale data interpretation

李雅普诺夫稳定性　Liapunov stability

理解　apprehension; comprehension; understand (v.); understanding

理解的决议　resolution of understanding

理解过程　understanding process

理解理论　understanding theory

理解力范畴　scope of understanding

理解系统　understanding system

理解现实　understood reality

理论　theorem; theory

理论的　theoretical

理论的证实　theoretical confirmation

理论构建　theory construction

理论建模框架　theoretical modeling framework

理论框架　theoretical framework

理论模型　theoretical model

理论数据　theoretical data

理论研究　theoretical study

理论验证　theory validation

理论有效性　theoretical validity
理论值　theoretical value
理想点模型　ideal point model
理想化仿真　gedanken simulation
理想化模型　idealized model
理想实验　gedanken experiment
理想搜索仿真　ideal-seeking simulation
理性模型　rational model
理性响应　rational response
理性知识　rational knowledge
力　effort
力学系统　mechanical system
历史　history
历史测试　historical test
历史仿真　historical simulation
历史数据　historic data; historical data
历史数据验证　historical-data validation
历史数据有效性　historical-data validity
历史有效性　historical validity
利他主义　altruism
利益相关者　stakeholder
例程　routine
隶属度　degree of membership
隶属度值　membership value
隶属关系　membership
粒度　granularity
连贯特性　coherent property
连贯性　coherence
连贯性框架　coherent framework
连接　connection; couple (v.); link; linking
连接到实时仿真　linkage to live simulation
连接模型　connectionist model
连接器　connector
连接误差　connection error
连接训练器　link trainer
连通度　interconnectedness
连通性　connectivity
连续/离散仿真语言（SL）　continuous/discrete SL
连续/离散混合　combined continuous/discrete model
连续/离散混合仿真　combined continuous/discrete simulation
连续变化的　continuous-change
连续变化的变量　continuous-change variable
连续变化仿真　continuous-change simulation
连续变化模型　continuous-change model
连续变量　continuous variable
连续变量仿真优化　continuous-variable simulation optimization
连续变量模型　continuous-variable model
连续的　continuous; sequential
连续对策　continuous game
连续多模型　continuous multimodel; sequential multimodel
连续仿真　continuous simulation
连续仿真语言　continuous SL
连续分布　continuous distribution
连续赋值变量　continuous-valued variable
连续轨迹　continuous trajectory

连续核心对策 continuous kernel game
连续活动 continuous activity
连续建模 continuous modeling
连续空间 continuous space
连续马尔可夫链 continuous Markov chain
连续模型 continuous model
连续片段 continuous segment
连续全局最优化 continuous global optimization
连续时间变量 continuous-time variable
连续时间仿真 continuous-time simulation
连续时间仿真语言 continuous-time SL
连续时间连续仿真 continuous-time continuous simulation
连续时间模型 continuous-time model
连续时间系统仿真语言 continuous-time system SL
连续时间线性系统 continuous-time linear system
连续时间状态空间模型 continuous-time state-space model
连续适应性 continuous adaptation
连续属性 continuous attribute
连续数据 continuous data
连续顺序变量 continuous ordinal variable
连续随机变量 continuous random variable
连续问题 continuous problem
连续系统 continuous system
连续系统仿真 continuous-system simulation
连续系统建模器 continuous-system modeler
连续现象 continuous phenomenon
连续相关变量 continuous dependent variable
连续性 continuity
连续性测试 continuity test
连续运行 continuous run
连续增长变量 continuously increasing variable
连续状态模型 continuous-state model
联邦 federation
联邦标记 federate markup
联邦成员 federate
联邦成员的互操作性 federate interoperability
联邦成员的外部克隆 external cloning of a federate
联邦成员复制 cloning of a federate
联邦成员集群 federate cluster
联邦成员模型 federate model
联邦成员资格认证 federate qualification
联邦成员组成 federate composition
联邦重用 federation reuse
联邦初始化 federation initialization
联邦的 federated
联邦对象模型 federation object model
联邦对象模型模版 federation object model template
联邦仿真 federate simulation; federated simulation
联邦仿真服务 federated simulation

联邦仿真环境 federated simulation environment
联邦仿真输出数据 federate simulation output data
联邦仿真体系结构 federated simulation architecture
联邦仿真系统 federated simulation system
联邦功能 federation functionality
联邦管理器 federation manager
联邦计算机 federate computer
联邦交换数据 federation exchange data
联邦接口规范 federate interface specification
联邦结构变换 federation structure transformation
联邦开发 federation development
联邦开发与执行过程 FEDEP
联邦可重用的 federation reusable
联邦可重用性 federation reusability
联邦联合 federation of federations
联邦模型 federation model
联邦目标 federation objective
联邦认证 federation certification
联邦设计 federation design
联邦设计方法学 federation design methodology
联邦时间 federate time; federation time
联邦时间轴 federation time axis
联邦实例化 federate instantiation
联邦所需的执行细节 federation required execution detail
联邦所有权管理 federate ownership management
联邦元素 federation element
联邦执行 federation execution
联邦执行发起者 federation execution sponsor
联邦执行控制 federation execution control
联邦执行数据 federation execution data
联邦组件 federation component
联动变量 yoked variable
联动的 yoked
联动仿真 yoked simulation
联合多分辨率建模 joint multiresolution modeling
联合仿真 combined simulation; joint simulation; cosimulation; conjoint simulation; co-simulation
联合仿真技术 cosimulation technique
联合仿真系统 joint simulation system (JSIMS); combined simulation system
联合仿真语言 combined simulation language
联合仿真组件 cosimulation component; co-simulation component
联合合成作战空间 joint synthetic battlespace
联合激活误差 associative activation error
联合建模与仿真 joint M&S
联合建模与仿真提议者 joint M&S proponent

联合建模与仿真投资计划　joint M&S investment plan
联合建模与仿真系统　JMASS (Joint M&S System)
联合建模与仿真执行程序面板　joint M&S executive panel
联合试验　joint experiment; joint experimentation
联合系统仿真　combined-system simulation
联合学习　combined learning
联合训练联盟　joint training confederation
联合作战系统　joint warfare system (JWARS)
联合作战训练　joint warfare training
联机模式　online mode
联络分支　tie branching
联盟　confederation
联想　association
联想理解　associative understanding
练习　exercise; training
链　chain
链反应系统　chain reaction system
链接　link
链路　link
两分法变量　dichotomous variable
两级仿真　two-level simulation
两级仿真方法学　two-level simulation methodology
两级析因设计　two-level factorial design
两阶段变换　two-stage transformation
两阶段理解　two-stage understanding
两用技术　dual use technology

量程　span
量程误差　span error
量出　measure (v.)
量化　quantification; quantization; quantizing; quantify (v.)
量化变量　quantified variable; quantized variable
量化仿真器　quantized simulator
量化函数　quantization function
量化积分器　quantized integrator
量化离散事件系统规范仿真器　quantized DEVS simulator
量化器　quantizer
量化输出变量　quantized output variable
量化输入　quantized input
量化输入变量　quantized input variable
量化数值　quantized value
量化系统　quantized system
量化滞后　quantization with hysteresis
量化状态　quantized state
量化状态变量　quantized state variable
量化状态系统　quantized state system
量化状态系统方法　quantized state system method
量化状态系统求解器　quantized state system solver
量子　quantum
量子变量　quantum variable
量子处理器　quantum processor
量子叠加　quantum superposition
量子对象　quantum object
量子仿真　quantum simulation
量子仿真器　quantum simulator

量子光子　quantum photonic
量子计算　quantum computing
量子计算机　quantum computer
量子连续变量　quantum continuous variable
量子逻辑　quantum logic
量子启发的智能系统　quantum-inspired intelligent system
量子算法　quantum algorithm
量子态　quantum state
量子特性　quantum property
量子系统　quantum system
量子系统结构　quantum system architecture
量子信息处理　quantum information processing
量子游走　quantum walk
猎食模型　predator-prey model
邻近值　adjacent value
林氏系统　Lindenmayer system（L-system）
林氏系统仿真　Lindenmayer system simulation（L-system simulation）
临界变量　critical variable
临界乘数　critical multiplism
临界刚性　marginally stiff
临界刚性系统　marginally stiff system
临界假设　critical assumption
临界稳定系统　marginally stable system
临界值　critical value
临界状态变换　critical state-transformation
临界状态转换　critical state-transition
临近事件　imminent event
临时变量　temporal variable; temporary variable
临时玩家　casual gamer
临时用户　casual user
灵活性　flexibility
灵敏度　sensitivity
灵敏度仿真　sensitivity simulation
灵敏度分析　sensitivity analysis
灵敏度分析的精确度　accuracy of sensitivity analysis
灵敏度模型　sensitivity model
灵敏度误差　sensitivity error
灵敏度研究　sensitivity study
零方差仿真　zero-variance simulation
零和博弈　zero-sum game
零和仿真　zero-sum simulation
零和信任博弈　zero-sum trust game
零假设　null hypothesis
零交点　zero-crossing
零刻度误差　zero-scale error
零图灵机仿真　null Turing machine simulation
零位误差　zero error
领域本体　domain ontology
领域边界　domain boundary
领域分析　domain analysis
领域模型　domain model
领域需求　domain requirements
领域应用　domain application
领域知识　domain knowledge
领域专家　domain expert
另一种现实　alternative reality
令牌　token
令人满意的行为　satisficing behavior
浏览　browsing; scan; scan (v.)

流　flow
流程模板　process template
流动代理　itinerant agent
流动智能体　itinerant agent
流量变化　flow variable
流片后的　postsilicon
流片后校核　postsilicon verification
流片后验证　postsilicon validation
流入　inflow
流体仿真　fluid simulation
六自由度　six degree of freedom
龙格－库塔积分　Runge-Kutta integration
龙格－库塔算法　Runge-Kutta algorithm
漏测值　missing value
漏检错误　undetected error
鲁棒的　robust
鲁棒仿真运行库　robust simulation runtime library
鲁棒决策　robust decision making
鲁棒控制　robust control; robustness control
鲁棒模型　robust model
鲁棒设计　robust design
鲁棒性　robustness
鲁棒性维持　preservation of robustness
鲁棒优化　robust optimization
鲁棒有效性　validity of robustness
录用　accepted
路德模拟　ludic simulation
路径　routing
路径测试　path testing
路径仿真　pathway simulation
路径规划器　path planner
路径空间　routing space
路由　routing
路由空间　routing space
伦理错误　ethical error
伦理的　ethical
伦理仿真　ethical simulation
伦理规则　ethical rule
伦理合理性　ethical rationality
伦理目标评价　ethical goal-assessment
伦理评价　ethical evaluation
伦理问题　ethical issue
伦理学　ethics
伦理原则　ethical principle
轮廓　profile
轮流　alternate (*v.*)
轮流的　alternate
论点　point of view; thesis
论文　thesis
论域　domain
论证　argument
论证系统　argumentation system
逻辑　logic
逻辑变量　logic variable; logical variable
逻辑错误　fallacies in logic; logic error
逻辑错误类型　types of fallacies in logic
逻辑倒错　paralogism
逻辑等价　logical equivalence
逻辑仿真　logic simulation; logical simulation
逻辑仿真器　logic simulator
逻辑仿真拓扑　logical simulation topology

逻辑分布　logistic distribution
逻辑关系　logical relation
逻辑互操作性　logical interoperability
逻辑环境　logical environment
逻辑技术　logical technique
逻辑架构　logical schema
逻辑接口　logical interface
逻辑结果　logical consequence
逻辑经验主义　logical empiricism
逻辑命名变量　logically named variable
逻辑模型　logical model
逻辑上的边界　logical boundary
逻辑上的谬论　logical fallacy
逻辑时间　logical time
逻辑时间管理　logical time management
逻辑时间轴　logical time axis
逻辑数据模型　logical data model
逻辑斯谛增长模型　logistic growth model
逻辑推理　logical inference
逻辑推理行为　logically deduced behavior
逻辑推论　logical deduction
逻辑误差　logical error
逻辑校核　logical verification
逻辑游戏　logic game
逻辑正确性　logical validity
逻辑知识　logical knowledge
逻辑值　logical value
滤波　filter (v.); filtering
滤波模型　filtering model
滤波器　filter
λ演算　lambda calculus

Mm

马尔可夫博弈　Markov game
马尔可夫仿真　Markov simulation
马尔可夫过程　Markov process
马尔可夫假设　Markov assumption
马尔可夫链　Markov chain
马尔可夫链仿真　Markov chain simulation
马尔可夫链模型　Markov chain model
马尔可夫模型　Markov model
马尔可夫特性　Markovian property
麦克风输入　microphone input
脉冲　impulse
脉冲变量　impulse variable
满量程误差　full-scale error
满意的　satisficing
慢时间常数　slow time constant
矛盾　contradiction
矛盾规则　contradictory rule
冒险游戏　adventure game
没有表现的特性　nonemergent property
媒体　media
美学计算　aesthetic computing
门户　portal
门级仿真　gate-level simulation
蒙特卡罗　Monte Carlo
蒙特卡罗测试　Monte Carlo test
蒙特卡罗方法　Monte Carlo method
蒙特卡罗仿真　Monte Carlo simulation
蒙特卡罗实验　Monte Carlo experiment
蒙特卡罗算法　Monte Carlo algorithm

米利机　Mealy machine
米利模型　Mealy model
米利自动机　Mealy automata；Mealy automaton
密度　density
密度估计　density estimation
密集的　intensive
密集连接　dense connection
密集输出　dense output
密码错误　password error
幂次法则　power law
幂律函数　power law function
面部动画标记语言　facial animation markup language
面对面的学习　face-to-face learning
面向……的　oriented
面向变量的　variable-oriented
面向变量的模型　variable-oriented model
面向表达式的仿真语言　expression-oriented SL
面向程序的仿真　program-oriented simulation
面向代理的　agent-oriented
面向代理的方法学　agent-oriented methodology
面向代理的模型　agent-oriented model
面向代数表达式的仿真语言　algebraic expression-oriented SL
面向动作的　action-oriented
面向动作的规则　action-oriented rule
面向对象的　object-oriented
面向对象的表示（法）　object-oriented representation
面向对象的程序设计　object-oriented programming
面向对象的仿真　object-oriented simulation
面向对象的建模　object-oriented modeling
面向对象的模型　object-oriented model
面向对象的佩特里网　object-oriented Petri net
面向对象的设计　object-oriented design
面向对象的实现　object-oriented implementation
面向对象的语言　object-oriented language
面向对象技术　object-oriented technology
面向多边形的　polygon-oriented
面向多边形的建模　polygon-oriented modeling
面向方程的仿真　equation-oriented simulation
面向方程的仿真系统　equation-oriented simulation system
面向方面的　aspect-oriented
面向方面的感知　aspect-oriented perception
面向方面的建模　aspect-oriented modeling
面向方面的转变　aspect-oriented transformation
面向仿真生命周期的系统工程　systems engineering for simulation life cycle

面向非方程式的仿真　non-equation-oriented simulation
面向服务的　service-oriented
面向服务的技术　service-oriented technology
面向服务的架构　service-oriented architecture
面向服务的架构方法　service-oriented architecture approach
面向关系的　concern-oriented
面向关系的建模　concern-oriented modeling
面向过程的　process-oriented
面向过程的离散事件系统规范　process-oriented DEVS
面向过程的模型　process-oriented model
面向过程的世界观　process-oriented world view
面向过程仿真　process-oriented simulation
面向过程建模　process-oriented modeling
面向行为的　behavior-oriented
面向活动的　activity-oriented
面向活动的模型　activity-oriented model
面向结果的仿真　outcome-oriented simulation
面向块的仿真　block-oriented simulation
面向块的仿真系统　block-oriented simulation system
面向块的仿真语言　block-oriented SL
面向领域的　domain-oriented

面向领域的灵敏度分析　domain-oriented sensitivity analysis
面向模块的　block-oriented
面向模式的　pattern-oriented
面向模式的建模　pattern-oriented modeling
面向模型的　model-oriented
面向目标的系统仿真　goal-oriented system simulation
面向区间的　interval-oriented
面向区间的仿真　interval-oriented simulation
面向任务的　task-oriented
面向任务的帮助　task-oriented help
面向任务的代理　task-oriented agent
面向任务的智能体　task-oriented agent
面向事件的　event-oriented
面向事件的仿真　event-oriented simulation
面向事件的规则　event-oriented rule
面向事件的模型　event-oriented model
面向图形的仿真语言　graphic-oriented SL
面向图形对象的模型　graphical object-oriented model
面向团队的　team-oriented
面向团队的多学科设计　team-oriented multidisciplinary design
面向团队的设计　team-oriented design
面向网络的　network-oriented
面向网络的仿真　network-oriented simulation
面向智能体的　agent-oriented

面向智能体的方法学 agent-oriented methodology
面向智能体的模型 agent-oriented model
面向自治的建模 autonomy-oriented modeling
面向自主性的 autonomy-oriented
面值 face value
描述 describe (v.); description; specify (v.)
描述本体 descriptive ontology
描述的 described; descriptive
描述的方法 descriptive approach
描述的复杂性 descriptive complexity
描述的模型 descriptive model
描述的评估 descriptive assessment
描述的数据 descriptive data
描述符 descriptor
描述格式 descriptive format
描述技术 description technique
描述误差 description error
描述性变量 descriptive variable
描述性仿真 descriptive simulation
描述性简化 descriptive simplicity
描述性建模 descriptive modeling
描述性决策 descriptive decision
描述性理解 descriptive understanding
描述性模型分析 descriptive model analysis
描述性知识 descriptive knowledge
描述语言 descriptive language
瞄准的 aimed
敏感的 sensitive
敏感度分析技术 sensitivity analysis technique
敏感数据 sensitive data
敏捷方法学 agile methodology
敏捷仿真 agile simulation
敏捷建模 agile modeling
明确的 specific; unambiguous; definite
明确的规则 unambiguous rule
明确的模型 unambiguous model
明确的输入 unambiguous input
明确目标 explicit goal
铭文标定误差 inscription error
命令 order
命令的 imperative
命令驱动的 imperative-driven
命令驱动的仿真 imperative-driven simulation
命令驱动的输入 command-driven input
命题 proposition; theorem
命题变量 proposition variable; propositional variable
命题模型 propositional model
命题性知识 propositional knowledge
谬论 fallacy; paralogism
谬误的 fallacious
模板 template
模板定位 model location
模仿 mimesis; mimetics
模仿表现 imitative representation
模仿的 mimetic
模仿视角 imitation perspective
模仿现实 imitated reality
模糊变量 fuzzy variable
模糊表示（法） ambiguous representation

模糊参数　fuzzy parameter
模糊初始条件　fuzzy initial condition
模糊的　fuzzy
模糊定性值　fuzzy qualitative value
模糊方程　fuzzy equation
模糊仿真　fuzzy simulation
模糊关系　fuzzy relation
模糊归纳推理　fuzzy inductive reasoning
模糊规则　ambiguous rule; fuzzy rule
模糊规则对象　fuzzy rule object
模糊规则库　fuzzy rule-base
模糊规则库类　fuzzy rule-base class
模糊规则类　fuzzy rule class
模糊函数　fuzzy function
模糊化　fuzzificate (*v.*); fuzzification
模糊化的　fuzzificated
模糊化方法　fuzzification method
模糊回归　fuzzy regression
模糊集合　fuzzy set
模糊集类　fuzzy set class
模糊建模　fuzzy modeling
模糊聚类去模糊化　fuzzy clustering defuzzification
模糊决策　fuzzy decision
模糊离散事件系统规范　fuzzy-DEVS
模糊隶属度函数　fuzzy membership function
模糊量子状态　fuzzy quantum state
模糊模型　ambiguous model; fuzzy model
模糊佩特里网　fuzzy Petri net
模糊器　fuzzifier
模糊认知图　fuzzy cognitive map
模糊输入　fuzzy input

模糊数　fuzzy number
模糊数据　fuzzy data
模糊推理　fuzzy inference
模糊推理算法　fuzzy inference algorithm
模糊推理网络　fuzzy inference network
模糊推理引擎　fuzzy inference engine
模糊微分方程　fuzzy differential equation
模糊误差　ambiguity error
模糊系统　fuzzy system
模糊系统仿真　fuzzy system simulation
模糊系统建模　fuzzy system modeling
模糊演算　fuzzy calculus
模糊因果有向图　fuzzy causal directed graph
模糊有限状态机　fuzzy finite-state-machine
模糊有向图　fuzzy digraph
模糊约束　fuzzy restriction
模糊真值　fuzzy truth value
模糊值　fuzzificated value; fuzzy value
模糊专家系统　fuzzy expert system
模糊状态迁移图　fuzzy state transition graph
模块　module
模块边界　module boundary
模块测试　module testing
模块化　modularity; modularization
模块化的　modular
模块化的离散事件系统规范形式　modular DEVS form

模块化仿真　modular simulation
模块化仿真程序　modular simulation program
模块化建模　modular modeling
模块化模型　modular model
模块化系统　modular system
模块化系统建模　modular system modeling
模块化形式　modular form
模块化组件　modular component
模块接口　module interface
模块性　modularity
模拟　analog; emulate (v.); imitation
模拟程序　emulation program
模拟的　analogous; emulative; simulant
模拟的故障　simulated mulfunction
模拟的伪观察　simulated pseudoobservation
模拟对象　simuland
模拟多水平仿真　analog multilevel simulation
模拟方法论　emulation methodology
模拟方法学规范　norms of simulation methodology
模拟仿真　analog simulation; mock simulation
模拟仿真软件　analog simulation software
模拟仿真研究　analog simulation study
模拟计算机　analog computer
模拟计算机仿真　analog-computer simulation
模拟控制器　simulation controller
模拟模型　analog model
模拟曝光　simulation exposure
模拟曝光效应　effects of simulation exposure
模拟器　emulator
模拟器参数　simulator parameter
模拟适应度　simulation fitness
模拟输入　analog input
模拟-数字转换器　analog-to-digital converter
模拟退火　simulated annealing
模拟退火过程　simulation annealing process
模拟退火算法　simulated annealing method
模拟物　simulacra; simulacre
模拟游戏　mimetic play
模拟状态　simulating state
模式　mode; pattern; schema
模式导向的多模型　pattern-directed multimodel
模式切换　mode switching
模式误差　mode error
模式效应　pattern effect
模式映射　schema mapping
模式语言　pattern language
模式值　modal value
模型　model
模型（存储）体　model bank
模型安全　model security
模型版本　model version
模型版本控制　model version control; model versioning
模型保障性　model supportability
模型本体　model ontology

模型逼真度　model fidelity
模型逼真性　model verisimilitude
模型比较　model comparison
模型边界　model boundary
模型编辑环境　model editing environment
模型编辑器　model editor
模型编译器　model compiler
模型变换　model transformation
模型变换的形式方面　formal aspect of model transformation
模型变换方法　model transformation approach
模型变换工具　model-transformation tool
模型辨识　model identification
模型表格　table model
模型表示（法）　model representation
模型表征　model characterization
模型部署　model deployment
模型裁剪　model pruning
模型参数　model parameter
模型参数的适当性　adequacy of model parameters
模型仓库管理　model repository management
模型操作　model manipulation
模型测试　model test; model testing
模型层规范　level of model specification
模型查询　model query; model querying
模型查询语言　model query language
模型差分化　model differencing
模型常数　model constant

模型成本　model cost
模型成熟度　model maturity
模型尺寸度量　model-size metric
模型充分性　model adequacy
模型抽象　model abstraction
模型抽象化技术　model abstracting technique
模型处理　model processing
模型处理环境　model processing environment
模型处理基础　basis for model processing
模型处理支持　model processing support
模型创作　model authoring
模型粗糙度　model coarseness
模型存档　model archive; model archiving
模型大小限制　model-size limit
模型代理　model brokering
模型代码变换　model-to-code transformation
模型导航　model navigation
模型导向的　model-directed
模型导向的系统　model-directed system
模型的被动实体　passive entity of a model
模型的参数评估　assessment of parameters of a model
模型的层次结构　hierarchy of models
模型的范围　scope of a model
模型的合理性　model plausibility
模型的活动实体　active entity of a model

模型的可接受性　acceptability of model; model acceptability
模型的可理解性　model intelligibility; model understandability
模型的可用性　model usability
模型的连续性　model continuity
模型的潜在方面　latent aspect of model
模型的似真性　model plausibility
模型的输入/输出　model input/output
模型的说明性部分　declarative part of a model
模型的同质性　model homogeneity
模型的休眠方面　dormant aspect of model
模型的异质性　model heterogeneity
模型的真实性　model veracity
模型的准用空间　admissible space of models
模型等价　model equivalencing
模型地貌　model topography
模型定位　model location
模型动力学　model dynamics
模型动态发现　dynamic model discovery
模型对接　model docking
模型对齐　model alignment
模型发现　model discovery
模型分辨率　model resolution
模型分割　model partitioning
模型分级　model staging
模型分解　model decomposition
模型分类　model classification; model taxonomy
模型分析　model analysis

模型封装　model packaging; model wrap; model wrapping
模型封装工具　model packaging tool
模型复杂性　model complexity
模型复杂性评估　model complexity assessment
模型复制　model replication
模型概念化　model conceptualization
模型更新　model update; model updating
模型工程　model engineering
模型工程本体　model engineering ontology
模型工程标准　model engineering standard
模型工程的知识体系　body of knowledge of model engineering
模型工程工具　model engineering tool
模型工程过程　model engineering process
模型工程支持　model engineering support
模型工程知识体系　model engineering body of knowledge
模型工具　model tool
模型共享　model sharing
模型构建　model construction
模型构建管理　model construction management
模型关联　model relationship; model relevance
模型关系　model relation
模型管理　model management
模型管理工具　model management tool

模型管理基础设施　model-management infrastructure
模型管理器　model manager
模型归并　model merging
模型规范　model specification
模型规范环境　model specification environment
模型规范语言　model specification language
模型规模　model size
模型过程管理　model process management
模型行为　model behavior; model's behavior
模型行为本体　model behavior ontology
模型行为不连续性　model behavior discontinuity
模型行为拟合　model behavior fitting
模型合格　model acceptability
模型互操作　model interoperation
模型互操作标准　model interoperability standard
模型互操作性　model interoperability
模型环境　model environment
模型获取　model acquisition
模型基准　model benchmarking
模型激活　model activation
模型级　model stage
模型集成　model integration
模型记忆　model memory
模型假定　model postulate; model assumption
模型间变换　model-to-model transformation
模型间映射　model-to-model mapping
模型检测技术　model checking technique
模型检测算法　model checking algorithm
模型检查　model checking
模型检索　model retrieval
模型简化　model simplification
模型建档　model documenting
模型建立　model building
模型建立形式　model building formalism
模型鉴定　model qualification
模型降阶　model reduction
模型降阶技术　model reduction technique
模型交互　model interaction
模型交换　model exchange
模型校核　model verification
模型校核、验证与确认　model verification, validation and certification
模型校验工具　model checking tool
模型校准　model calibration
模型接口测试　model interface testing
模型接口错误　model interface error
模型接口分析　model-interface analysis
模型结构　model structure
模型结构的适当性　adequacy of model structure
模型精度　model accuracy; model precision
模型精度范围　range of model accuracy

模型精化　model elaboration
模型精炼　model refinement
模型聚合　model aggregation
模型开发　model development
模型开发环境　model development environment
模型开发框架　model development framework
模型开发设施　model-development infrastructure
模型开发者　model developer
模型可传递性　model transferability
模型可复制性　model replicability
模型可共享性　model shareability
模型可交付性　model referability
模型可接受性标准　model acceptability standard
模型可靠性　model reliability
模型可扩展性　model scalability
模型可理解性　model comprehensibility
模型可维护性　model maintainability
模型可信度　model trustworthiness; model credibility
模型可修改性　model modifiability
模型可移植性　model portability
模型可用性　model availability
模型可支持行　model supportability
模型可重现性　model replicability
模型可重用的　model reusable
模型可重用性　model reusability
模型可追溯性　model traceability
模型可组合性　model composability
模型空间　model space
模型库　model base; model library;
model repository
模型库管理　model base management
模型库管理器　model base manager
模型类　model class
模型粒度　model granularity
模型联邦　model federation
模型灵活性　model flexibility
模型灵敏度　model sensitivity
模型鲁棒性　model robustness
模型论　model theory
模型描述　model description
模型描述语言　model description language
模型模板　model template
模型模块化　model modularity
模型内容　model content
模型拟合　model fitting
模型配置　model configuration
模型配准　model alignment
模型匹配　model matching
模型评估　assessment of the model; model appraisal; model assessment; model evaluation
模型切换　model switch; model switching
模型亲和性　model affinity
模型驱动　model-driven
模型驱动的代码生成　model-driven code generation
模型驱动的方法　model-driven approach
模型驱动的仿真工程　model-driven simulation engineering
模型驱动的可重复性　model-driven reproducibility

模型驱动的系统　model-driven system
模型驱动的系统工程　model-driven systems engineering
模型驱动的想定　model-driven scenario
模型驱动的用户界面（接口）　model-driven user interface
模型驱动方法的变异　mutation for model-driven approach
模型驱动方法学　model-driven methodology
模型驱动工程　model-driven engineering
模型驱动构架工具　model-driven architecture tool
模型驱动过程　model-driven process
模型驱动互操作性　model-driven interoperability
模型驱动技术　model-driven technique
模型驱动架构　model-driven architecture
模型驱动开发　model-driven development
模型驱动开发方法　model-driven development approach
模型驱动开发方法学　model-driven development methodology
模型驱动开发技术　model-driven development technique
模型驱动开发语言　model-driven development language
模型驱动器　model driver
模型驱动推理　model-driven reasoning
模型驱动语言　model-driven language
模型权衡算法　model trade-off algorithm
模型确认　model accreditation
模型认识论　model epistemology
模型认证　model certification
模型融合　model merging
模型软件　software model
模型设定误差　model specification error
模型设计　model design
模型生成　model generation
模型生成器　model generator
模型生成问题　model generation problem
模型生命周期　model life cycle
模型生命周期成本　model life cycle cost
模型生命周期工程　model lifecycle engineering
模型生命周期过程　model life cycle process
模型失活　model deactivation
模型实例　model instance
模型实例化　model instantiation
模型实体映射　model-to-entity mapping
模型实现　model implementation
模型实用性　model usefulness
模型实在论　model realism
模型使用　model use
模型使用者　model user
模型市场　model market place
模型试验　model test
模型适当性　model appropriateness

模型适应　model adaptation
模型适用性　model applicability
模型属性　model attribute
模型树　model tree
模型数据　model data
模型数据库　model database
模型数据描述　data representation model
模型算子　model operator
模型缩放　model scaling
模型提供方　model provider
模型提取　model extraction
模型替换　model replacement
模型同步　model synchronization
模型同构　model isomorphism
模型同态　model homomorphism
模型同型　model morphism
模型图　model graph
模型推荐人　model recommender
模型完整性　model integrity
模型完整性需求　model integrity requirements
模型维护　model maintenance
模型文本变换　model-to-text transformation
模型文档　model documentation
模型文件　model file
模型稳定性　model stability
模型误差　model error
模型误差统计　model error statistic
模型系谱　model pedigree
模型系统工程　model systems engineering
模型相关的　model relevant; model-related

模型效用　model utility
模型协同进化　model coevolution
模型形式　model formalism
模型需求　model requirements
模型需求管理　model requirements management
模型选择　model selection
模型压缩　model compression
模型延迟　model latency
模型研究　model research
模型演变　model evolution
模型验证　model validation
模型一致性　model consistency
模型依赖　model reliance
模型引导的过程　model-directed process
模型有效性　model validity
模型有效性程度　degree of model validity
模型与试验框架分离　separation of model and experimental frame
模型语义　model semantics
模型语用学　model pragmatics
模型元素　model element
模型约束　model constraint
模型在环　model-in-the-loop
模型正确性　model correctness
模型正确性分析　model correctness analysis
模型支持　model support
模型支架　model support
模型执行　model execution
模型执行属性　model execution property
模型质量　model quality

模型质量保证 model quality assurance
模型置信度 model confidence
模型中心 model centric
模型中心方法 model-centric approach
模型重构 model reconstruction; model refactoring
模型重构管理 model reconstruction management
模型重现 model replication
模型重用 model reuse
模型重用技术 model reuse technology
模型注册 model registry
模型专家 model expert
模型转换 model transition
模型转换的可扩展性 scalability of model transformation
模型转换的语义 semantics of model transformation
模型转换范式 model transformation paradigm
模型转换算子 model transformation operator
模型转换引擎 model transformation engine
模型转换语言 model transformation language
模型资源 model resource
模型自适应性 model adaptivity
模型综合 model synthesis
模型综合误差 model synthesis error
模型族 model family
模型组合 model composition
模型组合规则 model composition rule
模型组合器 model composer
模型组合实例 model composition instance
模型组和技术 model composition technology
模型组件 component of a model; model component
模型组件的语法 syntax of model component
模型组件模板 model component template
模型组件实用主义 model component pragmatism
模型组件语义 model component semantics
摩尔机 Moore machine
摩尔模型 Moore model
摩尔型邻域 Moore neighborhood
摩尔自动机 Moore automata; Moore automaton
末段 terminal section
默认 default
默认参数 default parameter
默认值 default value
木马仿真器 Trojan simulator
目标 goal; objective
目标变量 goal variable; objective variable
目标参数 goal parameter
目标处理系统仿真 goal-processing system simulation
目标代码 object code
目标导向的 goal-directed
目标导向的多模型 goal-directed

multimodel
目标导向的活动 goal-directed activity
目标导向的系统仿真 goal-directed system simulation
目标导向的系统模型 goal-directed system model
目标导向的知识处理 goal-directed knowledge processing
目标导向的智能体 goal-directed agent
目标导向行为 goal-directed behavior
目标导向实验 goal-directed experiment; goal-directed experimentation
目标的合规评估 ethical assessment of the goal
目标的语用评估 pragmatic assessment of the goal
目标仿真语言 target SL
目标分量的可还原性 reducibility of goal components
目标分量的矛盾 contradiction of components of a goal
目标分量的矛盾性 contradiction of goal components
目标分量的一致性 consistency of goal components
目标分量的正交性 orthogonality of goal components
目标规划 goal programming
目标规划方法 goal-programming method
目标函数 objective function
目标回归仿真 goal-regression simulation
目标计算机 object computer
目标结构 goal structure
目标控制的模型 target control model
目标模型 target model
目标配准误差 target registration error
目标驱动的 goal-driven
目标驱动的方法学 objectives-driven methodology
目标驱动的活动 goal-driven activity
目标驱动的决策 goal-driven decision
目标确定的 goal-determined
目标确定的仿真 goal-determined simulation
目标软化策略 goal softening strategy
目标设置系统 goal-setting system
目标设置系统仿真 goal-setting system simulation
目标生成系统仿真 goal-generating system simulation
目标适用性 fitness to purpose
目标数据 target data
目标搜索仿真 goal-seeking simulation
目标搜索系统 goal-seeking system
目标文件 object file
目标系统 target system
目标域 target domain
目标值 objective value
目标状态 goal state; target state
目标组件 goal component
目的 purpose

目的论　teleology
目的论模型　teleological model
目的论系统仿真　teleological system simulation
目的视角　purpose perspective
目的性的模型　teleonomic model
目的性的试验　purposeful experiment
目的性系统仿真　teleonomic system simulation
目录　directory

Nn

NP 困难　NP-hard
NP 难题　NP-hard problem
NP 完全　NP-complete
NP 完全问题　NP-complete problem
N 阶方案　nth order scheme
N 阶中心差分格式　nth order central difference scheme
n 路模型融合　n-way model merging
n 人博弈　n-person game
n 体仿真　n-body simulation
纳米仿真　nanosimulation
纳米级仿真　nanoscale simulation
纳米级系统　nanoscale system
纳米生物仿真　bio-nano simulation
奈斯特龙算法　Nyström algorithm
耐火期　refractory period
耐久状态　long-lived state
难解方程　intractable equation
难解仿真　intractable simulation
脑机接口　brain-computer interface; brain-machine interface
脑机接口技术　brain-computer interface technology
内变量　internal variable
内部变换　internal transition
内部参量　internal parameter
内部产生的假设　internally generated hypothesis
内部产生的目标　internally generated goal
内部产生的问题　internally generated question
内部代理　internal agent
内部的模型　internal model
内部的特性　internal characteristic
内部方案　internal scheme
内部仿真器　internal simulator
内部仿真数据　intrasimulation data
内部构件　internal structure
内部活动　internal activity
内部激活多模型　internally activated multimodel
内部激励　internal excitation
内部理解　internal understanding
内部模型　endomodel; endo-model
内部目标　internal goal
内部耦合　internal coupling
内部耦合的　internally coupled
内部生成的输入　internally generated input
内部时间管理　internal time management
内部事件　internal event
内部输入　internal input
内部完整性　internal completeness
内部误差　internal error
内部闲置　internal inactivity

内部性能演变　internal performance evolution
内部演化　internal evolution
内部一致性　internal consistency
内部有效的　internally valid
内部有效性　internal validity
内部语言　intralanguage
内部语言模型映射　intralanguage model mapping
内部语言映射　intralanguage mapping
内部智能体　internal agent
内部转移函数　internal transition function
内部状态　internal state
内存　memory
内存错误　memory error
内存泄漏　memory leak
内核游戏　kernel game
内聚性　cohesion
内联集成　inline integration
内联偏微分方程　inlining partial differential equation
内联隐式龙格-库塔算法　inlining implicit Runge-Kutta algorithm
内胚层　endomorph
内胚层体型　endomorphy
内容　content
内容聚合模型　content aggregation model
内容描述层　content description layer
内容耦合　content coupling
内容正确性　content accuracy
内生变量　endogenous variable
内生的　endogenous
内生假设　endogenous assumption
内生事件　endogenous event
内生输入　endogenous input
内生协变量　endogenous covariate
内省　introspection
内省仿真　introspective simulation
内省仿真模型　introspective simulation model
内省系统　introspective system
内隐模型　internal tacit model
内源的　endogenous
内源性　endogenous
内在仿真　intrinsic simulation; intrinsic simulism
内在互动　intra-action
内在价值　intrinsic value
内置的　built-in
内置仿真　built-in simulation
内置质量保证　built-in quality assurance
能触知的　tactile
能行性　effectiveness
能力　capability
能力成熟度模型　capability maturity model
能力分析　capability analysis
能量　energy
能量流　energy flow
能量守恒　conservation of energy
能量守恒定律　energy conservation law
能量转换器　energy transducer
拟合　fit; fit (v.); fitting
拟合指标　fit indicator
拟申请的域名注册　domain of intended application

逆　inverse; inversion
逆变换　inverse transformation
逆多体系统动力学　inverse multibody system dynamics
逆分对数　inverse logit
逆行为　inverse behavior
逆黑塞矩阵　inverse Hessian
逆模拟仿真　inverse ontomimetic simulation
逆三次多项式　inverse cubic polynomial
逆时因果关系　reverse time causality
逆问题　inverse problem
逆向仿真　reverse simulation
逆向工程　reverse engineering
逆向建模　inverse modeling
凝聚仿真　cohering simulation
牛顿迭代（法）　Newton iteration
牛顿-格雷戈里多项式　Newton-Gregory polynomial
牛顿-格雷戈里后向多项式　Newton-Gregory backward polynomial
牛顿-格雷戈里前向多项式　Newton-Gregory forward polynomial
扭曲的模型　distorted model
扭曲的现实　distorted reality
暖机时段　warm up period

Oo

欧拉积分　Euler integration
欧拉积分方法　Euler integration method
欧拉角　Euler angle
欧拉模型　Eulerian model

偶函数　even function
偶然（性）　contingency
偶然不确定性　aleatory uncertainty
偶然的（事件）　contingent
偶然的行为　aleatory behavior
偶然事件　contingent event
耦合　couple (v.); coupling
耦合闭包　coupling closure
耦合程度　degree of coupling
耦合程序　coupled program
耦合的　coupled
耦合的常微分方程　coupled ODE
耦合的连续模型　coupled continuous model
耦合的模型　coupled model
耦合多模型　coupled multimodel
耦合方法　coupling recipe
耦合仿真　coupled simulation
耦合仿真软件　coupled simulation software
耦合非线性模型　coupled nonlinear model
耦合函数　coupling function
耦合后继仿真研究　coupled successor simulation study
耦合机制　coupling scheme
耦合矩阵　coupling matrix
耦合可变成分模型　coupled variable-component model
耦合可变结构模型　coupled variable-structure model
耦合离散模型　coupled discrete model
耦合离散事件系统规范　coupled DEVS
耦合离散事件系统规范行为　coupled

DEVS behavior
耦合离散事件系统规范模型　coupled DEVS model
耦合模型　model coupling
耦合模型编辑器　coupled-model editor
耦合偏微分方程　coupled partial differential equation
耦合器　coupler
耦合强度　coupling strength
耦合人与材料系统　coupled human and material system
耦合算法　coupled algorithm
耦合特性下的闭包　closure under coupling property
耦合拓扑　coupling topology
耦合网络　coupled network
耦合系统　coupled system
耦合系统模型　coupled system model
耦合线性/非线性模型　coupled linear/nonlinear model
耦合线性模型　coupled linear model
耦合形式下的闭包　closure under coupling
耦合型偏微分方程　coupled PDE
耦合约束　coupling constraint

Pp

PC博弈　PC game
爬行　creep
帕德逼近　Padé approximation
帕德近似法　Padé approximation method
帕累托分布　Pareto distribution
帕累托前沿方法　Pareto frontier approach
帕累托随机变量　Pareto random variable
帕累托最优法　Pareto optimal approach
帕累托最优解　Pareto optimal solution
排比　alignment
排队规则　queuing discipline
排队论　queueing theory
排队模型　queueing model
排队网络模型　queueing network model
排队系统　queueing system
排练　rehearsal
排列关系　ranked relation
排列模型　ranked model
排列转换　ranked transformation
排序　ordering; sequencing
排序错误　sequencing error
派生　derivative
判别分析　discriminatory analysis
判别式变量　discriminant variable
判断数据　judgmental data
判断误差　judgment error
判据　criterion
庞泰利代斯算法　Pantelides algorithm
抛物线型偏微分方程　parabolic PDE
培训　training
佩特里网　Petri net
佩特里网仿真　Petri net simulation
佩特里网模型　Petri net model
配对　pair
配套工具　adjunct tool
配置　configuration

配置管理　configuration management
配置误差　configuration error
膨胀的时间　expanded time
碰撞　impact
碰撞建模　impact modeling
碰撞时间　impact time
批量仿真　batch simulation
批量输入　batch input
批准　authority
匹配　match (v.); matching
匹配算法　matchmaking algorithm
匹配引擎　matchmaking engine
偏差　bias; bias error
偏微分方程　partial differential equation; PDE (Partial Differential Equation)
偏微分方程仿真　partial differential equation simulation
偏微分方程建模　PDE modeling
偏微分方程建模与仿真　partial differential equation M&S
偏序　partial ordering
偏压　bias
偏移　bias
偏移误差　offset error
偏倚　bias
偏值　partial value
偏置　bias
偏转误差　deflection error
片　chip
片段　segment
片上系统　system on a chip
频带　band
频率　frequency
频率比　frequency ratio
频率图　frequency plot
频率误差　frequency error
频谱分析　spectral analysis
频谱分析技术　spectral analysis technique
频谱算法　spectral algorithm
品质度量　quality metric
品质问题　quality issue
平方误差　quadratic error
平行处理　parallel processing
平行过程　parallel process
平行扩展　parallel expansion
平行离散事件系统规范模型　PDEVS model
平行实境　alternate reality
平行实境游戏　alternate reality game; alternative reality game
平行世界　parallel world
平衡　equilibrium
平衡的　steady
平衡模型　equilibrium model
平衡输入　balanced input
平衡条件　equilibrium condition
平衡误差　balance error; balanced error
平滑　smooth; smooth (v.)
平滑参数　smoothing parameter
平滑的　smoothed; smoothing
平滑度仿真　smoothness simulation
平滑行为　smooth behavior
平滑数据　smooth data
平缓　smooth; smooth (v.)
平均的　average
平均恢复时间　mean time to restore
平均误差　average error; mean error

平均值　average value；averaged value
平视显示　head-up display
平视仪表　head-up instrument
平台　platform
平台独立性　platform independence
平台级　platform level
平台集成　platform integration
平台实现　platform implementation
平台特定的　platform-specific
平台特定的建模　platform-specific modeling
平台特定的模型　platform-specific model
平台无关的　platform independent
平台相关的　platform dependent
平台需求　platform requirements
平台依赖性　platform dependence
平太阳时　mean solar time
平稳变量　stationary variable
平稳过程　stationary process
平移变量　translational variable
评定　assess (*v.*)
评估　assessed；assessment；value
评估测度　assessment metric
评估工具　assesment tool
评估过程　assessment procedure
评估价值　assessed value
评估模型　assessment model；evaluative model
评估目标　evaluation goal
评估输入　evaluated input
评估输入源　evaluated source of input
评估数据　assessed data；evaluation data
评估数据集合　evaluation data set

评估约束　assessment constraint
评估者　evaluator
评价　appraisal；evaluation；assessing；assessment；assess (*v.*)；evaluate (*v.*) value
评价方法　assessment metric
评价模型分析　evaluative model analysis
评价认识　evaluated understanding
评价性理解　evaluative understanding
评价指标　evaluation metric
评审　review；review (*v.*)
评审技术　review technique
评注　document (*v.*)
屏幕　screen
屏幕可读性　screen readability
破坏性思维实验　destructive thought experiment
破碎　breaking
破译　break (*v.*)
剖面测量　profiling
剖面图　profile
朴素　simplicity
普遍的　pervasive
普遍的假设　common assumption
普拉克珀斯　Practopoiesis
普拉克珀斯的　Practopoietic
普拉克珀斯系统　Practopoietic system
普拉克珀斯周期　Practopoietic cycle
普利斯科识别　Prisk identification
普适仿真　pervasive simulation
普适化网格仿真　pervasive grid simulation
普适计算　pervasive computing
普适技术　pervasive technology

普适系统　pervasive system
普通克里格法　ordinary Kriging
普通克里格仿真　ordinary Kriging simulation

Qq

期望　anticipation; desirability; anticipate (v.)
期望的行为　desired behavior
期望驱动　expectation-driven
期望驱动型推理　expectation-driven reasoning
期望输入　anticipated input; expected input
期望值　expected value
期望值模型　expectancy-value model; expected-value model
期望状态　expected state
期限　term
齐次边界条件　homogeneous boundary condition
齐次方程　homogeneous equation
齐次模型　homogeneous model
齐次条件　homogeneous condition
其他条件相同的模型　ceteris paribus model
奇点　singular point; singularity
奇点耦合　singular coupling
奇点输入　singular input
奇怪吸引子　strange attractor
奇特问题　wicked problem
奇异性　singularity
棋牌游戏　board game
企业　business; enterprise
企业仿真模型　enterprise simulation model
企业模型　enterprise model
启动变量　activation variable
启动插值　start-up interpolation
启动规则　enabling rule
启动模型　starting model
启动条件　starting condition
启动周期　start-up period
启动状态　starting state
启发法　heuristics
启发过程　heuristic procedure
启发式的　heuristic
启发式的数值　heuristic value
启发式方法　heuristic method
启发式仿真　heuristic simulation
启发式评价　heuristic evaluation
启发式误差　heuristic error
启发式学习　heuristic learning
启发式知识　heuristic knowledge
启用实例　enabled instance
起始点　start place
起始段　initial section
起始模型　starting model
起源于　traceable
弃真错误模型　error of rejecting valid model
千兆级　petascale
千兆级仿真　petascale simulation
千兆级计算　petascale computing
千兆级计算机　petascale computer
前端　front-end
前端分析　front-end analysis
前端接口　front-end interface
前额视网膜　forehead retina

前额视网膜系统　forehead retina system
前进　advance (v.); advancement; advance
前期系统　precocial system
前人仿真　ancestor simulation
前哨事件　sentinel event
前摄的　proactive
前摄地　proactively
前摄行为　proactive behavior
前摄交互　proactive interaction
前摄决策　proactive decision
前摄模型维护　proactive model maintenance
前摄维护　proactive maintenance
前摄系统　proactive system
前提　antecedent; postulate
前向差分算子　forward difference operator
前向多仿真　forward multisimulation
前向欧拉积分　forward Euler integration
前向欧拉算法　forward Euler algorithm
前项　antecedent
前因变量　antecedent variable
前瞻时间　look-ahead time
前置变量　lead variable
潜在变量　latent variable
潜在的　latent
潜在的方面　latent aspect
潜在模型　latent model
潜在缺陷　potential defect
潜在误差　latent error
潜在值　latent value

浅层模型　shallow model
欠设计　poor design
欠实时时间　slower than real-time time
欠实时系统　slower than real-time system
嵌入　embedding
嵌入的　built-in
嵌入仿真器　built-in simulator
嵌入仿真中的线性规划　linear programming embedded within simulation
嵌入基于规则的系统中的仿真　simulation embedded within rule-based system
嵌入式　embedded
嵌入式仿真　embedded simulation
嵌入式仿真中的专家系统　expert system embedded within simulation
嵌入式交互　embedded interaction
嵌入式模型　embedded model
嵌入式数值　embedded value
嵌入式系统　embedded system
嵌入式训练　embedded training
嵌入式训练系统　embedded training system
嵌入式智能体　embedded agent
嵌入线性规划中的仿真　simulation embedded within linear programming
嵌入优化中的仿真　simulation embedded within optimization
嵌入专家系统中的仿真　simulation embedded within expert system
嵌套博弈　nested game
嵌套的模型　nested model
嵌套多仿真　nested multisimulation

嵌套仿真　nested simulation
嵌套分对数　nested logit
嵌套耦合　nested coupling
嵌套耦合多模型　nested coupled multimodel
强 NP 难题　strongly NP-hard problem
强度变量　intensive variable
强度测试　stress testing
强非线性系统　highly nonlinear system
强分布系统　highly distributed system
强互仿真　strong bisimulation
强互相似　strong bisimilarity
强加性输入　imposed input
强健仿真　strong simulation
强烈的期待　strong anticipation
强耦合算法　strongly coupled algorithm
强强仿真　strong-strong simulation
强生物相似性　strong biosimilarity
强双向耦合　strong two-way coupling
强双向耦合仿真　strong two-way coupling simulation
强涌现　strong emergence
强涌现行为　strong emergent behavior
强预期系统　strongly anticipatory system
强制　coerce (*v.*); coercion
强制的　coercive
强制仿真　coercion simulation
强制事件　forced event
强制输入　forced input
翘曲　warp
切换　switch; switching
切换模型　switching model

切片　slicing
切实稳定的　faithfully stable
侵入法　intrusive method
亲和度　affinity
轻量级仿真　portable simulation
轻量级仿真系统　portable simulation system
轻量级模型　portable model
倾向误差　tendency error
清算　reckoning
清算价值　abandonment value; disposal value
清晰度　clarity
情感标记语言　emotion markup language
情感参数　emotional parameter
情感触发　emotional trigger
情感计算　affective computing
情感建模　emotional modeling
情感交互　emotional interaction
情感界面　emotional interface
情感模型　emotion model; emotional model
情感渗透　emotional filter
情感用户界面　affective user interface
情感诱导　emotion-induced
情感智能交互系统　emotionally-intelligent interactive system
情感智能界面　emotionally intelligent interface
情感状态　emotional state
情景　scenario; situation
情景仿真　scenario simulation
情景活动　situated activity
情景计算　computed situation

情景交互　situative interaction
情景空间　scenario space
情景生成　scenario generation
情景演变　scenario evolution
情景自适应性　scenario adaptivity
情境（上下文）　context
情境帮助　contextual help
情境变量　context variable; situational variable
情境表达　context representation
情境沉浸　situated immersion
情境的(上下文的)　contextual
情境的复杂性　situational complexity
情境分析　context analysis; contextual analysis
情境概念　contextual concept
情境感知　context awareness; situation awareness; situational awareness
情境感知的　context-aware
情境感知输入　context-aware input
情境感知系统　context-aware system
情境管理　context management
情境-行动　situation-action
情境-行动规则　situation-action rule
情境兼容性　context compatibility; contextual compatibility
情境建模工具　context modeling tool
情境聚合　context aggregation
情境理解　context comprehension; contextual understanding
情境敏感的　context-sensitive
情境敏感的帮助　context-sensitive help
情境敏感的输入　context-sensitive input
情境敏感的应用　context-sensitive application
情境敏感系统　context-sensitive system
情境模型　context model
情境设计　context design
情境识别　context identification
情境属性　situational attribute
情境推理　context inference
情境无感的输入　context-unaware input
情境无关仿真　context-free simulation
情境无关模型　context-free model
情境相关　contextual dependency
情境相关的模型　context-dependent model
情境形式化　context formalization
情境验证　contextual validation
情境应急行为　contextualized emergent behavior
情况　status
情绪　emotion
情绪表现　emotional expression
情绪的行为　emotional behavior
情绪化输入　emotional input
情绪耦合　emotional coupling
情绪上耦合的　emotionally coupled
情绪实验　emotional experiment
情绪响应　emotional response
请求　request
囚徒困境　prisoner's dilemma
求解　solving
求解错误　solution error
求解技术　solution technique

求解器　solver
求值程序　evaluator
区分　partition (v.)
区间　interval
区间变量　interval variable
区间尺度　interval scale
区间尺度变量　interval-scale variable
区间公差　interval tolerance
区间值　interval value
区位　location
区域　domain; region
区域集合　range set
区组设计　blocked design
驱动　animate (v.); driven
驱动器　driver
驱动误差　driver error
趋势　trend
趋势校核　trend verification
曲面　surface
曲面建模　surface modeling
曲面模型　surface model
曲面细分　tessellation
曲线　curve
取景器　viewer
取消　cancel (v.)
取消的　cancelled
取消的活动　deselected action
取样与保持　sample and hold
去模糊　defuzzificate (v.)
去模糊的　defuzzificated
去模糊化　defuzzification
去模糊化方法　defuzzification method
去模糊化值　defuzzificated value
去模糊器　defuzzifier
趣味游戏　fun game

权标　token
权衡需求　trade-off requirements
权威　authoritative
权威模型　authoritative model
权威认证　certification authority
权威数据　authoritative data
权威数据源　authoritative data source
权限模型矩阵　right model matrix
全部状态　total state
全沉浸式游戏　total immersion gaming
全阶　full order
全阶模型　full-order model
全局变量　global variable
全局的　global
全局行为　global behavior
全局积分误差　global integration error
全局近似　global approximation
全局模型　global model
全局时间　global time
全局时间表　global schedular
全局时间管理器　global time manager
全局调度器　global state-transition function
全局误差　global error
全局误差界　global error bound
全局相对精度　global relative accuracy
全局相关误差　global relative error
全局信息　global information
全局优化　global optimization
全局约束　global constraint
全局知识库　global knowledge base
全局转移函数　global transition function

全局最优化问题　global optimization problem
全耦合仿真　fully coupled simulation
全球位置感知输入　global position sensing input
全软件仿真　all-software simulation
全生命周期支持　life cycle support
全数字仿真　all-digital simulation
全数字仿真软件　all-digital simulation software
全数字模拟仿真　all-digital analog simulation
全速仿真　as-fast-as-possible simulation
全息仿真　holographic simulation
全息合作　intra-holonic cooperation
全息化　holonization；holonize (v.)
全息化的　holonized
全息透镜系统　holacratic system
全息显示　holographic display
全系统仿真　full-system simulation
全系统行为　total system behavior
全协同仿真　full cosimulation
全隐式龙格-库塔算法　fully-implicit Runge-Kutta algorithm
全有序集合　totally ordered set
全域最小时间　global minimum time
劝导仿真　persuasive simulation
劝导技术　persuasive technology
劝导媒体仿真　persuasive media simulation
劝导模型　persuasion model
缺口　gap
缺口分析　gap analysis
缺失的数据　missing data
缺失值　missing value
缺陷　defect；flaw
缺陷密度度量　defect density metric
确定　certainty；identify (v.)
确定……的参数　parameterize (v.)
确定变迁　deterministic transition
确定的　certain；deterministic
确定的分布　deterministic distribution
确定的假设　deterministic assumption
确定的离散事件系统规范　deterministic DEVS
确定的上下文无关的林氏系统　deterministic context-free L-system
确定的信息　certain information
确定性定时转换　deterministic timed transition
确定性仿真　deterministic simulation
确定性函数　deterministic function
确定性模型　deterministic model
确定性实验　deterministic experiment
确定性算法　deterministic algorithm
确定性系统　deterministic system
确定性优化　deterministic optimization
确定值　certainty value
确切值　explicit value
确认　accreditation；verification；validation；validate (v.)
确认发起方　accreditation sponsor
确认管理　validation management
确认过程　accreditation process
确认计划　accreditation plan
确认提议者　accreditation proponent
确认推荐　accreditation recommendation

确认与认证 accreditation and certification
群 group
群体仿真 crowd simulation; swarm simulation
群体行为 crowd behavior; group behavior; swarm behavior
群体激发工作 stigmergy
群体激发工作的行为 stigmergic behavior
群体激发工作行为 stigmergy model
群体激发工作式交互 stigmergic interaction
群体激发工作式通信 stigmergic communication
群体激发工作式系统 stigmergic system
群体激发工作式协调 stigmergic coordination
群体激发工作式协作 stigmergic collaboration
群体激发工作系统 stigmergy system
群体激发工作现象 stigmergic
群体模型 swarm model
群体智能 swarm intelligence

Rr

扰动技术 distractive technology
热力学变量 thermodynamic variable
热值 calorific value
人 human
人的感知 human perception
人的功效测量 human performance measurement
人工变量 artificial variable
人工的 artificial
人工分类代码 labor classification code
人工辅助语言模型映射 interlanguage model mapping
人工辅助语言映射 interlanguage mapping
人工免疫系统 artificial-immune system
人工模拟 manual simulation
人工神经网络 artificial neural network
人工生命游戏 artificial life game
人工世界 artificial world
人工系统 artificial system
人工现实 artificial reality
人工因素 artifact
人工制品 artifact
人工智能 artificial intelligence
人工智能控制的仿真 AI-controlled simulation
人工智能控制的高级佩特里网 AI-controlled high-level Petri net
人工智能模型 AI model
人工智能体通信 human-agent communication
人工智能引擎 AI engine
人机仿真 man-machine simulation
人机工程 ergonomic
人机工程学 ergonomics
人机交互仿真 human-machine simulation
人机界面 human-computer interface; human-machine interface

人机系统仿真 man-machine system simulation
人际技巧 interpersonal skill
人类的行为 human behavior
人类行为表示（法） human behavior representation
人类行为建模 human behavior modeling
人类可读的模型 human-readable model
人类可解释的描述模型 human-interpretable descriptive model
人类视角 human view
人类转录错误 human transcription error
人体测量校正模型 anthropometrically correct model
人体动画标记语言 body animation markup language
人体建模 human modeling
人体建模器 human modeler
人为定义的活动 human-defined activity
人为仿真 human-initiated simulation
人为误差 human error; personal error
人为系统 anthropogenic system
人为因素 human factors
人为中断 artificial discontinuity
人与仿真交互 human-simulation interaction
人在环（回路） human-in-the-loop; man-in-the-loop
人在环（回路）仿真 human-in-the-loop simulation; man-in-the-loop simulation
人在环（回路）仿真器 human-in-the-loop simulator; man-in-the-loop simulator
人在系统 human-in-the system
人造的 synthetic
人造系统 human-made system; man-made system
人-智能体交互 human-agent interaction
人-智能体系统 human-agent system
认可 accept (v.); accreditation; certificate (v.)
认可的 accreditated
认识论 epistemology
认识论的 epistemological
认识论的涌现 epistemological emergence
认识论视角 epistemological perspective
认识论转变 epistemological shift
认证 certification; confirmation
认证代理 accreditation agent
认证方法 certification method
认证方法学 certification methodology
认证数据 authentication data; certification data
认证误差 confirmation error
认知 cognition
认知（系统）建模 cognitive modeling
认知傲慢 epistemic arrogance
认知变量 cognitive variable
认知不确定性 epistemic uncertainty
认知策略 cognitive strategy
认知的 cognitive
认知的不透明度 epistemic opacity

认知的复杂性　cognitive complexity
认知方面　cognitive aspect
认知仿真　cognitive simulation
认知仿真器　cognitive simulator
认知复杂性个体　cognitive-complexity individual
认知工程学　cognitive ergonomics
认知过程　cognitive process
认知监控　cognitive monitoring
认知建模策略　cognitive-modeling strategy
认知建模体系架构　cognitive-modeling architecture
认知结构　cognitive architecture
认知模式　cognitive schema
认知模型　cognitive model; cognitized model
认知耦合　cognitive coupling
认知偏差　cognitive bias
认知实验　cognitive experiment
认知系统　cognitive system
认知响应　cognitive response
认知信任　epistemic trust
认知信息论　cognitive informatics
认知映射　cognitive map
认知源　source of cognition
认知知觉　cognitive perception
认知主体　cognitive agent
认知状态　cognitive state
任命　accreditation
任期　administration
任务　task; mission
任务等级建模　mission-level modeling
任务共享　task sharing
任务规划器　task planner
任务级　mission level
任务级别　task level
任务级仿真　mission-level simulation
任务空间　mission space
任务空间概念模型　conceptual model of the mission space
任务空间实体　mission space entity
任务排演仿真　mission rehearsal simulation
任务切换　task switching
任务特定接口　task specific interface
任务系统　mission system
任务演练　mission rehearsal
任意变量　arbitrary variable
容差比较　tolerance on comparison
容错的　fault tolerant
容错仿真　fault-tolerant simulation
容错环境　mistake forgiving environment
容器模型　container model
容忍　tolerance
容忍的　tolerant
容忍误差　bearing error
容许　admission
容许的输入　admissible input
容许的阈值　acceptable threshold
容许空间　admissible space
容许区间　tolerance interval
容许条件　admissibility condition
容许误差　error containment
容许误差限　acceptance error
容许值　admissible value
融合　fusion
融合离散事件系统规范的转换函数　confluent DEVS transition-function

融合转换函数　confluent transition-function
冗余代码　redundant code
冗余系统　redundant system
冗余状态　redundant state
冗长方面　redundant aspect
柔点　flexible point
柔顺　compliance
柔体仿真　soft-body simulation
柔性　flexibility
柔性动态模型更新　flexible dynamic model updating
柔性系统　flexible system
入口　access; entry
软传感器输入　soft sensor input
软错误　soft error
软仿真　soft simulation
软计算　soft computing
软计算仿真　soft computing simulation
软件　software
软件错误　software error
软件代理　software agent
软件代码　software code
软件方法学　software methodology
软件仿真器　software simulator
软件工程　software engineering
软件工程过程　software engineering process
软件规范　software specification
软件环境　software environment
软件架构　software architecture
软件兼容的　software compatible
软件建模　software modeling
软件建模器　software modeler
软件建模器评估　assessment of software modeler
软件建模语言　software modeling language
软件开发　software development
软件密集系统　software-intensive system
软件密集型　software-intensive
软件认证　software certification
软件设计错误　software design error
软件实现　software implementation
软件套接字接口技术　software socket interface technique
软件完整性　software integrity
软件文档　software documentation
软件系统工程　software systems engineering
软件项目生命周期　software project life cycle
软件需求　software requirement
软件在环　software-in-the-loop
软件在回路仿真　software-in-the-loop simulation
软件资源　software resource
软件组成　software composition
软件组件　software component
软耦合　soft coupling
软实时约束　soft real-time constraint
软输出　soft output
软输入　soft input
软数据　soft data
软系统　soft system
软系统思考　soft system thinking
软约束　soft constraint

弱仿真　weak simulation
弱互仿真　weak bisimulation
弱互仿真等价　weak bisimulation equivalence
弱经典仿真　weak classical simulation
弱涌现　weak emergence
弱涌现行为　weak emergent behavior
弱预期　weak anticipation
弱预期系统　weakly anticipatory system
ρ-DEVS　rho-DEVS

Ss

Ⅲ型错误　type Ⅲ error
Ⅲ型错误概率　type Ⅲ error probability
赛博安全　Cybersecurity
赛博代理　Cyberagent
赛博空间　Cyberspace
赛博物理系统　Cyber-physical system
赛博物理系统仿真　Cyber-physical system simulation
赛博物理状态　Cyber-physical state
赛博系统　Cybersystem
赛车游戏　driving game
赛后分析　postgame analysis
三次插值　cubic interpolation
三次样条插值　cubic spline interpolation
三段论　syllogism
三极的　triple pole
三角分布　triangular distribution
三阶超隐式亚当斯公式　third-order overimplicit Adams formula
三阶精度　third-order accurate
三阶龙格-库塔积分　third-order Runge-Kutta integration
三维仿真　3-d simulation; third degree simulation
三维建模　3-d modeling
三维空间建模　3-dimensional modeling
三维模型　3-d model; 3-dimensional model
三维全息显示　3-d holographic display
三维位置输入　3-d positional input
三维显示　3-d display
扫描　scan; scan (v.); scanning
扫描活动　scan activity
沙箱式仿真　sandbox-style simulation
沙箱游戏　sandbox game
筛查　screening
筛选　selection
筛选阶段　selection phase
筛选准则　selection criterion
删除　pruning
删除错误　deletion error
伤害性刺激　nociceptive stimulus
伤害性输入　nociceptive input
商务　business
商务博弈　business gaming
商务博弈管理　business game management
商务仿真博弈　business simulation game
商务仿真博弈管理　business simulation game management
商务模型　business model
商业的现成仿真工具包　commercial

off-the-shelf simulation package
商业的现成仿真库　commercial off-the-shelf simulation library
商业的现成模型库　commercial off-the-shelf model library
商业游戏　business game
商用成品软件　COTS (Commercial Off-The-Shelf) software
上层本体　upper ontology
上层合并本体　upper merged ontology
上帝模拟游戏　god game
上界　upper bound
上确界　supremum
上下文仿真　in context simulation
上限　upper limit
上一个状态　last state
舍入误差　rounding error; round-off error
舍入值　rounded value
设备　device; equipment
设备误差　device error; equipment error
设备延迟　device latency
设定值　set value
设计　design
设计变更　design for change; designed for change
设计变量　design variable
设计不确定性　uncertainty in design
设计的模型　designed model
设计的收益系统　designed system of interest
设计方法学　design methodology
设计仿真器　design simulator
设计规范　design specification

设计阶段　design phase
设计可接受性　design acceptability
设计空间　design space
设计模式　design pattern
设计实验　designed experiment; designed experimentation
设计文档　design document
设计文件　design documentation
设计误差　design error
设计校核　design verification
设计需求　design requirement
设计选型　design selection
设计验证　design validation
设计一致性　design conformance
设计语言　design language
设计准则　design criterion; design guideline
设想　assumption
设想的　assumed
设想的现实　conceived reality
设置　set-up
设置误差　set-up error
社会博弈　social game
社会动力学模型　sociodynamic model
社会复杂性　societal complexity
社会合弄　social holon
社会媒体　social media
社会媒体环境　social media environment
社会媒体游戏　social media game
社会网络　social network
社会文化知识　socio-cultural knowledge
社会影响博弈　social impact game
社区　community

射击游戏　shooter game
摄动指数　perturbation index
申请　application
身份　identity；status
深度优先　depth first
深入的模型　deep model
深入理解　deep understanding
神经模糊推理算法　neuro-fuzzy inference algorithm
神经网络　neural network
神经网络元模型　neural network metamodel
神经系统级建模　neural-level modeling
神经信任游戏　neurotrust game
审级模型　trial-level model
审计　audit；audit (v.)
审计策略　audit policy
审计数据　auditable data
审计追踪　audit trail
审美经验　aesthetic experience
审议连贯　deliberative coherence
审议制度　deliberative system
审议主体　deliberative agent
生产代码　production code
生产力度量　productivity metric
生成　generated；produce (v.)
生成的　generative
生成仿真　generative simulation
生成行为　generative behavior
生成建模　generative modeling
生成模型　generative model
生成视差仿真　generative parallax simulation
生成现实　generated reality

生存博弈　game of life
生存性仿真　survivability simulation
生命周期　life cycle
生命周期成本　life cycle cost
生命周期成本建模　life cycle cost modeling
生命周期成本模型　life cycle cost model
生命周期管理　life cycle management
生命周期建模　life cycle modeling
生命周期模型　life cycle model
生态激励系统　biologically-inspired system
生物传感器　biosensor
生物仿真　biosimulation
生物仿制药　biosimilar
生物过程　biological process
生物激励　bio-inspired
生物启发的　bio-inspired
生物启发的通信　bio-inspired communication
生物启发式仿真　biologically-inspired simulation
生物启发式工程　biologically-inspired engineering
生物启发式计算　bio-inspired computation; bio-inspired computing; biologically-inspired computation; biologically-inspired computing
生物启发算法　biologically-inspired algorithm
生物输入　bio-input
生物相似物　biosimilar
生物相似性　biosimilarity
生物学建模器　biological modeler

声明 assert (*v*.); declaration; declare (*v*.); notice; predicate; statement
声明的变量 declared variable
声明的随机变量 declared random variable
声呐输入 sonar input
声音仿真 audio simulation
声音输入 acoustic input
省略 omission
剩余变量 surplus variable
剩余标准差 residual standard error
剩余故障 remaining fault
失败 failure
失活 deactivation
失效 failure
失效分析 failure analysis
失验 disconfirmation
施工 construction
施加的 imposed
施加的目标 imposed goal
湿实验室数据 wet-lab data
湿式实验室 wet laboratory
十进制数值 decimal value
十字形模型 cross model
时变模型 time-varying model
时变耦合 time-varying coupling
时变系统 time-variant system
时变系统仿真 time-varying system simulation
时不变 time invariance
时不变的 time invariant
时不变连续系统 time-invariant continuous system
时不变模型 time-invariant model
时不变耦合 time-invariant coupling
时不变系统 time-invariant system
时基 time base
时基发生器 time-base generator
时基精度 time-base accuracy
时间 time
时间边界条件 temporal boundary condition
时间变量 time variable
时间变迁 timed transition
时间变迁系统 timed transition system
时间表的 schedular
时间不确定性 temporal uncertainty
时间步进 time stepping
时间步进仿真 time-stepping simulation
时间步进式执行 time-stepped execution
时间步长 time step
时间步长模型 time step model
时间槽 time slot
时间常数 time constant
时间尺度 time scale
时间尺度变量 time-scale variable
时间触发的 time-triggered
时间触发的控制 time-triggered control
时间戳 time-stamp
时间戳上界 upper bound on the time stamp
时间戳顺序 time-stamp order
时间戳下界 lower bound on the time stamp
时间戳消息 time-stamped message
时间的 temporal
时间独立的 time independent

时间独立的协变量　time-independent covariate
时间独立的约束　time-independent constraint
时间对准　time alignment
时间方面　time aspect
时间分辨率　time resolution
时间分片　time slicing
时间分片的仿真　time-slicing simulation
时间管理　time management
时间管理模块　time-management module
时间管理数据　time-management data
时间管理算法　time management algorithm
时间管理透明度　time-management transparency
时间轨迹　time trajectory
时间集合　time set
时间假设　temporal assumption
时间间隔　time interval
时间间隔仿真　time-interval simulation
时间离散方法　discrete-time method
时间离散化　time discretization
时间粒度　time granularity
时间连续系统　continuous-time system
时间流　time flow
时间流机制　time flow mechanism
时间耦合　temporal coupling
时间佩特里网　timed Petri net
时间片段　time-slice
时间片仿真　time-slice simulation

时间翘曲　time warp
时间翘曲仿真　time-warp simulation
时间翘曲机理　time-warp mechanism
时间翘曲算法　time-warp algorithm
时间驱动　time-driven
时间驱动的仿真　time-driven simulation
时间事件　time-event
时间事件系统　time-event system
时间索引的　time-indexed
时间索引数据　time-indexed data
时间条件　temporal condition
时间推进　time advance; time advancement
时间推进函数　time advance function
时间推进机制　time advance mechanism
时间推进请求　time advance request
时间推进准许　time advance grant
时间误差　temporal error
时间相关　time dependent
时间相关的耦合　time-dependent coupling
时间相关的协变量　time-dependent covariate
时间相关的约束　time-dependent constraint
时间相关事件　time-dependent event
时间序列　time series
时间序列有效性　time-series validity
时间一致性　time consistency
时间约束　time constraint
时间约束的　time constrained
时间值　time value
时间周期　time period

时间轴　time axis
时间自动机　timed automata; timed automaton
时空方法　spatiotemporal method
时空仿真　spatiotemporal simulation
时空分析方法　spatiotemporal analytical method
时空连续体　space-time continuum
时空模型　spatiotemporal model
时空事件　spatiotemporal event
时期　period
时隙　time slot
时序仿真　timing simulation
时序分析　time series analysis
时序关系　temporal relation
时序规范　temporal specification
时序结构　temporal structure
时序模型　temporal model
时序时间　time series time
时序数据　time-series data
时延　latency; time delay
时域　time domain
时域仿真　time-domain simulation
时域设计　time-domain design
时钟　clock
时钟变量　clock variable; clocked variable
时钟输入　clock input
时钟同步　clock synchronization; time-synchronized
识别　identify (v.); recognition
实变量　real variable
实兵实抗训练　live training
实兵实抗训练环境　live training environment

实部　real-part
实际对象　real object
实际轨迹　actual trajectory
实际输出　realistic output
实际系统　real system
实际系统上的训练　training on the real-system
实际系统实验　real system experimentation
实际系统数据　real-system data
实际系统数据可接受性　real-system data acceptability
实际系统增强仿真　real-system-enriching simulation
实际系统支持仿真　real-system-support simulation
实际性能　actual performance
实际用途　practicality of use
实际值　actual value
实际状态　actual state
实践　practice
实践社区　community of practice
实践者　practitioner
实践知识　practical knowledge
实况仿真　live simulation
实况仿真、虚拟仿真与构造仿真间的联接　linkage of live, virtual, and constructive simulations
实况系统　live system
实况系统增强仿真　live system-enriching simulation
实况系统支撑仿真　live system-supporting simulation
实况训练仿真　live training simulation
实况训练系统　live training system

实例 instance
实例变量 instance variable
实例化 instantiated; instantiation
实例化变量 instantiated variable
实例化操作 instantiated operation; instantiation operation
实例化的模型 instantiated model
实例数据 instance data
实施 administration; implement (v.)
实施的 implemented
实施校验 implementation verification
实施平台 implementation platform
实施确认的机构 accreditation authority
实施确认的组织 accreditation organization
实时 real-time
实时程序设计 real-time programming
实时仿真 real-time simulation
实时仿真的 simulated real-time
实时服务 real-time service
实时环境 real-time environment
实时建模 real-time modeling
实时解决方案 real-time solution
实时决策仿真 real-time decision making simulation
实时可视化 real-time visualization
实时离散事件系统规范 RT-DEVS
实时连续仿真 real-time continuous simulation
实时面向对象的建模 real-time object-oriented modeling
实时平台参考 real-time platform reference
实时时钟 real-time clock
实时时钟同步脉冲 real-time clock synchronization impulse
实时输出 real-time output
实时输入 real-time input
实时输入测试 real-time input testing
实时数据 live data; real-time data
实时数据驱动仿真 real-time data-driven simulation
实时系统 real-time system
实时系统数据评估 assessment of real-system data
实时训练 real-time training
实时验证 real-time validation
实时预测 real-time prediction
实时域 real-time domain
实时约束 real-time constraint
实时执行 real-time execution
实体 entity
实体层次 entity level
实体关系模型 entity-relationship model
实体关系图 entity-relationship diagram
实体行为 entity behavior
实体级仿真 entity-level simulation
实体结构 entity structure
实体流 entity flow
实体模型 entity model
实体视角 entity perspective
实体坐标 entity coordinate
实现 implement (v.); implementation; realization
实现的对象 implemented object
实现的价值 realized value
实现的模型 archieved model

实现机制　implementation mechanism
实现阶段　implementation phase
实现目标的含意　implications of achieving the goal
实现情境　realization context
实验　experiment；experimentation
实验变量　experimentation variable; experimental variable
实验布置　experimental set-up
实验步骤　experimental procedure
实验参数　experimentation parameter
实验策略　experimentation policy
实验场景互操作性　experimental scenario interoperability
实验程序　experimental procedure
实验代理　experiment agent
实验的　experimental
实验的代码　experimental code
实验的概念　experimental concept
实验方法　experimental method
实验方法学　experimental methodology
实验仿真　experimental simulation
实验工具　experimental tool; experimentation tool
实验观察　experimental observation
实验管理　experiment management
实验管理器　experimentation manager
实验规范　experimentation specification
实验规划　experiment planning
实验互操作性　experimentation interoperability
实验技术　experimental technique
实验结果　experimental result

实验科学　experimental science
实验可信度　experimentation credibility
实验空间　experiment space
实验框架　experimental frame
实验框架的可接受性　experimental frame acceptability
实验框架的可用性　applicability of experimental frame
实验框架的可组合性　experimental frame composability
实验框架的实现　experimental frame realization; realization of experimental frame
实验框架对模型的适用性　applicability of an experimental frame to a model
实验框架对模型的适用性评估　assessment of the applicability of an experimental frame to a model
实验框架分解　experimental frame decomposition
实验框架互操作性　experimental frame interoperability
实验框架可导性　experimental frame derivability
实验框架库　experimental frame base
实验框架组合　experimental frame composition
实验论者　experimentalist
实验模型　experiment model; experimental model
实验偏差　experimentation bias; experimental bias
实验轻松　experimentation ease

实验情境　experimental context
实验上　experimentally
实验设计　design of experiment; experiment design; experimental design
实验设计方法学　experimental-design methodology
实验设计评估　assessment of experimental design
实验式地　experimentally
实验视角　experimentation perspective
实验室　laboratory
实验室仿真　laboratory simulation
实验室试验　laboratory experiment
实验数据　experimental data
实验挑战　experimentation challenge
实验条件　experimental condition
实验条件规范　specification of experimental condition
实验条件互操作性　experimental-conditions interoperability
实验条件评估　assessment of experimental conditions
实验调查　experimentation investigation
实验误差　experimentation error; experimental error
实验系统　experimental system
实验性的　experimentative
实验研究　experimental study
实验有效性　experimental validity
实验员　experimentator
实验约束　experiment constraint
实验者　experimenter
实验者偏见　experimenter bias
实验支持　experimentation support
实验知识　experimental knowledge
实验重现性　experiment reproducibility
实验主义　experimentalism
实验专家　experimentist
实用程序　utility; utility program
实用模型　practical model
实用主义　pragmatism
实用主义的　pragmatic
实在　reality
实在论　realism
实证方法论　positivist methodology
实证理论　positivist theory
实证论　positivism
实证效度　empirical validity
实证研究　empirical study
实证主义范式　positivist paradigm
实质　substance
实质互操作性　substantive interoperability
实轴　real axis
实装演习　live exercise
实装仪表化的仿真　live instrumented simulation
实装仪表化的实体　live instrumented entity
矢量　vector
使抽象化　abstract (*v.*)
使等价　equivalencing
使聚集　aggregate (*v.*)
使能　enabling
使能的　enabled
使能仿真　enabling simulation
使能仿真互操作性　enabling-

simulation interoperability
使能技术　enabling technology
使能器　enabler
使用　use；used
使用错误　usage error
使用的元模型　used metamodel
使用挖掘　usage mining
使有生气　animate (v.)
示范　demonstration
世界　world
世界观　world view
世界模型　world model
世界时　universal time
世界坐标　universal coordination
市场　market；market place
式样　style
似是而非的论点　sophism
似真场景　plausible scenario
似真轨迹　plausible trajectory
似真模型　plausible model
似真试验　plausible experiment
似真性　plausibility
似真值　plausible value
似真状态　plausible state
势场　potential field
事后变量　ex post facto variable
事后的　ex post；ex post facto
事后仿真　ex post simulation
事后计算　ex post calculation
事后预测　ex post prediction
事件　event
事件标志　event flag
事件撤回　event retraction
事件抽取　event extraction
事件处理　event processing

事件处理程序　event handler
事件处理器　event processor
事件触发　event-triggered
事件触发控制　event-triggered control
事件代码　event code
事件导向的程序设计　event-directed programming
事件的连续性　succession of events
事件的时间戳　time-stamp of an event
事件调度仿真　event-scheduling simulation
事件段　event segment
事件队列　event queue
事件发生　event occurrence
事件发生器　event generator
事件概率　probability of occurrence
事件跟踪　event trace；event following
事件跟踪仿真　event-following simulation
事件合理性测验　event-validity test
事件记录器　event recorder
事件交付　event delivery
事件交换　event exchange
事件交换格式　event-exchange format
事件节点　event node
事件解释器　event interpreter
事件可取性　event desirability
事件空间　event space
事件类　event class
事件类型　event type
事件历史　event history
事件历史方法　event-history approach
事件粒度　event granularity
事件链　chain of events
事件列表　event list
事件密度　event density

事件描述部分　event description section
事件描述符　event descriptor
事件排序　event sequencing
事件评估　event appraisal
事件评价模型　event-appraisal model
事件驱动　event-driven
事件驱动程序　event-driven program
事件驱动的程序设计　event-driven programming
事件驱动的仿真　event-driven simulation
事件驱动的应用程序　event-driven application
事件驱动架构　event-driven architecture
事件驱动逻辑　event-driven logic
事件扫描方法　activity-scanning method
事件时间　event time
事件实例化　event instantiation
事件视界　event horizon
事件条件　event condition
事件-条件-动作　event-condition-action
事件-条件-动作规则　event-condition-action rule
事件调度　event scheduling
事件调度法　event-scheduling approach
事件通信　event communication
事件通知　event notice
事件同步　event synchronization
事件图　event graph
事件文件　event file
事件线　event line
事件协调　event coordination
事件序列　event sequence
事件序列图　event-sequence diagram
事件选择　event selection
事件学习　event learning
事件因果关系　event causality
事件有效性　event validity
事件有效性测验　event-validity test
事件语言　event language
事件预期　event anticipation
事件追踪　event tracing
事前　ex ante
事前仿真　ex ante simulation
事前分析　ex ante analysis
事前价值　ex ante value
事前评估　ex ante evaluation
事前预测　ex ante prediction
事实　fact；verity
事实错误　factual error
事实探测　fact detection
事务　business
试错　trial-and-error
试错法　trial-and-error method
试探法　heuristic method；heuristics
试位法　regula falsi
试验　experiment；proofing；test
试验仿真　trial simulation
试验观测　test observance
试验规范　norms of experimentation
试验和训练使能架构　test and training enabling architecture
试验鉴定　test qualification
试验鉴定技术　test qualification technique

试验台　testbed
视差仿真　parallax simulation
视差图　parallax view
视场　field-of-view
视点　point of view
视角　perspective
视觉　vision
视觉的　visual
视觉感受　visual sensation
视觉感知　visual perception
视觉环境　visual environment
视觉监控　visual monitoring
视觉建模技术　visual modeling technique
视觉模型　visual model
视觉模型规范　visual model specification
视觉输入　vision input; visual input
视觉输入/输出　visual I/O
视觉特性　visual characteristic
视觉系统　visual system
视觉系统规范　visual system specification
视觉隐身　visual stealth
视景仿真器　visual simulator
视景系统　visual system
视频输入　video input
视频游戏　video game; video gaming
视频游戏仿真　video game simulation
视图　view
视网膜显示屏　retina display
视网膜显示游戏　retina-display game
视线　line-of-sight
视线仿真　line-of-sight simulation
适当　adequacy; appropriateness
适当的　adequate; appropriate
适定问题　well-posed problem
适度不稳定的涌现　moderated unstable emergence
适度的涌现　moderated emergence
适度稳定的涌现　moderated stable emergence
适合　appropriateness
适合的　suitable
适合的行为　adaptive behavior
适合的模型　suitable model
适配　adaptation
适配器　adaptor
适宜　adequate
适宜的仿真　appropriate simulation
适宜的激励　adequate stimulus
适应　adapt (*v.*); adaptation; adapting
适应的　adapted
适应度　fitness
适应度函数　fitness function
适应能力　adaptive capability
适应系统　adaptation system
适应性　adaptability; adaptiveness; applicability
适应性引擎　adaptation engine
适应自动化　adaptive automation
适用的实验框架　applicable experimental frame
适用范围　scope of applicability
适用性　applicability; suitability
适用性范围　range of applicability
适用性感知　perceived suitability
适用性检查　applicability check
释义　interpretation
收集　collect (*v.*); collecting;

collection
收集器　collector
收敛边界仿真　convergent boundary simulation
收敛的仿真　convergent simulation
收敛的有效性　convergent validity
收敛仿真　convergence simulation
收敛阶次　convergence order
收敛误差　convergence error
收缩性　contractivity
收益　interest
收益系统　system of interest
手动输入　manual input
手仿真　hand simulation
手机游戏　mobile game
手势　gesture
手势输入　gesture input; hand-gesture input
手写笔输入　stylus input
守恒系统　conservative system
守恒原理　conservation principle
首先　first
首选行为　preferred behavior
首选项变量　preference variable
首选状态变量　preferred state variable
受控的　controlled
受控佩特里网　controlled Petri net
受控实验　controlled experiment; controlled experimentation
受控系统　controlled system
受控状态　controlled state
受生物启发的　biologically-inspired
受限因变量　censored dependent variable; limited dependent variable
受影响的属性　affected attribute

授权　accreditation; authority
授权测试　authorization testing
授权的　authoring
授权系统　authoring system
书写错误　writing error
疏忽造成的误差　inadvertent error
输出　output
输出变量　output variable
输出测量　measure of outcome
输出层　output layer
输出插值　output interpolation
输出端　output terminal
输出端口　output port
输出段　output segment
输出方程　output equation
输出分析　output analysis
输出复制　output duplication
输出观测端口　output observation port
输出轨迹　output trajectory
输出过程　output process
输出矩阵　output matrix
输出可视化　visualization of output
输出空间　output space
输出理解　output understanding
输出模块管理器　output module manager
输出驱动的验证　output-driven validation
输出日志　output log
输出设备　output device
输出时间　output time
输出事件　output event
输出数据　output data
输出条件　output condition
输出通道　output channel

输出文件　output file
输出限制　output limited
输出限制器　output limiter
输出验证　output validation
输出值　output value
输出值集合　output value set
输出周期性　output periodicity
输入　input
输入/输出　I/O; input/output
输入/输出表　input/output table
输入/输出程序　input/output routine
输入/输出处理　input/output process
输入／输出处理器　input/output processor
输入/输出端口　input/output port
输入/输出端口规范　input/output port specification
输入/输出对　input/output pair
输入/输出分析　input/output analysis
输入／输出规范　input/output specification
输入/输出缓冲区　input/output buffer
输入／输出寄存器　input/output register
输入/输出矩阵　input/output matrix
输入/输出控制　input/output control
输入／输出控制器　input/output controller
输入／输出控制系统　input/output control system
输入/输出模型　input/output model
输入/输出耦合　input/output coupling
输入/输出设备　input/output device
输入／输出示意图　input/output diagram
输入/输出通道　input/output channel
输入/输出误差　input/output error
输入/输出系统　input/output system
输入／输出协处理器　input/output coprocessor
输入／输出要求　input/output requirements
输入／输出语句　input/output statement
输入／输出指令　input/output instruction
输入/输出中断　input/output interrupt
输入／输出转换　input/output switching
输入/输出总线　input/output bus
输入变量　input variable
输入程序　input program
输入存储　input storage
输入单元　input unit
输入的可接受性　input acceptability
输入端口　input port
输入段　input section
输入断言　input assertion
输入队列　input queue
输入范围　input range
输入方程　input equation
输入分布　input distribution
输入格式　input format
输入格式化　input formatting
输入功率　input power
输入观察端口　input observation port
输入规格　input specification
输入规格简易　input specification ease
输入轨迹　input trajectory
输入过程　input process

输入缓冲区　input buffer
输入缓冲区寄存器　input buffer register
输入记录　input log
输入加载　input loading
输入角　input angle
输入节点　input node
输入矩阵　input matrix
输入空间　input space
输入控制　input control
输入跨度　input span
输入块　input block
输入量化误差　input quantization error
输入灵敏度　input sensitivity
输入流　input stream
输入率　input rate
输入媒体　input media
输入逆变器　input inverter
输入片段　input segment
输入评估　evaluation of input; input assessment
输入区　input area
输入驱动的验证　input-driven validation
输入确认　input validation
输入设备　input equipment; input device
输入事件　input event
输入数据　input data
输入数据分析　input-data analysis
输入条件　input condition
输入通道　input channel
输入同步　input synchronization
输入文件　input file
输入限制　input limit; input restriction

输入限制的　input limited
输入限制器　input limiter
输入线性度　input linearity
输入向量　input vector
输入消息　input message
输入校核　input verification
输入信号　input signal
输入有效性　input validity
输入阈值　input threshold
输入源　source of input
输入源评估　evaluation of source of input
输入阅读器　input reader
输入站　input station
输入值　input value
输入值集合　input value set
输入指令　input instruction
输入终端　input terminal
输入轴　input axis
输入装置　input unit
输入准备　input preparation
输入字符　input characteristic
熟练的　warm up
属性　attribute
属性变换　attribute transformation
属性的基数　cardinality of property
属性更新　attribute update
属性过载　attribute overloading
属性实体　attributive entity
属性树　attribute tree
属性所有权　attribute ownership
属性误差　attribute error
属性相似度　attribute similarity
属性选择　attribute selection; property selection

属性值　attribute value
鼠标仿真器　mouse emulator; mouse simulator
术语　term; terminology
术语学　terminology
束缚的　yoked
树　tree
树模型　tree model
树形式　tree formalism
数　number
数据　data
数据安全　data security
数据逼真度　data fidelity
数据编码　data encryption
数据标准　data standard
数据标准化　data standardization
数据表达技术　data-representation technique
数据并行执行　data parallel execution
数据捕捞　data fishing
数据采购　data sourcing
数据采集　data acquisition
数据采集评估　data collection assessment
数据仓库　data repository; data warehouse
数据层　data layer
数据成本　data cost
数据传感器　data sensor
数据传送　data transfer
数据词典　data dictionary
数据词典系统　data dictionary system
数据单位　data unit
数据单元标准化　data element standardization
数据导向的　data-directed
数据导向的推论　data-directed inference
数据的可验证性　validatability of data
数据点　data point
数据发布　data issue
数据发生器　data generator
数据分辨率　data resolution
数据分析　data analysis
数据分析法　data analytics
数据分析技术　data-analysis technique
数据分享　data sharing
数据格式　data format
数据格式转换　data format converter
数据耕耘　data farming
数据更新　data update
数据关联　data association
数据管理　data administration; data management
数据管理员　data steward; data administrator
数据规划　data layout
数据过滤　data filtering
数据含义　data meaning
数据合格　data acceptability
数据集　dataset; data set
数据记录器　data logger
数据加密　data encryption
数据架构　data architecture
数据建档　data documenting
数据建模　data modeling
数据建模方法　data modeling method
数据交换　data interchange
数据交换标准　data exchange standard; data interchange standard

数据接口测试　data interface testing
数据结构　data structure
数据结构化　data structuring
数据结构耦合　data-structured coupling
数据精确性　data accuracy
数据聚集　data aggregation
数据科学家　data scientist
数据可接受性　data acceptability
数据可视化　data visualization
数据可用性　data availability
数据空间　data space
数据库　database
数据库错误　database error
数据库管理　database administration
数据库管理系统　database management system
数据库管理员　database administrator
数据库目录　database directory
数据库设计　database design
数据库系统　database system
数据流　data flow; data stream
数据流变量　data flow variable
数据流测试　data flow testing
数据流分析　data flow analysis
数据流机　data flow machine
数据流建模　data flow modeling
数据流通　data currency
数据流图　data flow diagram
数据滤波方法　data smoothing method
数据滤波机理　data-filtering mechanism
数据密度　data density
数据密集的　data-intensive
数据密集型仿真　data-intensive simulation
数据密集型计算　data-intensive computing
数据密集型系统　data-intensive system
数据描述　data representation
数据模板　data template
数据模型　data model
数据模型变换　data model transformation
数据耦合　data coupling
数据评估　assessment of data; data assessment
数据清理　data cleansing
数据驱动的　data-driven
数据驱动的建模　data-driven modeling
数据驱动的决策　data-driven decision
数据驱动方法　data-driven method
数据驱动方法论　data-driven methodology
数据驱动仿真　data-driven simulation
数据驱动过程　data-driven process
数据驱动推理　data-driven reasoning
数据驱动误差　data-driven error
数据驱动系统　data-driven system
数据驱动智能　data-driven intelligence
数据取证　data forensics
数据确认　data validation
数据认证　data authentication; data certification
数据融合　data fusion
数据入口　data entry
数据入口误差　data-entry error

数据栅格　data grid
数据生成器　data producer
数据生成器认证　data-producer certification
数据实体　data entity
数据适当性　data appropriateness
数据收集　data collection
数据收集阶段　data collection phase
数据收集器　data collector
数据输入　data input
数据属性　data attribute
数据缩放　data scaling
数据探测　data snooping
数据特性　data characteristic
数据提取　data extraction
数据调解　data mediation
数据同步　data synchronization
数据挖掘　data dredging; data mining
数据挖掘语言　data mining language
数据完整性　data completeness; data integrity
数据完整性需求　data integrity requirements
数据维护　data maintenance
数据误差　data error
数据相关性　data relevance
数据相关性分析　data dependency analysis
数据消费者　data consumer
数据校核　data verification
数据校核、验证与确认　data VV&A
数据校核、验证与认证　data verification, validation and certification; data VV&C
数据校核与验证　data V&V; data verification and validation
数据校准　data calibration
数据需求　data requirements
数据压缩　data compression
数据一致性　data consistency
数据仪器　data instrumentation
数据隐匿　data hiding
数据用户　data user
数据有效性　data validity
数据语义学　data semantics
数据元模型　data metamodel
数据元素　data element
数据源　data source
数据正确性　data correctness
数据支持者　data proponent
数据值　data value
数据质量　data quality
数据治理　data governance
数据治理保证　data quality assurance
数据中心　data center
数据中心变换　data center transformation
数据中心层　data center layer
数据中心的　data centric
数据专家知识　data expertise
数据转换　data transformation
数据转换技术　data-transformation technique
数据准备工具　data preparation tool
数据资源　data resource
数据组织　data organization
数量　quantity
数量度量　quantity metric
数学变量　mathematical variable
数学标记语言　mathematical markup

language
数学程序设计 mathematical programing
数学仿真 mathematical simulation
数学建模 mathematical modeling
数学模型 mathematical model
数学游戏 mathematical game
数学语言 mathematical language
数学知识 mathematical knowledge
数值 numerical value; value
数值变量 numerical variable
数值不稳定区域 numerically unstable region
数值不稳定性 numerical instability
数值常微分方程求解器 numerical ODE solver
数值的 numerical
数值范围验证 numeric range validation
数值方法 numerical approach; numerical method
数值仿真 numerical simulation
数值分析 numerical analysis
数值函数 numeric function
数值积分 numeric integration; numerical integration
数值积分法 numerical integration method
数值积分算法 numerical integration algorithm; numerical integration requirements
数值计算 numerical computation; numerical computing
数值解 numerical solution
数值模型 numerical model
数值佩特里网 numerical Petri net
数值实验 numerical experimentation; numerical experiment
数值输入 numerical input
数值微分方程求解器 numerical differential equation solver
数值稳定区域 numerically stable region
数值稳定性 numerical stability
数值状态变量 numerical state-variable
数字 digital
数字（化）双胞胎 digital twin
数字博弈 digital game
数字仿真 digital simulation
数字仿真程序 digital simulation program
数字仿真软件 digital simulation software
数字仿真研究 digital simulation study
数字仿真语言 digital SL
数字化数据 digital data
数字化误差 digitization error
数字计算机 digital computer
数字计算机仿真 digital computer simulation
数字离散系统仿真语言 digital discrete-system SL
数字连续系统仿真语言 digital continuous-system SL
数字量子仿真 digital quantum simulation
数字孪生 digital twin
数字媒体 digital media
数字模拟仿真 digital analog

simulation
数字模拟仿真语言　digital analog SL
数字-模拟转换器　digital-to-analog converter
数字人建模　digital human modeling
数字实现　digital realization
数字实验　digital experiment
数字输入　digital input
数字特性　digital property
数字微分分析仪　digital differential analyzer
数字硬件　digital hardware
数字源仿真语言　digital source SL
数字在线仿真语言　digital online SL
数字证书　digital certificate
数字转换器　quantizer
数字追踪　digital trace
数字足迹　digital footprint
双边博弈　two sided game
双变量　bi-variable
双变量的　bivariate
双变量关系　bi-variate relationship
双层元胞自动机　bi-layer cellular automaton
双动作纯协调博弈　two-action pure coordination game
双动作协调博弈　two-action coordination game
双峰的　bimodal
双极　double pole
双焦建模　bifocal modeling
双焦模型　bifocal model
双曲线偏微分方程　hyperbolic partial differential equation
双曲型偏微分方程　hyperbolic PDE

双人博弈　two person game
双人零和博弈　two person zero-sum game
双线性　bilinearization
双线性插值　bilinear interpolation
双线性的　bilinear
双线性系统　bilinear system
双向变化　bidirectional transformation
双向耦合　two-way coupling
双向输入　two-directional input
双向因果关系　two-directional causality
水平变量　level variable
水准点　benchmark
顺序变量　ordinal variable
顺序程序的并行化　parallelization of sequential program
顺序处理　sequential processing
顺序的　sequential
顺序仿真　sequential simulation
顺序仿真的并行化　parallelization of sequential simulation
顺序仿真器　sequential simulator
顺序仿真语言　sequential SL
顺序分叉　sequential bifurcation
顺序更新　sequential update
顺序行为　sequential behavior
顺序量表　ordinal scale
顺序算法　sequential algorithm
顺序映射　sequential mapping
顺序执行　sequential execution
顺序状态　sequential state
瞬变消除　transient elimination
瞬时变迁　immediate transition
瞬时行为　transient behavior

瞬态 transient state
瞬态灵敏度分析 transient sensitivity analysis
说明 prescription; description; help; interpret (*v.*); describe (*v.*)
说明性表示 declarative representation
说明性程序设计 declarative programming
说明性程序设计范式 declarative programming paradigm
说明性仿真语言 declarative SL
说明性建模 declarative modeling
说明性模型 prescriptive model
说明性模型变换语言 declarative model-transformation language
说明性语句 declarative statement
说明性语言 declarative language
说明性知识 prescriptive knowledge
说明性智能体通信语言 declarative agent communication language
说明性智能体协调语言 declarative agent coordination language
说明性智能体语言 declarative agent language
私密网络 privacy grid
私有事件 private event
私有状态事件 private state event
思考 thinking
思维 thinking
思维仿真 thought simulation
思维控制仿真 thought controlled simulation
思维实验仿真 thought experiment simulation
思维体验 thought experience

思想 thinking
撕裂变量 tearing variable
撕裂算法 tearing algorithm
死板式博弈 rigid game
死变量 dead variable
死代码 dead code
死锁 deadlock
死锁探测 deadlock detection
死锁状态 deadlock state
四极 quadruple pole
四阶精度 fourth-order accurate
松弛 relaxation
松弛变量 slack variable
松弛法 relaxation method
松弛算法 relaxation algorithm
松耦合 loose coupling
松耦合的 loosely coupled
松耦合的联合仿真 loosely coupled federated simulation
松耦合模型 loosely coupled model
松耦合系统 loosely coupled system
松散联系 loose association
松散时间耦合 loose temporal coupling
搜索 search
搜索空间 search space
搜索模型 search model
速度误差 velocity error
速率 rate; speed
速率变量 rate variable
速率数据 rate data
溯因推理 abductive reasoning
溯源性 traceability; tractability
算法 algorithm
算法不可计算的函数 algorithmically

incomputable function
算法程序设计范式　algorithmic programming paradigm
算法错误　arithmetic mistake
算法的检查　algorithmic check
算法等价　algorithm equivalence
算法分析　algorithm analysis
算法检查　algorithm check
算法模型　algorithmic model
算法评估　algorithm assessment
算法同步　algorithm synchronization
算法完整性检查　algorithmic check of completeness
算法稳定性　algorithm stability
算法误差　algorithm error; algorithmic error
算法效率　algorithmic efficiency
算法一致性检查　algorithmic check of consistency
算子　operator
随机　random
随机变量　chance variable; random variable; random variate; stochastic variable
随机变量生成　random variate generation
随机博弈　stochastic game
随机采样　random sampling
随机的　stochastic
随机的自动机　stochastic automaton
随机仿真　random simulation; stochastic simulation
随机仿真模型　stochastic simulation model
随机仿真优化　stochastic simulation optimization
随机分析　stochastic analysis
随机关系　stochastic dependence
随机过程　random process; stochastic process
随机函数　random function
随机化　randomization
随机建模　stochastic modeling
随机近似法　stochastic approximation method
随机矩阵　stochastic matrix
随机可变性　stochastic variability
随机克里格法　stochastic Kriging
随机模型　stochastic model
随机佩特里网　stochastic Petri net
随机平衡不完全区组设计　randomized balanced incomplete block design
随机设计　randomized design
随机失效　random failure
随机时间变迁　stochastic timed-transition
随机实验　random experiment
随机事件　stochastic event
随机输出过程　stochastic output process
随机输入过程　stochastic input process
随机数　random number
随机数发生器　random number generator
随机数种子　random number seed
随机数种子初始化　initialization of random number seed
随机搜索　random search
随机完全区组设计　randomized

complete block design
随机微分博弈 stochastic differential game
随机微分方程 stochastic differential equation
随机微分方程模型 stochastic differential equation model
随机误差 accidental error; random error; stochastic error
随机系统 stochastic system
随机样本 random sample
随机游走 random walk
随机游走仿真 random walk simulation
随机噪声 stochastic noise
随机值 stochastic value
随机转换 stochastic transition
随机状态 random state
随境游戏 pervasive game
碎片 fragmentation
缩比模型 scale model
缩放 scale; scaling
缩放的挂钟时间 scaled wallclock time
缩放的实时仿真 scaled real-time simulation
缩放规则 scaling law
缩小的副本 scaled-down replica
所有权 ownership
索引 index
索引错误 index error
索引的 indexed

Tt

t仿真 t-simulation
他生 allopoiesis
他生的 allopoietic
他生仿真 allopoietic simulation
他生模型 allopoietic model
他生系统 allopoietic system
塔尔简算法 Tarjan algorithm
台 station
太赫兹技术 Terahertz technology
态势 situation
态势感知 situation awareness; situational awareness
态势模型 situation model
泰勒级数 Taylor series
探测 detecting; detection
探测模型 exploration model
探测器 detector
探测误差 detection error
探索式多模型 exploratory multimodel
探索性测试 exploratory testing
探索性的多仿真方法学 exploratory-multisimulation methodology
探索性多仿真 exploratory multisimulation
探索性仿真 exploration simulation; exploratory simulation
探索性分析 exploratory analysis
探索性建模 exploratory modeling
探索性模型 exploratory model
探索性实验 exploratory experiment
逃避现实仿真 escapist simulation
讨价还价博弈 bargaining game

套件　suite
特定的　specific
特定仿真　simulation-specific
特定解　solution-specific
特定领域　domain-specific
特定领域代理　domain-specific agent
特定领域建模　domain-specific modeling
特定领域库　domain-specific library
特定领域模型　domain-specific model
特定领域知识　domain specific knowledge
特定领域智能体　domain-specific agent
特定模型　specific model
特定输入测试　special input testing
特定数据　specific data
特定问题知识　problem specific knowledge
特定系统　specific system
特定系统的　system-specific
特定系统库　system-specific library
特定项目的　project-specific
特定项目知识　project-specific knowledge
特定于应用的　application-specific
特定于应用的确认　application specific accreditation
特定域接口　domain-specific interface
特定值　specific value
特殊值　special value
特性　character; characteristic; quality; characterization
特性的涌现　emergence of a property
特性仿真　identity simulation

特性仿真游戏　identity simulation game
特性模型　speciality model
特性校核　property verification
特征　stamp; feature
特征波动性　feature volatility
特征向量　eigenvector
特征值　eigenvalue; characteristic value
梯度搜索法　gradient search method
梯形公式　trapezoidal formula
提供者　provider
提取　extract (v.); extraction
提议者　proponent
体内　in vivo
体内仿真　in vivo simulation
体内观察　in vivo observation
体内模拟　in vivo analog
体内模型　in vivo model
体内实验装置　in vivo experimental set-up
体内试验　in vivo experiment; in vivo experimentation
体内数据　in vivo data
体内体验　in vivo experience
体内装置　in vivo set-up
体外的　in vitro
体外仿真　in vitro simulation
体外观察　in vitro observation
体外模型　in vitro model
体外实验装置　in vitro experimental set-up
体外试验　in vitro experiment; in vitro experimentation
体外数据　in vitro data

体外体验　in vitro experience
体外装置　in vitro set-up
体系　system of systems
体系的　systematic
体系方法　system of systems approach
体系仿真　system of systems simulation
体系仿真语言　system of systems SL
体系分类　system of systems taxonomy
体系级　system of systems level
体系结构　architecture
体系结构开发　architecture development
体系模型　system of systems model
体系模型实例　system of systems model instance
体系生态学　system of systems ecology
体系应急行为　system of systems emergent behavior
体验　experience (v.)
体验规划　experiential planning
体验视角　experience perspective
体育博弈　sports game
体制　systems
体轴　axis
替代度　substitutability
替代仿真模型　alternative simulation model
替代经验　vicarious experience
替代模型　alternate model; surrogate model
替代实验　vicarious experiment
替代误差　substitution error
替换　replacement

替换键　alternate key
天气模型　synoptic model
天线输入　antenna input
挑战　challenge
挑战性假设　challenged assumption
条件　condition; qualification
条件变量　condition variable; conditional variable
条件测试　condition testing
条件刺激　conditioned stimulus
条件仿真　conditional simulation
条件蒙特卡罗仿真　conditional Monte Carlo simulation
条件模型　conditional model
条件事件　conditional event
条件语句　conditional sentence
条件值　conditional value
条款　term
调节　adjustment
调节变量　moderating variable; moderator variable
调节器　moderator; modifier
调节误差　adjustment error
调节装置　adjustment
调解代理　mediating agent
调试　debugging
调试阶段　debugging phase
调试运行　debugging run
调谐器误差　tuner error
调用结构分析　calling structure analysis
调整　adjust (v.)
调制　modulation
调准　alignment
跳跃不连续性　jump discontinuity

听觉的　auditory
听觉的特性　auditory characteristic
听觉反馈　auditory feedback
听觉感知　auditory sensation
听觉接口　auditory interface
听觉输入/输出　aural I/O
听知觉　auditory perception
停止准则　stopping criterion
通道　channel; gateway
通断　switching
通过　pass (v.)
通信　communication
通信仿真　communal simulation
通信技能增强　communication skill enhancing
通信技术　communication skill
通信间隔　communication interval
通信模型　communicative model
通信时间　communication time
通信网络　communication network
通信误差　communication error
通信效率　communication efficiency
通信性　communicativeness
通信智能体　communicative agent
通用本体　generic ontology
通用的　generic
通用的参考模型　common reference model
通用方法　versatile approach
通用方法论　generic methodology
通用仿真　common-use simulation; generalized simulation; generic simulation
通用仿真模型　generic simulation model
通用仿真系统　general purpose simulation system
通用仿真语言　general purpose SL
通用仿真知识　general simulation knowledge
通用分布仿真　general purpose distributed simulation
通用函数逼近器　universal function approximator
通用建模仿真　common-use M&S
通用建模环境　generic modeling environment
通用建模与仿真　general use M&S
通用建模与仿真应用　general use M&S application
通用库　generic library
通用框架　common framework
通用类型变量　generic-type variable
通用模型　generic model
通用模型仿真　generalized model simulation
通用模型模版　generic model template
通用模型组件模板　generic model component template
通用耦合模版　generic coupling template
通用属性　universal property
通用算法　generic algorithm
通用图灵机仿真　universal Turing machine simulation
通用语义图　general semantic graph
通用域　generic domain
通用元素　generic element
通用智能体框架　generic agent framework

同比例副本　exact-scale replica
同步　simultaneous；synchronization；synchronous
同步传输　synchronous transmission
同步的　synchronous
同步的模型　synchronous model
同步的状态　synchronous state
同步点　synchronization point
同步仿真　simultaneous simulation
同步分析　synchronous analysis
同步更新　simultaneous update
同步化的　synchronized
同步机制　synchronization mechanism
同步间隔技术　simultaneous interval technique
同步建模　synchronized modeling
同步控制　simultaneous control
同步输入　synchronized input；synchronous input
同步属性更新　simultaneous attribute update
同步协同建模　synchronized collaborative modeling
同步协议　synchronization protocol
同步执行　synchronized execution
同方差误差　homoscedastic error
同方差性　homoscedasticity
同构　isomorph；isomorphism
同构代理　homogeneous agent
同构的　isomorphic
同构的模型　isomorphic model
同构关系　isomorphic relation
同构网络　homogeneous network
同构现象　isomorphic phenomenon
同构型系统　homogeneous system
同构映射　isomorphic mapping
同化（产）物　assimilate (v.)
同化作用　assimilation
同时发生的　concurrent
同时发生的简单故障　simultaneous simple failure
同时发生的事件　simultaneous event
同态　homomorphy；homomorphism
同态的　homomorphic
同态模型　homomorphic model
同态映射　homomorphic mapping
同相输入　noninverting input
同形　homothety
同形的　homothetic
同型　morphism
同型的　morphic
同型关系　homomorphic relation
同意　accept (v.)
同余的　congruent
同源　homology
同质的　homogeneous
统计　statistics
统计变量　statistical variable
统计采样　statistical sampling
统计的　statistical
统计方法学　statistical methodology
统计仿真　statistical simulation
统计分析　statistical analysis
统计技术　statistical technique
统计检验　statistical test；statistical testing
统计建模　statistical modeling
统计决策　statistical decision
统计模型　statistical model
统计模型检验　statistical model

checking
统计确认　statistical confirmation
统计试验　statistical experiment
统计数据压缩　statistical data compression
统计显著性检验　statistical significance testing
统计学　statistics
统计学的　statistical
统计验证　statistical validation
统计验证技术　statistical validation technique
统计有效性　statistical validity
统一的度量标准　unified metric
统一的多重建模方法学　unified multimodeling methodology
统一的建模方法　unified modeling approach
统一的建模语言　unified modeling language
统一的离散合连续仿真　unified discrete and continuous simulation
统一的元模型　unified metamodel
统一仿真　unified simulation
统一建模　uniform modeling
统一建模框架　unifying modeling framework
统一建模语言　uniform modeling language
统一框架　unified framework; unifying framework
统一系统　uniform system
头戴显式　head-mounted display
头盔式显示器　head-mounted display
投放　delivery

投影误差　projection error
投资博弈　investment game
投资回报　return on investment
透明　transparency
透明度　transparency
透明度测试　alpha test; alpha testing
透明模型　transparent model
透明认知模型　transparent cognitive model
透明现实仿真　transparent reality simulation
透射　transmission
透视　perspective
透视式头盔显示器　see-through head-mounted display
凸对策　convex game
凸克里格　convex Kriging
突变　mutation
突变多模型　mutational multimodel
突变分析　mutation analysis
突变模型　mutational model
突发复杂性　emergent complexity
突发信号　burst
突发性不当行为　emergent misbehavior
突现　emergence
突现的实体　emergent entity
突现行为　emergent behavior
突现行为库　emergent behavior repository
突现论　emergentism
突现论者　emergentist
图　diagram; graph
图标　icon
图表　chart; graphing

图结构　graph structure
图灵测试　Turing test
图灵机仿真　Turing machine simulation
图灵机模型　Turing machine model
图灵机样模型　Turing machine-like model
图灵模型　Turing model
图模型　graph model
图驱动的　graph-driven
图驱动的方法　graph-driven approach
图驱动的方法论　graph-driven methodology
图式　schemata
图式变量　schematic variable
图式类型变量　schematic type variable
图数据库　graph database
图像　image
图像插值　image interpolation
图像模型　iconic model
图像生成　image generation
图形表达　graphic representation
图形处理单元　graphic processing unit
图形的　graphical
图形对象　graphical object
图形仿真　graphical simulation
图形工具　graphical tool
图形规范　graphical specification
图形化的比较　graphical comparison
图形建模工具　graphical modeling tool
图形界面　graphical interface
图形模型　graphical model
图形耦合　graphic coupling
图形失真　aliasing
图形学　graphics
图形引擎　graphics engine
图形用户界面　graphic user interface（GUI）
图形原语　graph primitive
团队　team
团队支持工具　team support tool
推测的　stochastic
推测的模型　conjectural model
推测数据　speculative data
推导　derivative；deduce (v.)
推动因素　push factor
推荐　recommendation
推荐人　recommender
推理　reasoning
推理的　a priori
推理仿真　reasoning simulation
推理规则　inference rule
推理模型　reasoning model
推理实验　a priori experiment；apriori experiment
推理误差　reasoning error
推理引擎　inference engine
推论　deduct (v.)；deduction；inference
推算　estimated
推演模型　deduction model
退化　regression
退化分布　degenerate distribution
退火　annealing
拖放式交互　drag-and-drop interaction
椭圆型偏微分方程　elliptic PDE
拓扑　topology
拓扑变换　topological transformation
拓扑的　topological

拓扑关系　topological relation
拓扑建模　topological modeling
拓扑交互　topological interaction
拓扑耦合　topological coupling
拓扑耦合的　topologically coupled
拓扑形状建模　topological shape modeling
拓扑学　topology
拓扑学的　topological

Ww

挖掘　mining
挖掘数据　mining data
瓦格纳模型　Wagner model
外表建模　aspect modeling
外部变换　external transition
外部变量　external variable; extraneous variable
外部参数　extrinsic parameter
外部错误　external error
外部代理　external agent
外部方案　external scheme
外部仿真　extrinsic simulation
外部仿真主义　extrinsic simulism
外部化知识　externalized knowledge
外部活动　external activity
外部激活多模型　externally activated multimodel
外部激励　external excitation
外部记忆　external memory
外部模式　external schema
外部模型　external model; extrinsic model
外部目标　external goal

外部耦合　external coupling
外部耦合的　externally coupled
外部生成的目标　externally generated goal
外部生成的输入　externally generated input
外部时间管理　external time management
外部事件　external event
外部输出耦合　external output coupling
外部输出耦合函数　external output coupling function
外部输入　external input
外部输入耦合　external input coupling
外部系统　external system
外部闲置　external inactivity
外部现实　external reality
外部性能演化　external performance evolution
外部演化　external evolution
外部有效性　external validity
外部智能体　external agent
外部转换函数　external transition-function
外部状态　external state
外观　guise
外观验证　face validation
外壳　shell
外生变量　exogenous variable
外生的　exogenous
外生假设　exogenous assumption
外生事件　exogenous event
外生协变量　exogenous covariate
外推法　extrapolation method;

extrapolation technique
外推误差　extrapolation error
外形建模　shape modeling
外延变量　extensive variable
外源　exogenous
外源故障检测　exogenous fault detection
外源性输入　exogenous input
外在误差　external error
弯曲　warp
完备的节点集　complete nodeset
完备的模型　complete model
完备信息　complete information
完备信息博弈　perfect information game
完美的模型　perfect model
完全　complete
完全沉浸式仿真　full-immersive simulation
完全错误的　completely false
完全仿真　complete simulation
完全可观工厂　fully observable plant
完全可观系统　fully observable system
完全可控系统　fully controllable system
完全可控装置　fully controllable plant
完全可区分系统　fully distinguishable system
完全理性　perfect rationality
完全随机化设计　completely randomized design
完全同步　complete synchronization
完全图　complete graph
完全析因设计　full factorial design
完全相同的标记　identical marking

完全有效性　full validity
完全正确　completely true
完善的模型维护　perfective model maintenance
完善性维护　perfective maintenance
完整代理仿真　holonic agent simulation
完整仿真　holonic simulation
完整建模　holonic modeling
完整模型　holonomic model
完整系统　holonic system; holonomic system
完整系统仿真　holonic system simulation
完整性　completeness; integrity
完整性检查　completeness check; integrity checking
完整性网格　integrity grid
完整约束　holonomic constraint
完整智能体仿真　holonic agent simulation
玩家　gamer
顽固期　refractory period
晚成系统　altricial system
万维网服务　web service
万维网可访问本体　web-accessible ontology
万维网使能的　web-enabled
万维网使能的仿真　web-enabled simulation
万维网使能的新趋势　web-enabled emerging trend
万维网使能技术　web-enabled technique
万维网语言　web language

万亿级仿真　terascale simulation
万亿级计算　terascale computing
万亿级计算机　terascale computer
网　net
网格　grid；mesh
网格仿真　grid simulation
网格仿真工具　grid simulation tool
网格仿真环境　grid simulation environment
网格仿真技术　grid simulation technology
网格计算　grid computing
网格宽度　grid width
网格联邦　grid federation
网格路由器　grid router
网格生成步骤　mesh generation step
网格拓扑　grid topology
网格细化　mesh refinement
网格与集群计算　grid and cluster computing
网格资源驱动器　grid resource driver
网关　gateway
网络　net；network
网络安全　network security
网络博弈论　network game theory
网络错误　network error
网络管理　network management
网络管理器　network manager
网络化仿真　networked simulation
网络化系统　networked system
网络化游戏　networked game
网络节点　network node
网络可访问的　web-accessible
网络理论　network theory
网络滤波器　network filter
网络模型　network model
网络权限错误　network permission error
网络设备故障　network device error
网络通信服务　network communication service
网络型超对策　network-type hypergame
网络游戏　network game；web-based game
网络游戏实验　network game experiment
网络中心　netcentric
网络中心博弈　netcentric gaming
网络中心仿真　netcentric simulation
网络中心系统　netcentric system
网络中心战争博弈　netcentric wargaming
危机　crisp
危机变量　crisp variable
危机参数　crisp parameter
危机管理对策　crisis management game
危机规则　crisp rule
危机规则的　crisp rule-base
危机性　crispness
危机值　crisp value
危险事件　dangerous event
威布尔随机变量　Weibull random variable
威胁表征　threat representation
微步长　microstep size
微分代数方程　differential algebraic equation
微分代数方程求解器　differential

algebraic equation solver
微分代数问题 diffential algebraic problem
微分对策 differential game
微分法 differentiation
微分方程 differential equation
微分方程的阶次 order of differential equation
微分方程模型 differential equation model
微分方程求解器 differential equation solver
微分分析器 differential analyzer
微分算子 differentiation operator
微观步长 micro step size
微观仿真 microsimulation
微观仿真模型 microsimulation model
微观分析仿真 microanalytic simulation
微观行为 microbehavior; micro-level behavior
微观模型 microscopic model
微机电系统 microelectromechanical system
微网格仿真 microgrid simulation
微小模型 micromodel
微小输入 paltry input
微型计算机仿真 microcomputer simulation
违反直觉的结果 counterintuitive result
唯相模型 phenomenological model
维度 dimension
维度分析 dimension analysis
维和博弈 peace support game

维护 maintenance
维护仿真 maintenance simulation
维护模型 maintenance model
维护训练 maintenance training
伪导数 pseudoderivative
伪仿真 pseudosimulation
伪仿真器 pseudosimulator
伪分析仿真 pseudoanalytical simulation
伪观测 pseudo-observation
伪码 pseudocode
伪随机数 pseudorandom number
伪随机数发生器 pseudorandom number generator
伪随机数评估发生器 assessment of pseudorandom number generator
伪造 counterfeit; falcification; falcify (v.); falsification; falsify (v.)
伪装 guise
伪自变量 pseudoindependent variable
尾段 endian
委托 accreditate (v.)
委托书 proxy
未初始化变量 uninitialized variable
未定义变量 undefined variable
未定义误差 undefined error
未发现的错误 undiscovered error
未公认的假设 unrecognized assumption
未观察到的变量 unobserved variable
未建模动力学 unmodeled dynamics
未经核实的理解 unverified understanding
未经确认的状态 unidentified state
未经授权的模型 unsolicited model

未来趋势　future trend
未来事件列表　future events list
未确认的错误　unacknowledged error
未设置的值　unset value
未说明的假设　unaccounted assumption
未完成的模型　incomplete model
未限定变量　unqualified variable
未预测数据　unpredicted data
未预料到的事件　unanticipated event
未知变量　unknown variable
未知错误　unknown error
未知现实　unknown reality
未知值　unknown value
未知状态　unknown state
未指明的错误　unspecified error
位错误　bit error
位移变量　displacement variable
位置跟踪　location tracing
位置输入　positional input
位置误差　position error
谓词变量　predicate variable
谓词演算　predicate calculus
谓词转换　predicate transformation
温度场问题　thermal field problem
文本规范　textual specification
文档　document
文档标准　documentation standard
文档的评估　assessment of documentation
文档管理　file management
文档检查　documentation checking
文档框架　documentation framework
文档模板　documentation template
文档频率　frequency of documentation
文档树　documentation tree
文档中心的　document centric
文化差异　cultural bias error
文化复杂性　cultural complexity
文化感知误差　cultural perception error
文化过滤　cultural filter
文化环境　cultural environment
文化特征　cultural feature
文化特征数据　cultural features data
文化值　cultural value
文件　file
文件编制　documentation
文件错误　file error
文件库　document repository
文件生成错误　file creation error
文字-模型转换　text-to-model transformation
纹理仿真　texture simulation
稳定变量　stabilized variable
稳定的　stable; stationary; steady
稳定的行为　stable behavior
稳定的环境　stable environment
稳定的离散化机制　stable discretization scheme
稳定的模型　stable model
稳定范围　stable region
稳定化　stabilization
稳定机制　stability mechanism
稳定积分算法　stable integration algorithm
稳定极限环　stable limit cycle
稳定价值　sound value
稳定结果　stable result
稳定聚合状态　stable clustering state

稳定区域　stability region
稳定特性　stability property
稳定系统　stable system
稳定线性时不变系统　stable linear time-invariant system
稳定性　stability
稳定性判据　stability criterion
稳定性条件　stability condition
稳定域　stability domain
稳定整合算法　faithfully stable integration algorithm
稳定状态　stable state
稳态　steady state
稳态变量　homeostatic variable
稳态的　steady
稳态仿真　steady-state simulation
稳态技术　steady-state technique
稳态阶段　steady-state phase
稳态解　stable solution
稳态模型　steady-state model
稳态双回波离散事件系统规范　DESS-DEVS
稳态误差　stable error；steady-state error
稳态相位　steady-state phase
稳态行为　steady behavior
稳态周期　steady-state period
问答游戏　quiz game
问卷　questionnaire
问卷式编程　programming by questionnaire
问题　issue；problem
问题分析　problem analysis
问题复杂性　problem complexity
问题公式化　problem formulation

问题-规范-时间的效率　problem-specification-time efficiency
问题规格说明阶段　problem specification phase
问题解决范式　problem solving paradigm
问题解决过程　problem solving process
问题空间　problem space
问题情境　problem context
问题求解　problem solving
问题求解环境　problem solving environment
问题说明　problem specification
问题文件　problem documentation
问题域　problem domain
问题域需求　problem domain requirements
涡流仿真　eddy simulation
沃罗诺伊边界　Voronoi boundary
沃罗诺伊单元　Voronoi cell
沃罗诺伊分解　Voronoi decomposition
沃罗诺伊曲面细分　Voronoi tessellation
沃罗诺伊图　Voronoi diagram
无标度复杂交互系统　scale-free complex interactive system
无标度拓扑　scale-free topology
无参随机变量生成　nonparametric random variate generation
无参照的变量　unreferenced variable
无法获取的预测　inaccessible prediction
无法解释的行为　unexplained behavior

无法解释的突现行为　unexplained emergent behavior
无法量化的关系　unquantifiable relationship
无分隔输入　delimiterless input
无缝的　seamless
无关数据　irrelevant data
无关随机变量　uncorrelated random variable
无关信息　irrelevant information
无规博弈　free game
无害化误差　harmless error
无环变结构多模型　acyclic metamorphic multimodel
无环多模型　acyclic multimodel
无环模型　acyclic model
无记忆变量　memoryless variable
无记忆的多模型　memoryless multimodel
无记忆的模型　memoryless model
无记忆量化　memoryless quantization
无记忆随机变量　memoryless random variable
无界变量　unbounded variable
无量纲的　dimensionless
无目标仿真　goal-free simulation
无偏估计　unbiased analysis
无偏估值器　unbiased estimator
无偏数据　unbiased data
无偏误差　unbiased error
无偏样本　unbiased sample
无歧义表示（法）　unambiguous representation
无歧义性　unambiguity
无趣变量　uninteresting variable
无私　altruism
无锁的　free-lock
无条件的仿真　unconditional simulation
无条件稳定　unconditionally stable
无网格仿真　meshfree simulation
无限对策　infinite game
无限刚度　infinite stiffness
无限状态系统　infinite-state system
无线电输入　wireless input
无效　invalidity
无效的　invalid
无效简化　invalid simplification
无效决策　ineffective decision
无效输入测试　invalid input testing
无异议的假设　unchallenged assumption
无意识状态　unintentional state
无用数据　useless data
无源系统　passive system
无源元件　passive component
无约束变量　unbound variable
无约束仿真　unconstrained simulation
物理逼真度　physical fidelity
物理变量　physical variable
物理方面　physical aspect
物理仿真　physical simulation
物理感知　physical perception
物理环境　physical environment
物理建模　physical modeling
物理接口　physical interface
物理模式　physical schema
物理模型　physical model
物理模型属性　physical model attribute

物理时间　physical time
物理实现　physical realization
物理试验　physical experiment
物理属性　physical attribute
物理数据模型　physical data model
物理网络　physical network
物理系统　physical system
物理系统仿真　physical system simulation
物理现实　physical reality
物理引擎　physics engine
物流　logistics
物质　substance
物质现实　material reality
误比特　bit error
误差　error
误差测量　error measurement
误差层　error layer
误差传播　error spread
误差带　error band
误差范围　margin of error
误差分析　error analysis
误差估计　error estimation
误差函数　error function
误差检测模型　error-detection model
误差减少　error reduction
误差界　error bound
误差控制　error control
误差控制仿真　error-controlled simulation
误差累积　error accumulation
误差模糊　error ambiguity
误差模型　error model
误差评估　error assessment
误差容限　error tolerance

误差挖掘　error mining
误差系数　error coefficient
误差显示　error indication
误差消除　error elimination
误差校验代码　error checking code
误差校正　error correcting; error correction
误差预测模型　error-prediction model
误差增长　error growth
误导　misleading
误动作　malfunction
误读　misinterpretation
误解　misunderstanding; misconception
误码增殖　error multiplication
误算　miscalculation
雾计算　fog computing

Xx

吸收　absorb (*v.*); absorbtion
吸收的　absorbing
吸收马尔可夫链　absorbing Markov chain
吸收状态　absorbing state
吸引子　attractor
希腊拉丁方设计　Graeco-Latin square design
析取耦合　disjunctive coupling
稀少事件　rare event
稀疏连接　sparse connection
稀疏数据　sparse data
稀疏线性系统求解器　sparse linear system solver
稀有事件采样　rare-event sampling

稀有事件仿真　rare-event simulation
稀有事件仿真方法　rare-event simulation methodology
稀有事件仿真技术　rare-event simulation technique
稀有事件分析　rare-event analysis
稀有事件概率　rare-event probability
习惯的　used
系谱　pedigree
系数　coefficient
系数矩阵　coefficient matrix
系数事件　coefficient event
系统　system; systems
系统安全标准　system safety standard
系统安全性　system safety
系统边界　system boundary
系统变量　system variable
系统辨识　system identification
系统表示（法）　system representation
系统采样　systematic sampling
系统测试系统　system test system
系统测试需求　system test requirements
系统层　system level
系统层规范　system-level specification
系统词典　systematic dictionary
系统的　systematic; systemic
系统的观测行为　system's observed behavior
系统的局部特征　local property of a system
系统的开始　launch of the system
系统的全局（综合）特性　global property of a system
系统等价性　equivalence of systems
系统动力学　system dynamics
系统动力学建模　system dynamics modeling
系统方法　system approach
系统仿真　system simulation
系统仿真理论　system simulation theory
系统分类　system classification
系统分析　system analysis
系统分析的仿真　simulation for system analysis
系统复杂性　system complexity
系统工程　systems engineering
系统工程化的系统　systems engineering system
系统工程生命周期　systems engineering life cycle
系统工程生命周期的仿真　simulation for systems engineering life cycle
系统故障　system failure
系统规范　system specification
系统规范层　level of system specification
系统行为　system behavior; system's behavior
系统行为等效性　behavioral equivalence of systems
系统化仿真　systematic simulation
系统化规范　systematic specification
系统化实验　systematic experimentation
系统活动　systemic activity
系统级不当行为　system-level misbehavior
系统级度量　system-level metric

系统级仿真　system-level simulation
系统级软错误　system-level soft error
系统级性能　system-level property
系统集成　system integration
系统集成服务　system integration service
系统家族　family of systems
系统架构　system architecture
系统架构规范　system architecture specification
系统建模　system modeling
系统建模语言　system modeling language
系统校核　system verification
系统交互　system interaction
系统解　systemic solution
系统开发　system development
系统开发过程　system development process
系统开发系统　system development system
系统可塑性　system plasticity
系统控制　system control
系统理论　systems theory
系统连通性　system connectivity
系统描述　system description
系统模型　system model; systemic model
系统目标　system goal
系统耦合　system coupling
系统耦合方法　system coupling recipe
系统配置　system layout
系统评估　system assessment
系统设计　system design
系统设计过程　system design process

系统设计理论　system design theory
系统生命周期　system life cycle
系统生物学标记语言　systems biology markup language
系统实体　system entity
系统实体结构　system entity structure
系统事件　system event
系统视图　system view
系统属性　system property
系统思维　system thinking
系统特性　system characteristic
系统调查中　system under investigation
系统同型　system morphism
系统完整性　system integrity
系统稳定性　system stability
系统响应　system response
系统性　systemness
系统性能　system-level performance
系统性误差　system error; systematic error; systemic error
系统原因　systemic cause
系统约束　system constraint
系统诊断学　system diagnostics
系统质量特性　system quality characteristic
系统质量特性的可预测性　predictability of system quality characteristic
系统质量预测　system quality prediction
系统转换中　system in transition
系统状态　system state
细胞的　cellular
细胞离散事件系统规范　cell-DEVS

细胞模型　cell model
细分　refinement
细分技术　refinement technique
细化网格　refined mesh
细节　detail
细节层次　level of detail
细节复杂性　detail complexity
细节量　amount of detail
细粒度　fine-grain granularity
细粒度的组件　fine-grained component; finer-grained component
细粒度方法　fine-grain approach
细粒度模型　fine-grain model
下标变量　subscripted variable
下层　lower level
下级　lower level
下界　lower bound
下拉式菜单　drop-down menu
下三角形式　lower-triangular form
下视显示　head-down display
下视仪表　head-down instrument
下限　lower limit
下一内部事件　next internal event
下一事件　next event
下一事件形式化　next event formalism
下一外部事件　next external event
先行　anticipatory
先行关系　anticipatory relationship
先进的概念　advanced concept
先进的理解系统　advanced understanding system
先进仿真　advanced simulation
先进仿真环境　advanced simulation environment
先进分布式仿真　advanced distributed simulation
先进环境　advanced environment
先进建模　advanced modeling
先进数字仿真　advanced numerical simulation
先决条件　prerequisite condition
先驱模型　ancestor model
先验的　a priori
先验模型　a priori model
先验知识　a priori knowledge
闲置　inactivity
闲置的模型　inactive model
显示　display
显示设备　display device
显式　explicit
显式单步法　explicit single-step method
显式的Adams-Bashforth方法　explicit Adams-Bashforth method
显式法　explicit method
显式积分　explicit integration
显式积分算法　explicit integration algorithm
显式模型检查　explicit model checking
显式奈斯特龙技术　explicit Nyström technique
显式声明　explicit declaration
显式事件　explicit event
显式数值积分　explicit numerical integration
显式数值积分公式　explicit numerical integration formula
显式算法　explicit algorithm
显式协调　explicit coordination

显式一阶算法　explicit first-order algorithm
显式依存关系　explicit dependency
显式约束　explicit constraint
显式中点法则　explicit midpoint rule
显式状态　explicit state
显型　phenotype
显性的假设　explicit assumption
显性知识　explicit knowledge
现场测试　field testing
现场经验　in situ experience
现场实验　field experiment; in situ experiment; in situ experimentation
现场实验装置　in situ experimental set-up
现场仪器　field instrumentation
现成的仿真程序包　off-the-shelf simulation package
现成的模型　off-the-shelf model
现成的模型库　off-the-shelf model library
现存条件　existing condition
现实　reality
现实博弈　realistic game
现实的模型　realistic model
现实视野　reality horizon
现实在环　reality-in-the-loop
现实主义　realism
现象　phenomenon
现象性经验　phenomenal experience
现象学的　phenomenological
现象学方法　phenomenological approach
现象学经验　phenomenological experience
现象学误差　phenomenological error
现有的收益系统　existing system of interest
限定变量　qualified variable
限定符条件　qualifier condition
限制　limit; limit ($v.$); qualification
限制耦合　limiting coupling
限制器　limiter
线程化游戏技术制作的游戏　threading game-technology game
线程技术　threading technology
线路　line
线性　linear
线性插值　linear interpolation
线性常系数系统　linear constant coefficient system
线性单输入/单输出模型　linear single input single-output model
线性的连接　linear connection
线性度　linearity
线性二阶导数模型　linear second derivative model
线性规划　linear programming
线性规划嵌入式仿真　linear programming embedded simulation
线性互相似系统　linear bisimilar system
线性化　linearization; linearize ($v.$)
线性化误差　linearization error
线性混合模型　linear mixed model
线性模型　linear model
线性耦合　linear coupling
线性耦合的　linearly coupled
线性时不变时间连续系统　linear time-invariant continuous-time system

线性守恒定律　linear conservation law
线性拓扑　linear topology
线性稳定性理论　linear stability theory
线性系统　linear system
线性系统仿真　linear system simulation
线性系统求解器　linear system solver
线性隐式方法　linearly implicit method
线性映射　linear mapping
线性优化　linear optimization
线性状态空间模型　linear state-space model
相变　phase transition
相称　appropriateness
相称的　appropriate
相等的　equivalent
相对的　relative
相对误差　fractional error; relative error; proportional error
相对误差控制　relative error control
相对真理　relative truth
相关　dependency
相关闭包　dependency closure
相关变量　relevant variable
相关处理　relevant processing
相关的　dependent; related; relevant
相关仿真　related simulation
相关激励　relevant stimulus
相关联的模型　associative model
相关模型　related model; relevant model
相关输入　relevant input
相关数据　relevant data
相关误差　correlated error
相关信息　relevant information
相关性　coherence
相关依赖的耦合　state-dependent coupling
相互关联的问题群　problematique
相量　phasor
相配的模型　suitable model
相平面图　phase plane plot
相容的　compliant
相容离散机制　consistent discretization scheme
相容性　compliancy
相容性条件　consistency condition
相容性证明　certificate of compliancy
相似　resemblance
相似变换　similarity transformation
相似的　similar
相似度　similarity degree
相似度测量　similarity measure
相似法　similarity method
相似关系　resemblance relation; mimetic relation
相似类型　similarity style
相似模型　analogy model; similar model
相似性　similarity
相似性关系　similarity relation
相似性理论　similarity theory
相似性准则　similarity criterion
相似值　resemblance value
相适应的　suitable
相位　phase
相位差　phase difference
相位模型　phase model

相位同步　phase synchronization
相像　similitude
相异　dissimilitude
相异的　dissimilar
相异点　dissimilarity
相异判据　dissimilarity criterion
详尽的规则　exhaustive discipline
详尽说明的变量　elaborated variable
详细　detailedness
详细的建模　detailed modeling
详细的模型　detailed model
详细的数据　detailed data
详细的系统　detailed system
详细的系统建模　detailed system modeling
响应　response
响应变量　response variable
响应函数　response function
响应空间　response space
响应量　responding variable
响应面　response surface
响应面法　response surface method
响应面设计　response surface design
响应曲线　response curve
响应特性　response characteristic
响应图　response diagram
想定　scenario
想定测试　scenario testing
想定规范　scenario specification
想定开发　scenario development
想定空间　scenario space
想定模块性　scenario modularity
想定模型　scenario model
想定生成　scenario generation
想定实例　scenario instantiation

想法　notion
想象的现实　imagined reality
向后　backward
向后插值　back interpolation
向后积分　backward integration
向后欧拉积分　backward Euler integration
向量　vector
向量值变量　vector-valued variable
向上因果　upward causation
向下因果关系　downward causation
项目管理　project management
项目规划　project planning
像素　pixel
消除　elimination；smooth (*v.*)
消费者　consumer
消费者数据　consumer data
消极思想实验　negative thought experiment
消极训练　negative training
消息　message
消息传递　message delivery；message passing
消息传递服务　message delivery service
消息误差　message error
消元法　elimination
小尺度　not-so-minimal
小尺度模型　not-so-minimal model
小尺度系统　small-scale system
小端字节　little endian
小应用程序　applet
效度　validity
效率　efficiency
效应　effect

效用　utility
效用仿真　utilitarian simulation
效用模型　utility model
协变量　covariate
协变量动力学　covariate dynamics
协处理器　coprocessor
协商　negotiation
协调　coordination
协调变量　coordination variable
协调博弈　coordination game
协调的　coordinated
协调的时间推进　coordinated time advancement
协调模型　coordination model
协调世界时间　coordinated universal time
协调系统　coordination system
协调语言　coordination language
协调者　coordinator
协同　collaboration；synergy
协同的　collaborative；coopetitive
协同仿真　collaborative simulation；coopetitive simulation
协同仿真的尝试　cooperative simulation effort
协同仿真对策　cooperative simulation game；coopetitive simulation game
协同仿真环境　collaborative simulation environment
协同仿真技术　collaborative simulation technology；co-simulation technique
协同仿真网格　collaborative simulation grid
协同仿真系统　collaborative simulation system
协同分布式仿真　collaborative distributed-simulation
协同规划　collaborative planning
协同过程　collaborative process
协同环境　collaborative environment
协同技术　collaborative technique；collaborative technology
协同建模　co-modeling；combined modeling；concerted model
协同建模方法学　co-modeling methodology
协同进化　coevolution
协同进化系统　coevolving system
协同克里格　Cokriging
协同克里格仿真　Cokriging simulation
协同离散事件系统规范仿真　collaborative DEVS simulation
协同模式　collaborative mode
协同模型　co-model；collaborative model
协同设计　codesign
协同式建模与仿真　collaborative M&S
协同适应　coadaptation
协同体系　collaborative system-of-systems
协同系统　collaborative system
协同想定　synergetic scenario
协同性关系　coopetitive relationship
协同性行为　coopetitive behavior
协同虚拟仿真　collaborative virtual simulation
协同虚拟和增强现实　collaborated virtual and augmented reality
协同需求建模　collaborative

requirements modeling
协同学　synergetics
协同智能体　collaborative agent
协议　protocol
协议实体　protocol entity
协议数据单元　protocol data unit
协议数据单元标准　protocol data unit standard
协议套组　protocol suite
协议转换器　protocol converter
协作　cooperation; cooperate (v.)
协作仿真　cooperation simulation
协作行为　cooperative behavior
协作基本结构　collaboration infrastructure
协作建模　collaborating modeling; collaboration modeling
协作图　collaboration diagram
协作型建模　collaborative modeling
写保护错误　write protect error
写入错误　write error; writing error
卸载程序错误　uninstaller error
心理逼真度　psychological fidelity
心理复杂性　mental complexity
心理偏差　mental bias
芯片级软错误　chip-level soft error
新继任者模型　emerging successor model
新目标博弈　repurposed game
新趋势　emerging trend
新闻游戏　news game
新型仿真方法　innovative simulation approach
新型仿真工具　innovative simulation tool
新型仿真技术　innovative simulation technique
新兴的　emerging
新兴的范式　emerging paradigm
新兴的技术　emerging technology
新兴的行为　emerging behavior
新兴的实践　emerging practice
新兴的特征　emerging feature
信号　signal
信念的形成过程　belief-forming process
信念-愿望-意图语言　belief-desire-intention language
信任　reliance
信任（置信度）评估　confidence assessment
信任博弈　trust game
信任博弈实验　trust game experiment
信任管理语言　trust management language
信息　information
信息不对称　information asymmetry
信息处理　information processing
信息处理模型　information processing model
信息动画　information animation
信息管理　information management
信息技术　information technology
信息交换　information exchange
信息居民　infohabitant
信息可视化　information visualization
信息论　informatics
信息模型　information model
信息数据　informative data
信息物理系统　cyber-physical system

信息物理系统仿真　cyber-physical system simulation
信息物理状态　cyber-physical state
信息系统　information system
信息系统建模　information systems modeling
信息隐藏　information hiding
信息战　information warfare
信息追踪　info trace
信息资源词典　information resource dictionary
信息资源词典系统　information resource dictionary system
信源　information source；source
形变模型　morphable model
行程生成　trip generation
行动　activity
行动变量　action variable
行动评估　action appraisal
行动评价模型　action appraisal model
行动者　actor
行为　behavior
行为(序列)脚本　behavioral script
行为比较　comparison of behavior
行为变量　behavior variable
行为表示(法)　behavior representation
行为不连续性　behavior discontinuity
行为层次结构　behavioral hierarchy
行为产生参数　behavior generation parameter
行为产生器参数　behavior generator parameter
行为抽象　behavior abstraction
行为处理　behavior processing
行为的　behavioral

行为的比较　behavioral comparison
行为的抽象　behavioral abstraction
行为的复杂性　behavioral complexity
行为的假设　behavioral assumption
行为的可替代性　behavioral substitutability
行为的特性　behavioral characteristic
行为的微观仿真模型　behavioral microsimulation model
行为的准确性　behavioral accuracy
行为等价系统　behaviorally equivalent systems
行为发生器　behavior generator
行为仿真器　behavioral simulator
行为分析　behavior analysis
行为风险因素　behavioral risk factor
行为复杂性　behavior complexity
行为关系　behavioral relation
行为管理　behavior management
行为规范　behavior specification
行为规则　behavior rule
行为合成器　behavior composer
行为集中　behavior concentration
行为建模　behavioral modeling
行为可信度　behavior credibility
行为空间　behavior space
行为库　behavior repository
行为粒度方法　course-grain approach
行为粒度模型　course-grain model
行为论域　behavioral domain
行为模型　comportmental model；behavioral model
行为模型比较　behavioral model comparison
行为模型的逼真度　behavioral model

fidelity
行为模型转换　behavioral model transformation
行为配置　behavior configuration
行为评价　behavior evaluation
行为上可替代的　behaviorally substitutable
行为生成　behavior generation
行为生成技术　behavior generation technique
行为生成技术评估　assessment of behavior generation technique
行为实验　behavioral experiment
行为适应性仿真　behaviorally adaptive simulation
行为数据库　behavior database
行为同型　behavior morphism
行为图　behavior graph
行为文档　behavior documentation
行为误差　course error
行为显示　behavior display
行为相似性　behavioral similarity
行为需求　behavioral requirement
行为依赖　behavioral dependency
行为因素　behavioral factor
行为映射　behavioral mapping
行为涌现,涌现行为　behavior emergence
行为预测多体仿真　behaviorally anticipatory multisimulation
行为预测仿真　behaviorally anticipatory simulation
行为预期　behaviorally anticipatory; behavioral anticipation
行为预期决策　behaviorally anticipatory decision
行为预期模型　behaviorally anticipatory model
行为预期系统　behaviorally anticipatory system
行为元模型　behavioral metamodel
行为质量　behavior quality
行为重置　behavior reconfiguration
行为自组织　behavioral self-organization
行政　administration
形貌测量　profiling
形式本原　formalism primitive
形式变量　formal variable
形式的可扩展性　formalism extensibility
形式定义　formal definition
形式方法　formal method
形式概念模型　formal conceptual model
形式规范　formal specification
形式化　formalization
形式化本体　formal ontology
形式化表现力　formalism's expressive power
形式化的适当性　adequacy of formalism
形式化方式方法　formal-method approach
形式化技术　formalization technique; formal technique
形式化特性　formal property
形式化特性校核　formal property verification
形式化校核　formal verification

形式化校核、验证与测试技术　formal VV&T technique
形式化校核技术　formal verification technique
形式建模　formal modeling
形式建模方法　formal modeling approach
形式控制　formal control
形式扩展　formalism extension
形式模型　formal model
形式模型评价　formal model evaluation
形式评价　formal evaluation
形式评审　formal review
形式评审过程　formal review process
形式情境模型　formal contextual model
形式上的　formal
形式上的组合性　formal composability
形式上下文模型　formal context model
形式系统　formal system
形式下的闭包　closure of formalism
形式语言　formal language
形式语义系统　formal semantic system
形式主义　formalism
形式主义比较　formalism comparison
形式主义表现力　formalism expressiveness
形式主义的世界观　world view of a formalism
形式主义力量　formalism's strength
形式主义粒度　formalism's granularity
形态理解　morphological understanding
形态学　morphology
形态学的　morphic
形态学评估　morphological assessment
形象　profile
形象化　visualize (v.)
形象化的　visualized
形状仿真　shape simulation
兴奋　stimulation
兴趣变量　interesting variable
兴趣管理　interest management
性能　performance；property
性能参数　performance parameter
性能测量　measure of performance；performance measurement；performance measure
性能测试　performance testing
性能度量　performance metric
性能分析　performance analysis
性能跟踪　performance tracking
性能模型　performance model
性能评估　performance assessment
性能属性　performance attribute
性能需求　performance requirements
性能演化　performance evolution
性能预测　performance prediction
性能指标　performance indicator
性质分类　property classification
性质上的　qualitative
休眠的方面　dormant aspect
休眠模型　dormant model
休眠状态　dormant state
休闲游戏　leisure game
修复　restore (v.)

修改　modify (v.)
修改的数据　revised data
修剪　pruning
修正　update；update (v.)；correct (v.)
修正对策　modified game
修正假设　correct assumption
修正结果　correct result
修正模型　correct model；modification model；modified model
修正屏幕　modified screen
修正输出　correct output
修正系数　auxiliary value
修正性模型维护　corrective model maintenance
修正性维护　corrective maintenance
修正预测　correct prediction
修正自模型　correct self-model
宿主变量　host variable
宿主模型　host model
虚部　imaginary part
虚构的　imaginary
虚构的变量　dummy variable
虚构的经验　imaginary experience
虚假　falsity
虚假程度　degree of falsity
虚假错误　false error
虚假的现实　pretended reality
虚假的自模型　pretended self-model
虚假行为　spurious behavior
虚假知识　false knowledge
虚拟　virtual
虚拟表达　virtual representation
虚拟参赛者　virtual player
虚拟呈现　virtual presence
虚拟传感器输入　virtual sensor input

虚拟的　virtual
虚拟仿真　virtual simulation
虚拟仿真环境　virtual simulation environment
虚拟仿真器　virtual simulator
虚拟化　virtualization；virtualize (v.)
虚拟化仿真　virtualization simulation
虚拟化系统　virtualized system
虚拟化现实　virtualized reality
虚拟环境　virtual environment
虚拟计算机　virtual computer
虚拟价值　virtual value
虚拟建模显示　virtual modeling display
虚拟模拟计算机　virtual analog computer
虚拟内容　virtual content
虚拟人　virtual human
虚拟人体标记语言　virtual human markup language
虚拟设备错误　virtual device error
虚拟社区　virtual community
虚拟时间　virtual time
虚拟时间仿真　virtual time simulation
虚拟时钟　virtual clock
虚拟实体　virtual entity
虚拟实验　virtual experiment
虚拟实验室　virtual laboratory
虚拟世界　artificial world；virtual world
虚拟视网膜显示器　virtual retina display
虚拟体验　virtual experience
虚拟图像　virtual image
虚拟完整约束　virtual holonomic

constraint
虚拟网络　virtual network
虚拟网络设备错误　virtual network device error
虚拟系统仿真　virtual system simulation
虚拟系统体系　virtual system-of-systems
虚拟现实　fictitious reality; virtual reality
虚拟现实仿真　virtual reality simulation
虚拟现实仿真方法学　virtual reality simulation methodology
虚拟现实工具　virtual reality tool
虚拟现实技术　virtual reality technology
虚拟现实建模语言　virtual reality modeling language
虚拟协同　virtual collaboration
虚拟性　virtuality
虚拟训练　virtual training
虚拟训练仿真　virtual training simulation
虚拟训练环境　virtual training environment
虚拟训练系统　virtual training system
虚拟研究环境　virtual research environment
虚拟演播室　virtual actor
虚拟样机　virtual prototype; virtual prototyping
虚拟战场空间　virtual battlespace
虚拟知识　virtual knowledge
虚拟状态　virtual state

虚态　imaginary state
虚轴　imaginary axis
需求　requirement; requirements
需求测定　requirements determination
需求错误　requirement error
需求分解　requirement decomposition
需求分析　requirement analysis
需求分析阶段　requirement analysis phase
需求管理　requirements management
需求规范　requirements specification
需求合规　requirements compliance
需求建模　requirements modeling
需求可信性　requirements credibility
需求可追溯性矩阵　requirements traceability matrix
需求驱动的目标　need-driven goal
需求说明阶段　requirement specification phase
需求文档　requirements document
需求验证　requirements validation
需求质量　requirements quality
需要　require (v.)
需要的特性　desirable feature
许可　authority
许可的模型　licenced model
许可控制　admission control
序贯博弈　sequential game
序贯仿真的自动并行化　automated parallelization of sequential simulation
序贯高斯仿真　sequential Gaussian simulation
序贯算法　sequential algorithm
序列　sequence
序列式　sequenced

序列图　sequence diagram
序列误差　sequence error
序数尺度变量　ordinal scale-variable
序数值　ordinal value
序优化　ordinal optimization
序优化理论　ordinal optimization theory
序优化算法　ordinal optimization algorithm
叙述仿真　narrative simulation
叙述模型　narrative model
旋转变量　rotational variable
选单　menu
选单驱动　menu-driven
选定的动作　selected action
选手简介　player profile
选通规则　gated discipline
选择　selection
选择的输出　selected output
选择器　selector
选择文档　selective documentation
选择性注意　selective attention
学科　discipline
学科交叉建模　interdisciplinary modeling
学期　term
学术博弈　academic gaming
学术仿真　academic simulation
学术知识　scholarly knowledge
学说　theory
学习　learning
学习博弈　learning game
学习经验　learning experience
学习理论　learning theory
学习启发式　learning heuristic

学习数据　learning data
学习数据集合　learning data set
学习算法　learning algorithm
学习系统　learning system
学习自动机　learning automaton
循环　loop
循环变量　circular variable
循环测试　loop testing
循环的因果关系　circular causality
循环多模型　cyclic multimodel
循环算法　cyclic algorithm
循环突变多模型　cyclic metamorphic multimodel
训练　training
训练仿真　training simulation
训练仿真器　training simulator
训练仿真游戏　training simulation game
训练分析　training analysis
训练环境　training environment
训练技术　education technique; training technology
训练联盟　training confederation
训练平台　training platform
训练器　trainer
训练器频谱　spectrum of trainers
训练人员　training staff
训练任务分析　training task analysis
训练数据　training data
训练数据集合　training data-set
训练系统　training system
训练效能　training effectiveness
训练效能评价　training effectiveness evaluation
训练游戏　training game

训练装置　training device

Yy

压力测试　stress testing
压缩　compress (*v.*); compression
压缩的时间　compressed time
压缩方法　compression method
压缩时间序列　compressed time-series
压缩实时仿真　compressed-time simulation
压型　profiling
雅可比线性化　Jacobi linearization
雅克比矩阵　Jacobian matrix
亚当斯-莫尔顿算法　Adams-Moulton algorithm
亚稳态　metastable state
延迟变量　lag variable
延时　delay
延时转换　delayed transition
延拓条件　continuation condition
严格有效性　strict validity
严肃仿真　serious simulation
严肃玩家　serious gamer
严肃游戏　serious game
严肃游戏倡议　serious game initiative
言语模型　verbal model
研究　research; study
研究、开发与采办　research, development and acquisition
研究变量　research variable
研究重现性　research reproducibility
研究的合规评估　ethical assessment of the study
研究后分析　poststudy analysis
研究后活动　poststudy activity
研究后计算　poststudy computation
研究后输出　poststudy output
研究环境　research environment
研究模型　research model
研究目标　goal of the study
研究目标的合规评估　ethical assessment of the goal of the study
研究目标的语用评估　pragmatic assessment of the goal of the study
研究目标评估　assessment of the goal of the study
研究之后的　poststudy
衍生部分　derivative section
衍生的　derived
衍生多种仿真　generative multisimulation
衍生假设　derivative assumption
掩饰　dissimulation; dissimulated; dissimulating; dissimulate (*v.*)
掩饰的　dissimulative
掩饰器　dissimulator
演变　evolution
演变系统　evolving system
演化环境　evolving environment
演化模型　evolving model
演化现象　evolving phenomenon
演练　exercise
演练管理　exercise management
演练管理器　exercise manager
演示管理器　presentation manager
演绎　deduct (*v.*); deduction
演绎的　deductive
演绎的模型　deductive model
演绎法建模　deductive modeling

演绎方法　deductive method
演绎逻辑上的谬论　deductive logical fallacy
演绎推理　deductive reasoning
演绎误差　deductive error
演绎一致性　deductive coherence
演员　player
演奏者　player
厌恶刺激　aversive stimulus
验收　acceptance
验收测试　acceptance testing
验收公差　acceptable tolerance
验收准则　acceptance criterion
验算　proofing
验证　verification; validation; verified; validate (v.); verify (v.)
验证操作者　validation operator
验证错误　authentication error; validation error
验证代理　validation agent
验证度量　validation metric
验证方法　validation method
验证计划　validation plan
验证技术　validation technique
验证检查　validation check
验证评估　validation assessment
验证试验　compliance test; compliance testing
验证数据　validation data
验证数据需求　validation data requirements
验证算法　validation algorithm
验证系统　validation system
验证需求　validation requirement
验证者　verifier
验证准则　validation criterion
样本　sample
样机　mock-up; prototype
样机模型　mock-up model
样机研究　prototyping
要求　application; requirements; require (v.)
要求的模型　solicited model
业务需求　business requirements
业务训练博弈　business training game
一般边界条件　general boundary condition
一般错误　generic error
一般的　generic
一般故障　generic failure
一般互操作性　general interoperability
一般平衡模型　general equilibrium model
一般认证　general accreditation
一般性错误　general error
一般性建模与仿真知识　general M&S knowledge
一般性模型　general model
一定的耦合　certain coupling
一对　pair
一对多关系　one-to-many relation
一对一关系　one-to-one relation
一个仿真研究的生命周期　life cycle of a simulation study
一级仿真　first-degree simulation
一阶段理解　one-stage understanding
一阶精度　first-order accurate
一阶控制　first-order control
一阶控制论　first-order cybernetics
一阶模型　first-order model

一阶数字转换器　first-order quantizer
一阶效应　first-order effect
一阶协同作用　first-order synergy
一体化多重模型　integrative multimodeling
一体化仿真　integrated simulation
一体化建模　integrative modeling
一体化建模方法学　integrative modeling methodology
一维波动方程　one-dimensional wave equation
一元模型　monadic model
一直紧急行为　consistent emergent behavior
一致　conformity
一致初始条件　consistent initial condition
一致的　congruent; consistent
一致的行为　congruent behavior
一致互仿真等价　uniform bisimulation equivalence
一致输入　uniform input
一致性　coherence; conformance; consistency
一致性的模型　consistent model
一致性关系　coherence relation
一致性管理　consistency management
一致性检查　consistency check; consistency checking
一致性检验器　consistency checker
一致性建模　consistent modeling
一致性耦合　consistency of coupling
一致性驱动　coherence-driven
一致性数据　consistent data
一致性条件　consistency condition

一致性误差　consistency error
一致性现实　consensus reality
一致性准则　consistency criterion
一种电路仿真软件　multisim
一种因果关系　A causality
医疗培训　medical training
依存　dependency
依靠的假设　dependent assumption
依赖　depend (v.); dependence
依赖的　dependent
依赖实验的　experimentarian
依赖于技术的模型　technology-dependent model
依赖于实现　implementation dependent
仪表化的　instrumented
仪表化的模型　instrumented model
仪表化的实体　instrumented entity
仪表误差　instrument error
仪器　instrument; instrumentation
仪器误差　instrumental error; instrumentation error
移动边界仿真　moving-boundary simulation
移动仿真　mobile simulation
移动计算　mobile computing
移动模型　mobility model
移动设备触发的仿真　mobile device-triggered simulation
移动设备发起的仿真　mobile device initiated simulation
移动设备激活的仿真　mobile device activated simulation
移动协同　mobile collaboration
移位算子　shift operator

移位误差　shift error
遗传算法　genetic algorithm
遗传算法仿真　genetic-algorithm simulation
遗留　legacy
遗漏　omission
遗漏变量　omitted variable
遗漏误差　error of omission; omission error
遗赠　legacy
疑义度　equivocation
已分配变量　allocated variable
已核实的协定　verified understanding
已认证　authenticated
已扫描的　scanned
已识别的系统　identified system
已逝时间　elapsed time
已有变换　legacy transformation
已有仿真　legacy simulation
已有仿真系统　legacy simulation system
已有工具　legacy tool
已有建模与仿真　legacy M&S
已有模型　legacy model
已有数据　legacy data
已有系统　legacy system
已有应用　legacy application
已知变量　known variable
已知的对象　known object
已知的值　known value
已知的状态　known state
以代理为中心的　agent centric
以机器为中心的仿真　machine-centered simulation
以计算机为中性的模型　computer-centric model
以理论为中心　theory centric
以人为中心的　human-centered; human-centric
以人为中心的方法　human-centric approach
以人为中心的仿真　human-centered simulation; man-centered simulation
以人为中心的过程　human-centered process
以人为中心的环境　human-centered environment
以人为中心的活动　activity human centered; human-centered activity
以人为中心的模型　human-centric model
以人为中心的系统　human-centered system
以事件为中心　event-centered
以事件为中心的方法　event-centered approach
以数据为中心的方法　data-centric approach
以数据为中心的计算　data-centric computing
以万维网为中心　web-centric
以万维网为中心的仿真　web-centric simulation
以为　understand (v.)
以文档为中心的方法　document-centric approach
以细胞为中心的　cell-centered
以细胞为中心的建模　cell-centered modeling
以下　following

以艺术为中心的仿真　art-directed simulation
以用户为中心　user-centered
以用户为中心的设计　user-centered design
以用户中心的实验　user-centered experimentation
以智能体为中心的　agent centric
艺术仿真　artistic simulation
议程上的游戏　game with agenda
议事日程　agenda
异步传输　asynchronous transmission
异步传输模式　asynchronous transfer mode
异步的　asynchronous
异步仿真　asynchronous simulation
异步建模　asynchronized modeling
异步离散事件仿真模型　asynchronous discrete-event simulation model
异步输入　asynchronous input
异步协调建模　asynchronized collaborative modeling
异常的行为　aberrant behavior
异常事件　abnormal event
异地备份　off-site backup
异地实验　ex situ experiment
异方差误差　heteroscedastic error
异方差性　heteroscedasticity
异构代理　heterogeneous agent
异构多尺度方法　heterogenous multiscale method
异构仿真　heterogeneous simulation
异构仿真技术　heterogeneous simulation technique
异构仿真器　heterogeneous simulator
异构分布系统　heterogeneous distributed system
异构建模形式　heterogenous modeling formalism
异构数据　heterogeneous data
异构网络　heterogeneous network
异构型模型　heterogeneous model
异构型系统　heterogeneous system
异构智能体　heterogeneous agent
异化　dissimilating; dissimilation; dissimilate (v.)
异化的　dissimilative; dissimilatory
异态检测　anomaly detection
异源　antithetic
异质变化　alloplastic-change
异质的　heterogeneous
异质适配　alloplastic adaptation
异质条件　heterogeneous condition
异质系统　alloplastic system
异质性　heterogeneity
抑制变量　suppressor variable
抑制弧　inhibitory arc
抑制佩特里网　inhibitor Petri net
易处理　tractability
易处理的　tractable
易处理方程　tractable equation
易处理模型　tractable model
易懂性　understandability
易失变量　volatile variable
易失性　volatility
易于驾驭的　tractable
益智游戏　puzzle game
意识　awareness
意识到的　aware
意识读取头盔　mind-reading helmet

意识实验　conscious experiment
意图建模　intent modeling
意图模型　intent model
意图状态　intentional state
意外出现的行为　unintended emerging behavior
意外的行为　abrupt behavior; unexpected behavior
意外的输入　unexpected input
意外行为　unintended behavior
意外后果　unintended consequence
意外结果　unintended outcome
意外事件模型　contingency model
意象　imagery
意象仿真　imagistic simulation
溢出错误　overflow error
因变量　dependent variable
因果　causal; cause-effect
因果的　causalized
因果断言　causal assertion
因果对称博姆模型　causally symmetric Bohm model
因果方程　causal equation
因果方程系统　causalized equation system
因果分配　causal assignment
因果关系　causal relation; causality; causation; cause-and-effect
因果关系分析　causality analysis
因果关系条　causality bar
因果化　causalization
因果建模　causal modeling
因果键合图　causal bond graph
因果键合图模型　causal bond graph model
因果结构　causal structure
因果描述建模　causal-descriptive modeling
因果描述模型　causal-descriptive model
因果模型　causal model
因果顺序　causal order
因果图　causal diagram; cause-and-effect diagram
因果图示　cause-effect graphing
因果推理　causal reasoning
因果系统　causal system
因果循环　causal loop
因果语言　cause-and-effect language
因果原理　causality principle
因果追溯　causal tracing
因果追溯性　causal traceability
因素　factor
因素筛选　factor screening
因子设计　factorial design
音频输入　audio input
引导机制　anticipatory mechanism
引道　approach
引擎　engine
引入阶段　introductory phase
引用　reference
隐边界条件　implicit boundary condition
隐藏期　latent phase
隐含变量　hidden variable; lurking variable
隐含层　hidden layer
隐含值　hidden value; implicit value
隐马尔可夫链　hidden Markov chain
隐式单步法　implicit single-step

method
隐式的　implicit
隐式方法　implicit method
隐式积分　implicit integration
隐式积分算法　implicit integration algorithm
隐式交互　implicit interaction
隐式解　implicit solution
隐式密尔恩法　implicit Milne method
隐式事件　implicit event
隐式外推法　implicit extrapolation method
隐式协调　implicit coordination
隐式约束　implicit constraint
隐式中点规则　implicit midpoint rule
隐式状态　implicit state
隐属性　hidden attribute
隐形观察者　stealth viewer
隐性的假设　implicit assumption
隐性假设　tacit hypothesis
隐性目标　implicit goal
隐性知识　implicit knowledge; tacit knowledge
隐喻　metaphor
隐喻表示　metaphorical expression
隐喻模型　metaphorical model
影响　impact
影响变量　affecting variable
影响建模　impact modeling
影响图　influence diagram
应急对策　emergent game
应急对策平台　emergent gaming platform
应急范式　emergent paradigm
应急仿真　emergent simulation

应急仿真平台　emergent simulation platform
应急行为　emergent behavior
应急结构　emergent structure
应急平台　emergent platform
应急系统特性　emergent system property
应急系统特征　emergent system characteristic
应急自模型　emergent self-model
应用　application
应用层　application layer
应用程序　application
应用程序错误　application error
应用开发框架　application development framework
应用误差　application error
映射　map; mapping
映射模型　model mapping
硬错误　hard error
硬仿真　hard simulation
硬核玩家　hardcore gamer
硬件　hardware
硬件安装错误　hardware installation error
硬件错误　hardware error
硬件仿真器　hardware simulator
硬件环境　hardware environment
硬件开发　hardware development
硬件模型　hardware model
硬件实现　hardware implementation
硬件套接字接口技术　hardware socket interface technique
硬件完整性　hardware integrity
硬接线的　hard-wired

硬连线数据　hard-wired data
硬耦合　hard coupling
硬耦合的　hard coupled
硬实时系统　hard real-time system
硬实时约束　hard real-time constraint
硬挺度　stiffness
硬系统思维　hard system thinking
涌现　emergence
涌现本体　ontology of emergence
涌现的　emergent
涌现的集群　emergent cluster
涌现的特性　emergent property
涌现的问题　emergent problem; emerging problem
涌现动力学　emergent dynamics
涌现仿真　emergence simulation
涌现仿真方法　emergence simulation approach
涌现仿真系统　emerging simulation system
涌现复杂性　emergence complexity
涌现建模　emergence modeling
涌现特性　emerging property
涌现特征　emergent characteristic; emergent feature
涌现条件　emergence condition
涌现系统　emerging system
涌现现象　emergent phenomenon; emerging phenomenon
涌现预测　emergence prediction
涌现主义　emergentism
涌现状态　emergent state
用参数表示　parameterize (v.)
用法　use
用户　consumer; user

用户/系统接口　user/system interface
用户创造的　user-created
用户错误　user error
用户的模型　model of user; user's model
用户概念模型　user's conceptual model
用户行为　user behavior
用户行为模型　user behavior model
用户建模　user modeling
用户接口　user interface
用户界面（接口）的模型驱动开发　model-driven development of user interface
用户界面测试　user interface testing
用户界面错误　user interface error
用户界面分析　user interface analysis
用户界面设计　user interface design
用户领域需求　user domain requirements
用户输入　user input
用户为中心的分析　user-centered analysis
用户文档　user documentation
用户心理模型　user's mental model
用户验收　user acceptance
用户域　user domain
用户支持　user support
用例图　use case diagram
用途　use
用于模型转换的语义网　semantic web for model transformation
用于全局逼近的克里格建模　Kriging modeling for global approximation
用于训练的仿真　simulation for

training
用于训练的游戏　gaming for training
用于验证的转换　transformation for validation
用于遗留转换的语义网　semantic web for legacy transformation
优化仿真　optimizing simulation
优化嵌入仿真　optimization embedded within simulation
优化中的仿真　simulation within optimization
优势模型　dominance model
优先队列数据结构　priority queue data structure
优先佩特里网　priority Petri net
"幽灵"　demon
幽灵般的突现行为　spooky emergent behavior
幽灵般的涌现　spooky emergence
"幽灵"控制的仿真　demon-controlled simulation
由建模与仿真产生的数据　data produced by M&S
由模拟主义者所做的评估　assessment of a simulationist
由人工智能指导的仿真　AI-directed simulation
由误解引起的错误　error due to misunderstanding
游戏　game; playing
游戏比赛策略　game-playing strategy
游戏程序　game program
游戏程序设计　game programming
游戏更新　game update
游戏化　gamification; gamify (v.)
游戏化的　gamified
游戏化模块　gamified module
游戏化训练　gamified training
游戏技术　game technology
游戏架构　game architecture
游戏教育　gaming education
游戏开发　game development
游戏开发工具　game development tool
游戏开发路线　game development curriculum
游戏开发者　game developer
游戏软件　game software; gameware
游戏设备　gaming device
游戏设计　game design
游戏社区　gaming community
游戏式广告　advergaming
游戏式学习　game-based learning
游戏室　play room
游戏体验　gaming experience
游戏外壳　game shell
游戏项目管理　game project management
游戏引擎　game engine; gaming engine
游戏引擎设计　game-engine design
游戏硬件　gaming hardware
游戏者　player
游戏者空间　player room
游戏资产管理　game asset management
游戏资产生成　game asset generation
有帮助的训练系统　constructive training system
有参随机变量生成　parametric random variate generation

有故障的　faulty
有害误差　harmful error
有机体系　organic system-of-systems
有界变量　bounded variable
有界的　bounded
有界合理性　bounded rationality
有界佩特里网　bounded Petri net
有界时域自动机　bounded time domain automaton
有界输出有界输入　bounded-input bounded-output
有界输入　bounded input
有界输入有界输出系统　bounded-input bounded-output system
有界增长　bounded growth
有目的仿真　purposeful simulation
有偏测量　biased measurement
有偏估值器　biased estimator
有偏数据　biased data
有偏误差　biased error
有偏样本　biased sample
有缺陷的　defective
有缺陷的模型　imperfect model
有上界的　bounded above
有生气的　animated
有受约束的分对数　conditional logit
有条件的　conditional
有瑕疵的断言　flawed assertion
有瑕疵的模型　flawed model
有下界的　bounded below
有限　finite
有限差分　finite difference
有限差分法　finite-difference method
有限差分近似　finite-difference approximation
有限差分模型　finite-difference model
有限存储离散事件系统规范　finite-memory DEVS
有限的　limited
有限的环境交互　limited environmental interaction
有限的模型　finite model
有限的输入　limited input
有限对策　finite game
有限规则　limited discipline
有限互动　limited interaction
有限解空间　limited solution space
有限确定性离散事件系统规范　FD (finite deterministic) DEVS
有限实时离散事件系统规范　FRT (finite and real-time) DEVS
有限元法　finite-element method
有限元分析　finite-element analysis
有限元近似　finite-element approximation
有限元模型　finite-element model
有限元素　finite element
有限状态机　finite-state-machine
有限状态机仿真　finite-state machine simulation
有限状态系统　finite-state system
有限状态自动机　finite-state automaton
有限状态自动机模型　finite-state automaton model
有向的系统的系统　directed system-of-systems
有向图　digraph; directed graph; oriented graph
有向图模型　digraph model

有效的　active；valid
有效的模型　valid model
有效的误差　active error
有效的组成　valid composition
有效范围　validity range
有效简化　valid simplification
有效决策　effective decision
有效理解　valid understanding
有效期　validity period
有效性　effectiveness；usefulness；validity
有效性测量　validity measure；measure of effectiveness
有效性的等级　level of validity
有效性度量　effectiveness metric；validity metric
有效性范围　scope of validity
有效性检查　validity checking
有效性条件　validity condition
有效值　virtual value
有形的价值　tangible value
有序的　ordered
有序分对数　ordered logit
有序集合　ordered set
有意的耦合　intentional coupling
有意图的知识处理　intentional knowledge processing
有用　usefulness
有用性　usefulness
有源系统　active system
有噪声的模型　noisy model
有助于　stimulate (v.)
有组织的　organized
右特征向量　right eigenvector
诱导　induced

诱导出现　induced emergence
诱导的　inductive
诱发　induction
娱乐　entertainment
娱乐仿真　entertainment simulation；simulation in entertainment
娱乐模型　entertainment model
娱乐游戏　entertainment game；gaming in entertainment
娱乐游戏仿真　simulation in entertainment game
与平台无关的建模　platform-independent modeling
与平台无关的模型　platform-independent model
与实验（试验）相关的参数　experiment-related parameter
"与"输入　AND-input
与智能实体进行的博弈　game with intelligent entity
语法　syntax
语法错误　grammatical error；syntactic error；syntactical error；syntax error
语法的评估　syntactic assessment
语法分析　syntax analysis
语法互操作水平　syntactical interoperability level
语法检查　syntax check
语法可变性　syntactic variability
语法水平　syntactic level
语法验证　syntactic validation
语句　statement
语句变量　statement variable
语句测试　statement testing

语言　language
语言变量　linguistic variable
语言错误　language error
语言的语义　semantics of language
语言规范　language specification; verbal specification
语言化数据　verbalized data
语言值　linguistic value
语义变量　linguistic variable
语义标记　semantic markup
语义层　semantic layer
语义层次　semantic level
语义错误　semantic error
语义代理　semantic agent
语义导向的　semantic-directed
语义的可组合性　semantic composability
语义的相容性　semantic compatibility
语义发现　semantic discovery
语义分析　semantic analysis
语义丰富的元数据　semantically rich metadata
语义改进　semantic improvement
语义互操作水平　semantic interoperability level
语义互操作性　semantic interoperability
语义机制　semantic scheme
语义记忆模型　semantic memory model
语义间隙　semantic gap
语义建模　semantic modeling
语义建模范式的适用性　suitability of a paradigm for semantic modeling
语义建模语言的适用性　suitability of a language for semantic modeling
语义结构　semantic structure
语义可操作性　semantic operability
语义理解　semantic understanding
语义论域　semantic domain
语义模型　semantic model
语义模型比较　semantic model comparison
语义模型差分　semantic model differencing
语义模型评估　semantic model assessment
语义模型评价　semantic model evaluation
语义评估　semantic assessment
语义评价　semantic evaluation
语义上下文　semantic context
语义实体　semantic entity
语义数据模型　semantic data model
语义通信　semantic communication
语义突现行为　semantic emergent behavior
语义图　semantic graph
语义网　semantic web
语义网络　semantic network
语义学　semantics
语义涌现　semantic emergence
语义-语用模型　semantic-pragmatic model
语义元数据　semantic metadata
语义约束　semantic constraint
语义增强的　semantically augmented
语义增强元数据　semantically augmented metadata
语义真理论　semantic theory of truth

语义值　semantic value
语义智能体　semantic agent
语音标记语言　speech markup language
语音输入　speech input; voice input
语音压缩　speech compression
语用层次　pragmatic level
语用方法　pragmatic approach
语用互操作性　pragmatic interoperability
语用互操作性水平　pragmatic interoperability level
语用理解　pragmatic understanding
语用论　pragmatics
语用模型　pragmatic model
语用评估　pragmatic assessment
语用评价　pragmatic evaluation
语用视角　pragmatic perspective
预测　forcast (v.); forecast; look ahead; predict (v.); prediction; predictor
预测变量　predicted variable; predictor variable
预测的　forcasting; forecasting
预测的轨迹　predicted trajectory
预测的行为　predicted behavior
预测的数据　predicted data
预测仿真　prediction simulation; predictive simulation
预测技术　predictive technique
预测建模　predictive modeling
预测模型　forecasting model; prediction model; predictive model; predictor model
预测模型的可维护性　prediction-model maintainability
预测模型的可追溯性　prediction-model traceability
预测模型有效性　predictive model validity
预测器　predictor
预测误差　prediction error
预测系统　prediction system; predictive system
预测校正　predictor-corrector
预测校正法　predictor-corrector method
预测效度　predictive validity
预测性　predictive
预测性仿生仿真　predictive biosimulation
预测性决策　predictive decision
预测性显示　predictive display
预测性验证　predictive validation
预测有效的　predictively valid
预测值　predicted value; prognosed value
预处理　preprocessing
预订的活动　scheduled activity
预订的消息　scheduled message
预定的　scheduled
预定义活动　predefined activity
预定义假设　predefined hypothesis
预防　avoidance; prevention; prevent (v.)
预防性模型维护　preventive model-maintenance
预防性维护　preventive maintenance
预仿真阶段　presimulation phase
预估　predict (v.)

预后变量　prognostic variable
预后分析　postprognostic analysis；prognostic analysis
预后模型　prognostic model
预见的现实　anticipated reality
预见性分析　anticipatory analysis
预期出现　anticipated emergence
预期出现的行为　intended emerging behavior
预期代理　anticipatory agent
预期的　anticipative
预期的行为　anticipatory behavior；expected behavior
预期的事实　anticipated fact
预期的形势　anticipated situation
预期多仿真　anticipatory multisimulation
预期仿真　anticipatory simulation
预期感知仿真　anticipatory perceptual simulation
预期行为　intended behavior
预期后果　intended consequence
预期机制　anticipation mechanism
预期价值　anticipatory value
预期结果　intended outcome
预期事件　anticipated event
预期数据　anticipated data
预期误差　anticipation error；anticipated error
预期系统　anticipatory system；anticipative system；desirable system
预期现实　intended reality
预期性能　desired performance
预期值　desired value
预期中的模型　anticipative model

预热的　warm up
预热时间　warm up time
预设值　assigned value
预先的　anticipated
预想中的模型　anticipatory model
预研究活动　prestudy activity
预演　walkthrough
预运行活动　prerun activity
预执行状态　preexecution state
域　domain；field
域变量　domain variable
域名注册　domain application
域相关　domain dependent
阈　threshold
阈下刺激　subliminal stimulus
阈值　threshold；threshold value
阈值模型　threshold model
寓教于乐的仿真游戏　educational simulation game
元胞的　cellular
元胞仿真语言　cellular SL
元胞空间模型　cellular space model
元胞离散事件系统规范模型　cellular DEVS model
元胞模型　cellular model
元胞自动机　cellular automata；cellular automaton
元胞自动机仿真　cellular automaton simulation
元胞自动机模型　cellular automaton model
元胞自动机形式　cellular automaton formalism
元编程　metaprogramming
元变量　metavariable；metavariate

元对策　metagame
元仿真　metasimulation
元仿真研究　metasimulation study
元分析　meta-analysis
元构件　metamechanism
元活动　primitive activity
元级的组件　metalevel component
元级概念　metalevel concept
元计算　metacomputing
元件级　component level
元件库　component library
元建模　metamodeling
元建模基础　metamodeling foundation
元建模理论　metamodeling theory
元类　metaclass
元模型　metamodel
元模型构建技术　metamodeling technique
元模型校核　metamodel verification
元模型演变　metamodel evolution
元模型验证　metamodel validation
元数据　metadata
元数据标准　metadata standard
元数据模板　metadata template
元素　element
元素对齐　element alignment
元素分组　element grouping
元调度器　metascheduler
元语法变量　metasyntactic variable; metasyntactical variable
元语言的　metalinguistic
元元模型　meta-metamodel
元知识　metaknowledge
原理　principle; theory
原始的　original; primitive

原始的模型　crude model; primitive model
原始概念　original notion
原始假设　primary assumption
原始建模　primitive modeling
原始数据　original data
原始信息　raw information
原位仿真　in situ simulation
原位模拟　in situ analog
原位模型　in situ model
原位实验　in situ experiment; in situ experimentation
原位数据　in situ data
原型　ancestor; archetype; prototype
原型的　archetypical
原型地　archetypally; archetypically
原型仿真　ancestor simulation
原型仿真语言　prototype SL
原型模型　ancestor model; archetypal model; prototypical model
原型设计　prototyping
原因　cause
原语　primitive
原语转换　primitive transformation
原子多模型　atomic multimodel
原子级模拟　atomistic simulation
原子离散事件系统规范　atomic DEVS
原子模型　atomic model
原子模型编辑器　atomic model editor
原子模型规范　atomic model specification
原子实体　atomic entity
原子型离散事件系统规范模型　atomic DEVS model

圆　circle
源　source
源代码　source code
源点　source
源仿真程序　source simulation program
源仿真语言　source SL
源数据　source data
源文件　source documentation
源系统　source system
源域　source domain
远程沉浸　teleimmersion
远程错误　remote error
远程仿真　remote simulation
远传仿真　teleozetic simulation
远期多仿真博弈　forward multisimulation gaming
远生系统　teleogenetic system
远源模型　teleogenic model
约定的价值　aggred value
约束　constraint; restriction
约束变量　constrained variable
约束的　constrained
约束方程　constraint equation
约束仿真　constrained simulation
约束分析　constraint analysis
约束建模　constraint modeling
约束决策逆模糊化　constraint decision defuzzification
约束模型　constrained model; constraint model
约束驱动　constraint-driven
约束驱动的多模型　constraint-driven multimodel
约束条件模型　constrained conditional model
约束误差　constraint error
约束系统　constraint system
约束因子模型　constrained factor model
约束优化模型　constrained optimization model
约束主动外观模型　constrained active appearance model
跃迁选择规则　transition selection rule
云博弈　cloud gaming
云仿真　cloud simulation
云仿真工具　cloud simulation tool
云仿真环境　cloud simulation environment
云和赛博物理系统　cloud and Cyberphysical system
云计算　cloud computing
云计算环境　cloud computing environment
云际　intercloud
云间联邦　inter-cloud federation
云托管仿真　cloud-hosted simulation
云网格　cloud grid
允许的　acceptable
允许的输出　permissible output
允许值　allowable value
运筹博弈　operational gaming
运筹学　operations research
运动技术增强　motor skill enhancing
运动控制　motor skill
运动控制技术　motor control skill
运动模型　kinetic model
运动游戏　activism game
运行　operation; run; run (*v.*)

运行次数　number of runs
运行分析　operational analysis
运行后的　postrun
运行后分析　postrun analysis
运行后活动　postrun activity
运行后计算　postrun computation
运行后输出　postrun output
运行阶段　operation phase; run phase
运行可视化　execution visualization
运行控制变量　run-control variable
运行日志　log
运行时　runtime
运行时仿真更新　runtime simulation update
运行时仿真连接　runtime simulation linking
运行时仿真重置　runtime simulation reconfiguration
运行时更新　runtime update; runtime updating
运行时环境　runtime environment
运行时活动　runtime activity
运行时基础结构　runtime infrastructure
运行时基础结构消息　runtime infrastructure-message
运行时间的　runtime
运行时间效率　execution-time efficiency
运行时接口　runtime interface
运行时可扩展性　runtime extensibility
运行时连接的数据　runtime linked data
运行时联邦成员　runtime federate
运行时联邦成员发现　runtime federate discovery
运行时联邦成员互操作性　runtime federate interoperation
运行时联邦成员实例　runtime federate instantiation
运行时联邦成员组合　runtime federate composition
运行时模型更新　runtime model update
运行时模型鉴定　runtime model qualification
运行时模型切换　runtime model switching
运行时模型替换　runtime model replacement
运行时模型推荐　runtime model recommendation
运行时耦合　runtime coupling
运行时输出　runtime output
运行时数据　runtime data
运行时推荐　runtime recommendation
运行时执行程序　runtime executive
运行长度　run length
运输　transport
运算　operation

Zz

Z变换　Z-transform
杂波抑制　clutter reduction
灾变行为不连续性　catastrophic behavior-discontinuity
灾难性的仿真语言　catastrophic SL
灾难性的行为　catastrophic behavior
灾难性故障　catastrophic failure

灾难性事件　catastrophic event
再创造　anapoiesis
再评估　re-evaluate
再生　anapoietic
再生仿真　regenerative simulation
再实例化　re-instantiation
再现性　reproducibility
在大型仿真中　in the large simulation
在其他条件相同的规律下　ceteris paribus law
在其他条件相同假设条件下　ceteris paribus assumption
在其他条件相同情况下　ceteris paribus
在体外模拟　in vitro analog
在线帮助　online help
在线存储设备　online storage device
在线的　online
在线仿真　online simulation
在线仿真更新　online simulation update
在线仿真语言　online SL
在线角色扮演仿真　online role-play simulation
在线决策支持　online decision support
在线模型推荐器　online model recommender
在线模型文档　online model documentation
在线体验　online experience
在线学习　online learning
在线游戏　online game
在线诊断　online diagnosis
在小型仿真中　in the small simulation
在原位的　in situ

在职培训　on-the-job training
暂时的　temporal
暂时的行为　temporal behavior
暂态　temporal state; transitory state
赞助商　sponsor
早期应用　early application
早现　anticipation
噪声　noise
噪声输入　noisy input
噪声数据　noisy data
噪声值　noisy value
增广　augmented
增加的　incremental
增加的能量　incremental energy
增量　increased value
增量变换　incremental transformation
增量仿真　incremental simulation
增强　augmented; enhancing; enhance (v.)
增强的　enhanced
增强仿真　augmented simulation
增强实况仿真　augmented live simulation
增强视觉　augmented vision
增强物理仿真　augmented physical simulation
增强现实　augmented reality; enhanced reality
增强现实仿真　augmented-reality simulation
增强现实游戏　augmented-reality game
增强型仿真　simulation-enhanced
增强型先进分布式仿真　ADS-enhanced

增强型先进分布式仿真系统工程　ADS-enhanced systems engineering
增强虚境　augmented virtuality
增益误差　gain error
增长　growth
增长误差　growth error
摘要　abstract; abstract (v.)
栈　repository
战场　battlefield
战场可视化　battlefield visualization
战场视图　battlefield view
战斗　action
战斗游戏　combat game
战略变量　strategic variable
战略博弈　strategic game; strategy game
战略电子游戏　strategy video game
战略仿真　strategic simulation; strategy simulation
战略观点　strategic view
战略管理知识　strategic management knowledge
战略决策仿真　strategic-decision simulation
战术仿真　tactical simulation
战术决策仿真　tactical decision simulation
战术训练　tactics training
战争　war; warfare
战争博弈　war game; wargaming
战争模拟　war simulation
战争演示厅　theater of war
战争游戏　war gaming
战争综合演示室　synthetic theater of war

站　station
掌上游戏　handheld game; handheld gaming
遮挡　occlusion
折叠佩特里网　folded Petri net
折叠状态　collapsed state
折中算法　trade-off algorithm
哲学基础　philosophical foundation
着色结构有向图　colored structure digraph
着色佩特里网　colored Petri net
着色佩特里网仿真器　colored Petri net simulator
针对经验的　experience-aimed
针对经验的仿真　experience-aimed simulation
针对实验的　experiment-aimed
针对实验的仿真　experiment-aimed simulation
侦查误差　detected error
帧　frame
帧速率　frame rate
真理　truth; verity
真理论　theory of truth
真实　truth; verity
真实程度　degree of truth
真实的　real; true
真实的战场　real battlefield
真实仿真　real-life simulation
真实感虚拟现实　realistic virtual reality
真实环境　real environment
真实经验　real experience
真实生活相似的经验　real-life-like experience

真实实体　live entity
真实实验　actual experiment; real experiment
真实世界　real world
真实世界的时间　real-world time
真实世界数据　real-world data
真实事件学习　live event learning
真实数据　actual data
真实体验机会　real-life experience opportunity
真实系统　real system
真实系统数据的可接受性　acceptability of real-system data
真实性　veracity; verity
真实知识　real knowledge
真知　true knowledge
真值　real value; true value; truth; truth value
真值变量　real-valued variable
诊断　assertion; diagnosis
诊断建模　modeling for diagnosis
诊断模型　diagnostic model
诊断误差　diagnostic error
诊断系统　diagnostic system
阵列仿真　array simulation
振荡　oscillation
振荡状态　oscillating state
震动触觉演示　vibrotactile display
整合费用　integration cost
整数值　integer value
整数值变量　integer-valued variable
整体变量　holistic variable
整体的　holonic
整体仿真　holistic simulation
整体概念　holistic concept

整体行为　holistic behavior
整体宏观行为　holistic macro-level behavior
整体特性　global property
整体效应　holistic effect
整体性模型　holistic model
整体训练　holistic training
整型变量　integer variable
正常的　normative
正常优先级佩特里网　normal priority Petri net
正当的　valid
正当信赖　justified reliance
正当信念　justified belief
正断法规则　modus pollens rule
正反馈环　positive-feedback loop
正规佩特里网　normal Petri net
正交的　orthogonal
正交多智能体模型　orthogonal multiagent model
正交矩阵　orthogonal matrix
正交性　orthogonality
正空间误差　positive spatial error
正确　correct; exactness
正确的　true
正确的数据　correct data
正确性　correctness
正确性测定　correctness determination
正确性分析　correctness analysis
正确性校核　correctness verification
正确性证明　correctness proof; proof of correctness
正式的　formal
正式行为　formal behavior
正式授权　accreditate (*v.*)

正态分布　normal distribution
正向工程　forward engineering
正向推理　forward reasoning
正向推理模型　forward-reasoning model
正约束　positive constraint
正在经历的　experiencing
正在收敛的仿真　converging simulation
正在运行的　running
正则错误　regular error
正则模型　canonical model
正则形式　canonical form
正直的　true
证明　certification; proofing; proof; certify (v.)
证明(鉴定)过程　certification process
证明是真实的、可靠的或有效的　authenticate (v.)
证实　authentication; verification
证书　certificate
证书授权　certification commission
证伪　falsification
政府成品仿真库　government off-the-shelf simulation library
政府成品模型库　government off-the-shelf model library
政府成品软件　GOTS (Government Off-the-Shelf) software
政权　regime
政体　regime
政治变数　political variable
政治博弈　political game
支　branched
支撑工具　support tool

支撑论域　supporting domain
支撑软件　support software
支持　support
支持的　supported
支持者　proponent
支付　payoff
支付函数　payoff function
支付矩阵　payoff matrix
知道的　aware
知觉偏差　perception bias
知识　knowledge
知识产生活动　knowledge-generation activity
知识处理活动　knowledge-processing activity
知识处理视角类型　types of knowledge processing perspective
知识的基本结构　knowledge infrastructure
知识发现　knowledge discovery
知识概要工具　knowledge compendium tool
知识管理视角　knowledge management perspective
知识核心体　core body of knowledge
知识获取　knowledge acquisition
知识获取阶段　knowledge capture stage
知识库　knowledge base
知识领域　knowledge area
知识密集型多主体系统　knowledge-intensive MAS
知识密集型系统　knowledge-intensive system
知识面　knowledge horizon

知识模型　epistemic model
知识视野　knowledge horizon
知识属性　knowledge attributed
知识提炼　knowledge elaboration
知识体系　body of knowledge; BOK (Body of Knowledge)
知识体系规范　body of knowledge specification
知识体系元素　BOK element
知识渊博的行为　knowledgeable behavior
知识重构　knowledge reconstruction
知识转换器　knowledge transducer
执行　execute (*v.*); execution; run (*v.*)
执行测试　execution testing
执行程序　executive
执行错误　execution error
执行的　executive
执行分析　execution profiling
执行跟踪　execution tracing
执行环境　execution environment
执行机构　actuator
执行阶段　execution phase
执行控制　execution control
执行模型　execution model
执行软件　executive software
执行时间管理　execution-time management
执行细节　execution detail
执行效率　execution efficiency
直观的　visualized
直觉行为　counterintuitive behavior
直接操纵接口　direct manipulation interface
直接的通信　direct communication

直接的问题　direct problem
直接多体系统动力学　direct multibody system dynamics
直接仿真　direct simulation
直接交互　direct interaction
直接解决方法　direct solution method
直接量测技术　direct measurement technique
直接耦合　direct coupling
直接输入　direct input
直接输入设备　direct input device
直接数字仿真　direct numerical simulation
直接协调　direct coordination
直通变量　through variable
直线法　method-of-lines
直线法技术　method-of-lines technique
直线法解　method-of-lines solution
直证的　deictic
直证设备　deictic device
职业　career
指标驱动　metric-driven
指定　specify (*v.*)
指定模型　specified model
指定域仿真语言　domain-specific simulation language
指定域建模方法　domain-specific modeling approach
指定域建模语言　domain-specific modeling language
指定域语言　domain-specific language
指令　command; instruction
指令控制　order control
指令误差　command error

指南 guide; guideline
指派 assignment
指示变量 indicator variable
指示模型 denotational model
指示器 indicator
指示物 referent
指数 index
指数分布 exponential distribution
指数随机变量 exponential random variable
制图 graphing
制造模型者 modeler
制造者 producer
制作 tailor (v.)
质量 quality
质量保证 quality assurance
质量保证范式 quality assurance paradigm
质量测量 quality measure
质量的 qualitative
质量估计 quality estimation
质量矩阵 mass matrix
质量控制 quality control
质量模型 quality model
质量评估 qualitative assessment
质量守恒 conservation of mass
质量特征 quality characteristic
质量预测 quality prediction
质量指标 quality indicator
致命误差 fatal error
智力建模 mental modeling
智力模型 mental model
智能 intelligence
智能博弈 intelligent game
智能代理 intelligent agent; smart agent
智能的复杂适应系统 intelligent complex adaptive system
智能电话激活仿真 smart phone activated simulation
智能仿真 intelligent simulation
智能仿真环境 intelligent simulation environment
智能化模型 intelligent model
智能混合系统解 intelligent hybrid systemic solution
智能交互系统 intelligent interactive system
智能接口 intelligent interface
智能控制 intelligent control
智能社会协调群 intelligence community coordinating group
智能实体 intelligent entity
智能实验代理 smart experiment agent
智能体 agent; agenthood
智能体触发式仿真 agent-triggered simulation
智能体导向的 agent-directed
智能体导向的仿真 agent-directed simulation
智能体导向的框架 agent-directed framework
智能体对智能体 agent-to-agent
智能体仿真 agent simulation
智能体仿真器 agent simulator
智能体合理性 agent rationality
智能体化 agentify (v.)
智能体技术 agent technology
智能体间 inter-agent
智能体间的交互 agent-to-agent

interaction
智能体间的通信　inter-agent communication
智能体监控的　agent-monitored
智能体监控的多仿真　agent-monitored multisimulation
智能体监控的仿真　agent-monitored simulation
智能体监控的环境　agent-monitored environment
智能体监控的耦合　agent-monitored coupling
智能体监控的预期多仿真　agent-monitored anticipatory multisimulation
智能体建模　agent modeling
智能体交互　agent interaction
智能体交互协议　agent-interaction protocol
智能体控制的仿真　agent-controlled simulation
智能体控制的耦合　agent-controlled coupling
智能体框架　agent framework
智能体模型　agent model
智能体耦合　agent coupling
智能体平台　agent platform
智能体评估准则　agent evaluation criterion
智能体启动的仿真　agent-initiated simulation
智能体设计方法学　agent design methodology
智能体设计工具　agent design tool
智能体特征　agent characterization

智能体调和的仿真　agent-mediated simulation
智能体调和的仿真游戏　agent-mediated simulation game
智能体调和的环境　agent-mediated environment
智能体系统　agent system
智能体系统建模　agent system modeling
智能体协调的　agent-coordinated
智能体协调的仿真　agent-coordinated simulation
智能体行为　agent behavior
智能体行为图　agent-behavior diagram
智能体支持的　agent supported
智能体支持的多种仿真　agent-supported multisimulation
智能体支持的仿真　agent-supported simulation
智能体支持的框架　agent-supported framework
智能体仲裁的　agent-mediated
智能体仲裁解决方案　agent-mediated solution
智能系统　intelligent system; smart system
智能系统仿真　intelligent-system simulation
智能系统解　intelligent systemic solution
智能行为　intelligent behavior
智能用户接口　intelligent user interface
滞后　delay; hysteresis; lag; lag (*v.*)

滞后的　hysteretic；lagged
滞后的变量　lagged variable
滞后量化　hysteretic quantization
滞后内生变量　lagged endogenous variable
滞后微分方程　delay-differential equation
滞后误差　hysteresis error
滞环宽度　hysteresis width
置换技术　replacement technology
置换矩阵　permutation matrix
置换值　replacement value
置信测度　confidence measure
置信度　confidence；verifiability
置信度评价　confidence value
置信区间　confidence interval
置信区间技术　confidence interval technique
置信区域　confidence region
置信限制　confidence limit
置信值　belief value
中尺度仿真　mesoscale simulation
中尺度建模　mesoscale modeling
中尺度系统　mesoscale system
中等强度噪声　moderate noise
中度建模噪声　moderate modeling noise
中断　interrupt；interrupt (v.)
中断处理算法　discontinuity-handling algorithm
中断探测　discontinuity detection
中规模系统　medium-scale system
中间变量　intermediate variable；mediator variable；temporary variable
中间件　middleware
中间模型　intermediate model
中间输入　intermediate input
中间误差　intermediate error
中间值　intermediate value
中间状态　intermediate state
中介　mediated；mediation
中介变量　intervening variable
中介语　interlanguage
中期可预测性　medium-term predictability
中心　center
中心差分公式　central difference formula
中心差分机制　central difference scheme
中心差分近似法　central difference approximation
中心的　centric
中心极限定理　central limit theorem
中心涌现　neutral emergence
中心站　central station
中性输入　neutral input
中轴　axis
终端　back-end；terminal
终端用户定制工具　end-user customization tool
终结模型　final model
终态　final state
终值　final value；terminal value
终止仿真　terminating simulation
终止规则　dead-end rule
终止条件　termination condition
终止准则　termination criterion
种类　category
种子　seed

种子选择　seed selection
仲裁　mediated
仲裁员　moderator
众包数据　crowdsourced data
众包数据分析　crowdsourced data analysis
众包数据收集　crowdsourced data collection
众包数据输入　crowdsourced data entry
众包行为　crowdsourced behavior
重要事件　significant event
重要数据　significant data
重要特征　significant feature
重要性采样　importance sampling
重要性采样估计器　importance-sampling estimator
重要性术语　significance term
周期　cycle; period
周期边界条件　periodic boundary condition
周期的行为　cyclic behavior
周期函数　periodic function
周期时间尺度变量　cyclical time-scale variable
周期事件　cyclic event
周期系统　periodic system
周期性　periodicity
周期性的　periodic
周期性行为　periodic behavior
周期性环境　recurrent environment
周期性输出　periodic output
轴　axis
轴线　axis
逐步的　step-by-step

主导变量　dominant variable
主导动力学　dominant dynamics
主导极点　dominant pole
主动的　active
主动的外观模型　active appearance model
主动的组件　active component
主动地　proactively
主动感知的输入　actively perceived input
主动机制　proactive mechanism
主动式　proactive
主动式方法　proactive approach
主动式系统　active system
主动外激励　active stigmergy
主动外激励系统　active stigmergic system
主仿真计算机　host simulation computer
主仿真系统　host simulation system
主观错误　subjective error
主观度量　subjective metric
主观价值　subjective value
主观经验　subjective experience
主观数据　subjective data
主观误差　subjective error
主观现实　subjective reality
主观验证　subjective validation
主机　host
主机游戏　console game
主计划　master plan
主计算机　host computer
主题　topic
主题变量　subject variable
主题范围　subject area

主题模型　topic model
主题相关知识　subject-related knowledge
主题专家　subject matter expert
主要任务　primary task
主要效果　main effect
主要元素　essential element
贮存　storage
注册　register
注册表错误　registry error
注册错误　registration error
注销　cancellation
注意　attention
注意交互　attentive interaction
专家　expert
专家建模　expert modeling
专家模型　expert model
专家系统　expert system
专家系统外壳　expert system shell
专门的　specific
专门的方法学　ad hoc methodology
专门的仿真驱动实验　ad hoc simulation-driven experimentation
专门的分布式仿真　ad hoc distributed simulation
专门的建模　ad hoc modeling
专门的模型　ad hoc model
专门的实验　ad hoc experiment
专门的实验法　ad hoc experimentation
专题　topic
专题词典　thematic dictionary
专业的　professional
专业仿真　professional simulation
专业仿真工程师　professional simulation engineer
专业仿真系统工程师　professional simulation systems engineer
专业仿真者　professional simulationist
专业认证　professional certification
专业认证委员会　professional certification commission
专业知识驱动的　expertise-driven
专业知识驱动的决策　expertise-driven decision
专用仿真硬件　special-purpose simulation hardware
专用仿真语言　special purpose SL
专用模型　dedicated model
专用游戏硬件　special-purpose gaming hardware
转存错误　dumping error
转存错误函数　dumping error function
转化　transformation；translate (*v.*)
转换　conversion；transition；transform (*v.*)
转换操作　transition operation
转换策略　transformation strategy
转换范式　transformation paradigm
转换函数　transition function；translation function
转换精度　transformational accuracy
转换器　converter；transducer
转换使能　transition enabling
转换使能触发　transition-enabling firing
转换输出　transition output
转换系统　transition system
转换语言　transformation language
转换值　conversion value
转录错误　transcription error

转向带仿真　turning bands simulation
转向带条件仿真　turning bands conditional simulation
转移　transfer; transfer (*v.*)
转移误差　transfer error
转移值　transfer value
转运　transfer; transfer (*v.*)
转运体模型　transporter model
转置　transpose
装载　docking
装置　unit
状态　state; status
状态保持仿真　state maintaining simulation
状态变化　state change
状态变量　state variable
状态变量辨识　state-variable identification
状态变量不连续性　state-variable discontinuity
状态变量初始化　initialization of state variable
状态变量导数不连续性　state-variable derivative discontinuity
状态变量的量化　quantization of state variable
状态变量量化　state-variable quantization
状态表　state chart
状态超定　overdetermination of state
状态触发的控制　state-triggered control
状态触发的　state-triggered
状态叠加　state superposition
状态独立　state independent

状态独立的耦合　state-independent coupling
状态反馈　state feedback
状态方程　state equation
状态改变　state changing
状态轨迹　state trajectory
状态机　state machine
状态矩阵　state matrix
状态空间　state space
状态空间表示法　state-space notation
状态空间描述　state-space description
状态空间模型　state-space model
状态空间图　state-space graph
状态空间形式　state-space form
状态控制　state control
状态历史向量　state history vector
状态量化　state quantization
状态流　state flow; stateflow
状态密度　state density
状态模型　state model
状态耦合　state coupling
状态迁移图　state transition diagram
状态确定的　state-determined
状态事件　state event
状态事件定位　state event localization
状态事件检测　state event detection
状态图　state chart diagram; state diagram; state graph
状态向量　state vector
状态依赖　state dependent
状态依赖的事件　state-dependent event
状态值　state value
状态值集合　state value set
状态转换　state transformation; state

transition
状态转换机　state-transition machine
状态转换模型　state-transition model
状态转移分析　state transition analysis
状态转移函数　state-transition function
状态转移矩阵　state-transition matrix
状态转移系统　state-transition system
追加　append (v.)
追溯性　retrodiction
追踪　tracing
准备　preparation
准备工作　warm up (v.)
准分析（解析）仿真　quasi-analytic simulation
准分析（解析）仿真技术　quasi-analytic simulation technique
准连续仿真　quasi-continuous simulation
准蒙特卡罗　quasi-Monte Carlo
准蒙特卡罗仿真　quasi-Monte Carlo simulation
准确的　true
准确的世界时间　true global time
准确度　accuracy; accuracy rating
准确性　veracity
准入　admission
准入控制　admission control
准入资格　admissibility
准实验　quasi experiment; quasi experimentation
准线性偏微分方程　quasi-linear PDE
准优化　quasi optimization
准自然实验　quasi-natural experiment
桌面检查　desk checking

咨询者　consultant
姿态　gesture
姿态标记语言　gesture markup language
资产　asset
资产价值　asset value
资格　qualification
资格标准　qualification criterion
资格证书　qualification certificate
资源　resource
资源错误　resource error
资源库　resource repository
资源利用率　resource use efficiency
资源描述层　resource description layer
资源模型　resource model
资源驱动器　resource driver
资源约束　resource constraint
滋扰变量　nuisance variable
子变量　subvariable
子博弈　subgame
子分量　subcomponent
子活动　subactivity
子结构　substructure
子句　clause
子类　subclass
子类化　subclassing
子类耦合　subclass coupling
子模块　submodule
子模块性　submodularity
子模型　submodel
子模型测试　submodel testing
子模型激活　submodel activation
子模型失活　submodel deactivation
子模型替换　submodel replacement
子模型有效性　submodel validity

子目标　subgoal
子耦合　subcoupling
子上下文　subcontext
子事件　subevent
子调制性　submodularity
子网　subnet
子网隶属度函数　subnet membership function
子系统目标　subsystem goal
子系统外推　subsystem extrapolation
子系统执行　subsystem execution
子形式　subformalism
子整体　holon
子整体进化　evolutionary holon
子组件　subcomponent
自保护系统　self-protective system
自变量　argument; argument variable; independent variable
自变量传递　argument passing
自变量值　argument value
自测试输入　self-test input
自创生　autopoiesis
自创生的　autopoietic
自创生的仿真　autopoietic simulation
自创生的模型　autopoietic model
自存档模型　self-archiving model
自底向上的推理　bottom-up reasoning
自定义选项　customization option
自动半代码生成　automatic semicode generation
自动被试变量　automated subject variable
自动变量　automatic variable
自动并行化　automatic parallelization
自动测试工具　automated test tool
自动程序生成　automatic program generation
自动错误检测　automatic error detection
自动代码生成　automatic code generation
自动定理分析　automatic theorem analysis
自动仿真　autosimulation
自动仿真的　autosimulative
自动仿真模型生成　automatic simulation model generation
自动分析　automated analysis
自动更新　automatic update
自动化　automation
自动化过程　automatization process
自动化实验　automated experiment
自动化信息系统　automated information system
自动机　automata; automaton
自动机模型　automata model
自动机网络　automata network
自动模态　auto-modal
自动模型转换　automated model transformation
自动配置　automatic configuration
自动生成　automatic generation
自动生成的　autogenerated
自动态仿真　endomorphic simulation
自动误差修正　automatic error correction
自仿真　self-simulation
自复制系统仿真　self-replicating system simulation
自管理系统　self-managing system

自回归建模技术　autoregressive modeling technique
自回归模型　autoregressive model
自激励　self-excitation
自监控系统　self-monitoring system
自进化　self-evolution
自配置系统　self-configuration system; self-configuring system
自驱动的　self-driven
自驱动仿真　self-driven simulation
自驱动模型　self-driven model
自驱动输入测试　self-driven input testing
自然复杂度　natural complexity
自然计算　natural computing
自然模型　natural model
自然启发　nature-inspired
自然启发范式　nature-inspired paradigm
自然启发算法　nature-inspired algorithm
自然实验　natural experiment
自然系统　natural system
自然主义实验　naturalistic experiment
自然组变量　natural group variable
自认知模型　autoepistemic model
自上而下测试　top-down testing
自上而下理解　top-down understanding
自上而下涌现　top-down emergence
自上向下构建　top-down construction
自身　self
自身具有目的的仿真　autotelic simulation
自身具有目的的系统仿真　autotelic system simulation
自身具有目的的游戏　autotelic game
自身模型　model of self
自省模型　introspective model
自适应　self-adaptation
自适应变形的模型　adaptive deformable model
自适应步长　adaptive stepsize
自适应的　adaptive; self-adaptive
自适应的多模型　adaptive multimodel
自适应多重仿真方法学　adaptive multisimulation methodology
自适应仿真　adaptive simulation; self-adaptive simulation
自适应仿真基本结构　adaptive simulation-infrastructure
自适应共生的多种仿真　adaptive symbiotic multisimulation
自适应计划　adaptive plan
自适应技术　adaptation technology; adaptive technique
自适应界面(接口)　adaptive interface
自适应控制　adaptive control; self-adaptive control
自适应模型　adaptive model
自适应模型维护　adaptive model maintenance
自适应网格细化　adaptive mesh refinement
自适应维护　adaptive maintenance
自适应系统　adaptive system; self-adaptable system
自适应系统仿真　adaptive system simulation; self-adaptive system simulation

自适应现象　adaptive phenomenon
自适应性　adaptivity
自适应用户界面　adaptive user interface
自适应优化　adaptive optimization
自适应值　adaptive value
自适应最优化技术　adaptive optimization technique
自体变化　autoplastic-change
自体刺激的　autostimulant
自体适应　autoplastic adaptation
自体移植的　autoplastic
自调节仿真　self-regulating simulation
自同态　endomorphism
自同态的　endomorphic
自同态仿真模型　endomorphic simulation model
自同态仿真器　endomorphic simulator
自同态模型　endomorphic model
自同态系统　endomorphic system
自稳定　self-stabilization
自稳定系统仿真　self-stabilizing system simulation
自我管理　self-managing
自我模型　self-model
自我意识系统　self-aware system
自我再生的　autopoietic
自我再制　autopoiesis
自下而上测试　bottom-up testing
自下而上构建　bottom-up construction
自下而上设计　bottom-up design
自相似的　self-similar
自相似过程　self-similar process
自相似性　self-similarity
自校正系统　self-tuning system

自修复技术　self-healing technique
自修复系统　self-healing system
自修理模型　self-repair system
自学习仿真　self-learning simulation
自优化行为　self-optimized behavior
自优化系统　self-optimizing system
自由　freedom
自由变量　free variable
自由度　degree of freedom
自由形式博弈　free-form game
自有属性　owned attribute
自治　autonomy
自治常数微分方程　autonomous ordinary differential equation
自治的　autonomous
自治的行为　autonomous behavior
自治模型　autonomous model
自治时不变线性系统　autonomous time-invariant linear system
自治系统　autonomic system; autonomous system
自治主体　autonomous agent
自重置　self-reconfiguration
自主　autonomous; autonomy
自主仿真模式　autonomous simulation mode
自主理解　autonomous understanding
自主实现　implementation independent
自主体　agent
自主系统　autonomic system
自主协商　autonomous negotiation
自主型计算　autonomy-oriented computing
自组织　self-organization
自组织的　self-organized; self-

organizing
自组织仿真　self-organizing simulation
自组织模型　self-organizing model
自组织系统　self-organizing system
自组织系统仿真　self-organizing system simulation
自组织智能　self-organized intelligence
字符　character
字符串变量　string variable
字母输入　alphabetical input
字母数字输入　alphanumerical input
宗旨　purpose
综合　synthesis
综合的　comprehensive; synthetic
综合仿真系统　comprehensive simulation system
综合核心知识体系　comprehensive core body of knowledge
综合环境建模　synthetic environment modeling
综合建模　integrative modeling; synthetic modeling
综合建模方法学　integrative modeling methodology
综合模型　synthesized model; synthetic model
综合评价　comprehensive evaluation
综合系统　comprehensive system
综合相似性理论　comprehensive similarity theory
踪迹分析　trace analysis
踪迹驱动的仿真　trace-driven simulation

总是犯错却自认为正确的人　errorist
总线　bus
总线控制器　bus controller
纵距　advance
足够好的解决方案　good enough solution
阻尼　damping
阻尼矩阵　damping matrix
阻尼图　damping plot
阻尼误差　damping error
阻尼误差符号　damping error sign
阻塞　occlusion
组　group; suite
组播　multicast
组成　compose (v.)
组合博弈论　combinatorial game theory
组合错误　combined error
组合仿真语言　combined SL
组合服务　composition service
组合复杂性　combinatorial complexity
组合公理　composition axiom
组合化　modularization
组合化离散事件连续变化仿真语言　combined discrete-event continuous-change SL
组合技术　composition technique
组合模型　compositional model; assembled model; combined model
组合属性　composite attribute
组合条件　combinatorial condition
组合外观模型　combined appearance model
组合系统　combined system
组合形式下的闭包　closure under

composition
组合状态　combinational state
组和复杂性　composition complexity
组间变量　between-subject variable
组件　component; composant; module
组件的模型　componential model
组件对象模型　component object model
组件仿真　component simulation
组件分配　component allocation
组件封装　component encapsulation
组件化　componentized
组件化对象模型接口　component object model interface
组件化模型　componentized model
组件级别的不当行为　component-level misbehavior
组件级属性　component-level property
组件级性能　component-level performance
组件集成标准　component integration standard
组件交互　component interaction
组件类　component class
组件类型　component type
组件模型　component model
组件模型库　component model repository
组件模型模板　component model template
组件认证　component certification
组件失效　component failure
组件图　component diagram
组件系统　component system
组件之间的耦合　intermodular coupling
组内变量　within subject variable
组织　organization; set-up; structure
组织行为　organizational behavior
组织输入　organizational input
组装　packaging
祖先　ancestor
最大步长　largest step size
最大化　maximization
最大化可能性模型　maximum likelihood model
最大收益　maximum payoff
最大值　maximum value
最佳降阶　optimal reduced order
最佳实践　best practice
最佳应用守则　code of best practice
最小的　minimal
最小规模的训练　least training
最小化　minimization
最小模型　minimal model
最小上界　least upper bound
最小影响　minimal effect
最小影响模型　minimal effect model
最小值　minimal value
最新的模型　up-to-date model
最优　optimum
最优的　optimal
最优方法　best practice
最优化　optimization
最优化方法　optimization approach
最优化技术　optimization technique
最优化理论　optimization theory
最优化问题　optimization problem
最优降阶模型　optimal reduced-order

  model
最优解 optimal solution
最优条件 optimal condition
最优性 optimality
最优性的有效性 validity of optimality
最终条件 final condition
遵守 observance
作为结果而发生的 resultant
作为云服务的建模与仿真 M&S as a cloud service
作业阶段 working phase

作用 action
作用域 scope
作战对抗推演 wargame；wargaming
作战仿真 warfare simulation
作战建模 combat modeling
作战模型 combat model
作战训练 warfare training
坐标 coordinate
坐标变换 coordinate transformation
做实验 experimentalize (*v.*); experiment (*v.*)

第二部分　英汉建模与仿真术语集

3-d display 三维显示
3-d holographic display 三维全息显示
3-d model 三维模型
3-d modeling 三维建模
3-d positional input 三维位置输入
3-d simulation 三维仿真
3-dimensional model 三维模型
3-dimensional modeling 三维空间建模

Aa

a posteriori 后验
a posteriori error estimate 后验误差估计
a posteriori error estimation 后验误差估计
a posteriori experiment 后验实验
a posteriori knowledge 后验知识
a posteriori model 后验模型
a priori 先验的；推理的
a priori experiment 推理实验
a priori knowledge 先验知识
a priori model 先验模型
A-stable A-稳定
ab initio 从开始起；自始
ab initio simulation 从头仿真
abandonment value 清算价值
abandonware game 过时的软件游戏
abductive reasoning 溯因推理，反绎推理
Abelian sandpile model 阿贝尔沙堆模型
aberrant behavior 异常的行为
abnormal event 异常事件
abrupt behavior 意外的行为

absolute error 绝对误差
absolute error control 绝对误差控制
absolute error of measurement 测量的绝对误差
absolute measurement-error 绝对测量误差
absolute reality 绝对现实
absolute stability 绝对稳定性，A 稳定性
absolute truth 绝对事实
absolute validity 绝对有效性
absolute value 绝对值
absorb (v.) 吸收
absorbing 吸收的
absorbing Markov chain 吸收马尔可夫链
absorbing state 吸收状态
absorbtion 吸收
abstract 抽象；摘要
abstract (v.) 使抽象化；提炼；摘要
abstract architecture 抽象结构
abstract DEVS simulator 抽象离散事件系统规范仿真器
abstract domain 抽象（论）域
abstract entity 抽象实体
abstract layer 抽象层
abstract model 抽象模型
abstract model structure 抽象模型结构
abstract object class 抽象对象类
abstract primitive 抽象本原
abstract representation 抽象表示（法）
abstract sequential mapping 抽象顺序映射
abstract sequential simulator 抽象顺序仿真器

abstract simulation 抽象仿真
abstract simulation protocol 抽象仿真协议
abstract simulator 抽象仿真器
abstract specification 抽象规范
abstract state machine 抽象状态机
abstract-system model 抽象系统模型
abstract threaded simulator 抽象线程仿真器
abstracting 抽象(注重过程);编写摘要
abstracting in simulation 仿真中的抽象化
abstraction 抽象;抽象概念
abstraction criterion 抽象准则
abstraction level 抽象级别
abstraction mapping 抽象映射
abstraction refinement 抽象精化
abstraction technique 抽象化技术
abusive behavior 滥用的行为
academic gaming 学术博弈
academic simulation 学术仿真
acausal 非因果关系的
acausal equation 非因果方程
A causality 一种因果关系
accelerated simulation 加速仿真
acceleration error 加速度误差
acceleration of simulation 仿真加速
accept (v.) 接受;认可;同意
acceptability 可接受性;合格率
acceptability assessment 可接受性评估
acceptability criterion 可接受准则
acceptability of model 模型的可接受性
acceptability of real-system data 真实系统数据的可接受性

acceptability of simulated data 仿真数据可接受性
acceptability of simulation study 仿真研究可接受性;仿真研究合格性
acceptability parameter 可接受性参数
acceptability threshold 可接受性阈值
acceptable 可接受的;允许的
acceptable data 可接受数据
acceptable threshold 容许的阈值,可接受的阈值
acceptable tolerance 验收公差
acceptable validity range 可接受的有效范围
acceptable value 可接受值
acceptance 接受;验收
acceptance criterion 验收准则
acceptance error 容许误差限
acceptance probability 接受概率
acceptance testing 验收测试
accepted 接受;录用
accepted data 接受的数据
accepted state 接受的状态
accepted value 接受值
accepting 接受
accepting state 接受状态
acceptor 接受者;接收器
access 访问;接近;入口
access error 访问错误
access reliability 接入可靠性
accessibility 可访问性
accessibility error 可访问性错误
accessible 可接近的;可访问的;可及,可达
accessible data 可访问数据
accessible model 可访问模型

accessible web　可访问的万维网
accidental error　随机误差
accreditate (v.)　正式授权, 批准; 委托
accreditated　认可的
accreditation　认可; 任命; 授权, 批准; 确认
accreditation agent　认证代理
accreditation and certification　确认与认证
accreditation authority　实施确认的机构
accreditation organization　实施确认的组织
accreditation plan　确认计划
accreditation process　确认过程
accreditation proponent　确认提议者
accreditation recommendation　确认推荐
accreditation sponsor　确认发起方
accumulated error　累积的误差
accumulated value　累加值
accumulation error　累积误差
accumulation model　累积模型
accuracy　准确度; 精度
accuracy cost　精度代价
accuracy domain　精确论域
accuracy of sensitivity analysis　灵敏度分析的精确度
accuracy rating　准确度
accurate　精确的
accurate discretization scheme　精准离散化方案
accurate measurement　精确测量
accurate model　精确模型
accurate modeling　精确建模
accurate parameter　精确参数
accurate simulation　精确仿真

accurate solution　精确解法; 精确解
acknowledged error　公认误差
acknowledged system-of-systems　公认的体系
acoustic input　声音输入
acquisition　采办
acquisition model　采办模型
acquisition view　采办视图
across variable　跨变; 交叉变量
action　动作; 战斗; 作用; 活动
action appraisal　行动评估
action appraisal model　行动评价模型
action-based error detection　基于动作的错误探测
action game　动作游戏
action-oriented　面向动作的
action-oriented rule　面向动作的规则
action variable　行动变量
actionable　可操作的
actionable data　可操作数据
actionable information　可操作信息
actionable knowledge　可操作的知识
activate (v.)　激活; 建成
activated model　激活的模型
activation　活化, 激活
activation criterion　激活准则
activation energy　激活能量; 活化能
activation error　活化误差
activation function　活化作用
activation model　激活模型
activation value　激活值
activation variable　激活变量; 启动变量
active　积极的; 主动的; 有效的
active appearance model　主动的外观模型

active behavior 积极行为
active component 主动的组件
active entity 活动实体
active entity of a model 模型的活动实体
active error 有效的误差
active multimodel 活跃的多模型
active shape model 活动形状模型
active state 活动状态
active stigmergic system 主动外激励系统
active stigmergy 主动外激励
active system 有源系统；主动式系统
active value 活跃值
actively perceived input 主动感知的输入
activism game 运动游戏
activity 活动；行动；活动力；活力
activity-based 基于活动的
activity-based model 基于活动的模型
activity-based modeling 基于活动的建模
activity-based simulation 基于活动的仿真
activity cycle diagram 活动周期图
activity diagram 活动图
activity human centered 以人为中心的活动
activity language 活动性语言
activity model 活动模型
activity-oriented 面向活动的
activity-oriented model 面向活动的模型
activity rule 活动规则
activity scan 活动扫描
activity scanning 活动扫描
activity-scanning approach 活动扫描法
activity-scanning method 事件扫描方法
activity SL 活动仿真语言
actor 角色；行动者；参与者
actor-based simulation 基于角色的仿真
actor-based simulation engine 基于角色的仿真引擎
actor model 角色模型
actor-programming model 角色程序设计模型
actual data 真实数据
actual experiment 真实实验
actual performance 实际性能
actual state 实际状态
actual trajectory 实际轨迹
actual value 实际值
actuator 操动件；执行器，执行机构
acyclic metamorphic multimodel 无环变结构多模型
acyclic model 无环模型
acyclic multimodel 无环多模型
ad hoc distributed simulation 专门的分布式仿真
ad hoc experiment 专门的实验
ad hoc experimentation 专门的实验法
ad hoc methodology 专门的方法学
ad hoc model 专门的模型
ad hoc modeling 专门的建模
ad hoc simulation-driven experimentation 专门的仿真驱动实验

Adams-Moulton algorithm 亚当斯-莫尔顿算法
adapt (v.) 适应
adaptability 适应性
adaptable 可适应的
adaptable interface 可适应界面（接口）
adaptation 适应；改编；适配
adaptation engine 适应性引擎
adaptation system 适应系统
adaptation technology 自适应技术
adapted 适应的
adapting 适应
adaptive 适应的，自适应的
adaptive automation 适应自动化
adaptive behavior 适合的行为
adaptive capability 适应能力
adaptive control 自适应控制
adaptive deformable model 自适应变形的模型
adaptive interface 自适应界面（接口）
adaptive maintenance 自适应维护
adaptive mesh refinement 自适应网格细化
adaptive model 自适应模型
adaptive model maintenance 自适应模型维护
adaptive multimodel 自适应的多模型
adaptive multisimulation methodology 自适应多重仿真方法学
adaptive optimization 自适应优化
adaptive optimization technique 自适应最优化技术
adaptive phenomenon 自适应现象
adaptive plan 自适应计划
adaptive simulation 自适应仿真
adaptive simulation-infrastructure 自适应仿真基本结构
adaptive stepsize 自适应步长
adaptive symbiotic multisimulation 自适应共生的多种仿真
adaptive system 自适应系统
adaptive-system simulation 自适应系统仿真
adaptive technique 自适应技术
adaptive user interface 自适应用户界面
adaptive value 自适应值
adaptiveness 适应性
adaptivity 自适应性，适应性
adaptor 适配器
added value 附加值
additional value 附加价值
additional variable 附加的变量
additive game 附加游戏
additive trust game 附加信任游戏
adequacy 充分；适当
adequacy of dynamic model structure 动态模型结构的适当性
adequacy of formalism 形式化的适当性
adequacy of model parameters 模型参数的适当性
adequacy of model structure 模型结构的适当性
adequacy of static model structure 静态模型结构的适当性
adequate 适宜；充足的；适当的
adequate data 充足的数据
adequate model 充分的模型
adequate stimulus 适宜的激励

adiabatic-system simulation 绝热系统仿真
adjacent value 邻近值
adjunct tool 辅助工具，配套工具
adjust (*v.*) 调整
adjustable parameter 可调节的参数
adjusted data 校正后数据
adjusted value 校正值
adjustment 调节；校正；调节装置
adjustment error 调节误差
administration 实施；管理；行政；任期
administrator 管理员
admissibility 准入资格
admissibility condition 容许条件
admissible 可容许的；合理的
admissible input 容许的输入
admissible segment 可采纳的片段
admissible space 容许空间
admissible space of models 模型的准用空间
admissible value 容许值
admission 容许；准入
admission control 许可控制；准入控制
ADS-enhanced 增强型先进分布式仿真
ADS-enhanced systems engineering 增强型先进分布式仿真系统工程
advance 纵距；朝前；进度；前进
advance (*v.*) 前进，推进
advanced concept 先进的概念
advanced distributed simulation 先进分布式仿真
advanced environment 先进环境
advanced modeling 先进建模
advanced numerical simulation 先进数字仿真
advanced simulation 先进仿真
advanced simulation environment 先进仿真环境
advanced understanding system 先进的理解系统
advancement 前进，进步；晋升
adventure game 冒险游戏
advergame 广告游戏
advergaming 游戏式广告
adverse event 不良事件
aesthetic computing 美学计算
aesthetic computing for M&S 建模与仿真的美学计算
aesthetic experience 审美经验，审美体验
affected attribute 受影响的属性
affecting variable 影响变量
affective computing 情感计算
affective user interface 情感用户界面
affinity 亲和度，亲和力
agenda 议事日程
agent 智能体；自主体；代理
agent-based 基于智能体的；基于代理的
agent-based application 基于智能体的应用程序；基于代理的应用程序
agent-based approach 基于智能体的方法；基于代理的方法
agent-based data mining 基于智能体的数据挖掘；基于代理的数据挖掘
agent-based distributed data mining 基于智能体的分布式数据挖掘；基于

代理的分布式数据挖掘

agent-based game 基于智能体的博弈；基于代理的博弈

agent-based gamification 基于智能体的博弈；基于代理的博弈

agent-based gaming 基于智能体的博弈；基于代理的博弈

agent-based grid computing 基于智能体的网格计算；基于代理的网格计算

agent-based mechanism 基于智能体的机制；代理机制

agent-based methodology 基于智能体的方法学；基于代理的方法学

agent-based model 基于智能体的模型；基于代理的模型

agent-based modeling 基于智能体的建模；基于代理的建模

agent-based multisimulation 基于智能体的多仿真；基于代理的多仿真

agent-based participatory simulation 基于智能体的参与性仿真；基于代理的参与性仿真

agent-based Petri net 基于智能体的佩特里网；基于代理的佩特里网

agent-based programming paradigm 基于智能体的程序设计范式；基于代理的程序设计范式

agent-based simulation 基于智能体的仿真；基于代理的仿真

agent-based simulation game 基于智能体的仿真游戏；基于代理的仿真游戏

agent-based solution 基于智能体的解决方案；基于代理的解决方案

agent-based system 基于智能体的系统；基于代理的系统

agent behavior 智能体行为；代理行为

agent-behavior diagram 智能体行为图；代理行为图

agent centric 以智能体为中心的；以代理为中心的

agent characterization 智能体特征；代理特征

agent class 智能体类；代理类

agent-controlled coupling 智能体控制的耦合；代理控制的耦合

agent-controlled simulation 智能体控制的仿真；代理控制的仿真

agent-coordinated 智能体协调的；代理协调的

agent-coordinated simulation 智能体协调的仿真；代理协调的仿真

agent coupling 智能体耦合；代理耦合

agent design methodology 智能体设计方法学；代理设计方法学

agent design tool 智能体设计工具；代理设计工具

agent-directed 智能体导向的；代理导向的

agent-directed framework 智能体导向的框架；代理定向的框架

agent-directed simulation 智能体导向的仿真；代理导向的仿真

agent evaluation criterion 智能体评估准则；代理评估准则

agent framework 智能体框架；代理框架

agent-initiated simulation 智能体启动的仿真；代理启动的仿真

agent interaction 智能体交互；代理交互

agent-interaction protocol 智能体交互协议；代理交互协议

agent-mediated 智能体仲裁的；代理仲裁的

agent-mediated environment 智能体调和的环境；代理调和的环境

agent-mediated simulation 智能体调和的仿真；代理调和的仿真

agent-mediated simulation game 智能体调和的仿真游戏；代理调和的仿真游戏

agent-mediated solution 智能体仲裁的解决方案；代理仲裁的解决方案

agent model 智能体模型；代理模型

agent modeling 智能体建模；代理建模

agent-monitored 智能体监控的；代理监控的

agent-monitored anticipatory multisimulation 智能体监控的预期多仿真；代理监控的预期多仿真

agent-monitored coupling 智能体监控的耦合；代理监控的耦合

agent-monitored environment 智能体监控的环境；代理监控的环境

agent-monitored multisimulation 智能体监控的多仿真；代理监控的多仿真

agent-monitored simulation 智能体监控的仿真；代理监控的仿真

agent-oriented 面向智能体的；面向代理的

agent-oriented methodology 面向智能体的方法学；面向代理的方法学

agent-oriented model 面向智能体的模型；面向代理的模型

agent platform 智能体平台；代理平台

agent rationality 智能体合理性；代理合理性

agent simulation 智能体仿真；代理仿真

agent simulator 智能体仿真器；代理仿真器

agent supported 智能体支持的；代理支持的

agent-supported framework 智能体支持的框架；代理支持的框架

agent-supported multisimulation 智能体支持的多种仿真；代理支持的多种仿真

agent-supported simulation 智能体支持的仿真；代理支持的仿真

agent-supported simulation framework 智能体支持的仿真框架；代理支持的仿真框架

agent system 智能体系统；代理系统

agent system modeling 智能体系统建模；代理系统建模

agent technology 智能体技术；代理技术

agent-to-agent 智能体对智能体；代理对代理

agent-to-agent interaction 智能体间的交互；代理间的交互

agent-triggered simulation 智能体触发式仿真；代理触发式仿真

agenthood 智能体

agentified 被智能体化的

agentify (*v.*) 智能体化
agentifying 智能体
aggreed value 约定的价值
aggregate (*v.*) 使聚集
aggregate-level simulation 聚合级仿真
aggregate-level simulation protocol 聚合级仿真协议
aggregate model 聚合模型
aggregated data 聚合的数据
aggregated model 聚合的模型
aggregated modeling 聚合形式建模
aggregation 聚集
aggression level 攻击级别
agile methodology 敏捷方法学
agile modeling 敏捷建模
agile simulation 敏捷仿真
AI-based modeling 基于人工智能的建模
AI-based programming paradigm 基于人工智能的程序设计范式
AI-controlled high-level Petri net 人工智能控制的高级佩特里网
AI-controlled simulation 人工智能控制的仿真
AI-directed simulation 由人工智能指导的仿真
AI engine 人工智能引擎
AI model 人工智能模型
aided 辅助
aimed 瞄准的
aleatory behavior 偶然的行为
aleatory uncertainty 偶然不确定性
algebraic 代数的
algebraic differentiation 代数微分
algebraic equation model 代数方程模型
algebraic expression-oriented SL 面向代数表达式的仿真语言
algebraic loop 代数环
algebraic system of equations 代数方程组
algebraic variable 代数变量
algorithm 算法
algorithm analysis 算法分析
algorithm assessment 算法评估
algorithm-based 基于算法的
algorithm check 算法检查
algorithm equivalence 算法等价
algorithm error 算法误差
algorithm stability 算法稳定性
algorithm synchronization 算法同步
algorithmic check 算法的检查
algorithmic check of completeness 算法完整性检查
algorithmic check of consistency 算法一致性检查
algorithmic efficiency 算法效率
algorithmic error 算法误差
algorithmic model 算法模型
algorithmic programming paradigm 算法程序设计范式
algorithmically incomputable function 算法不可计算的函数
aliasing 混叠；图形失真
alignment 排比；对准；调准
all-digital analog simulation 全数字模拟仿真
all-digital simulation 全数字仿真
all-digital simulation software 全数字

仿真软件
all-software simulation　全软件仿真
allocate (*v.*)　分派，分配；放在一边
allocated　配属，分配
allocated variable　已分配变量
allocation　分派，分配；安置
allocation error　分配误差
alloplastic adaptation　异质适配
alloplastic-change　异质变化
alloplastic system　异质系统，对变更适应的系统
allopoiesis　他生
allopoietic　他生的
allopoietic model　他生模型
allopoietic simulation　他生仿真
allopoietic system　他生系统
allosteric transition　变构转变
allostelic game　变构游戏
allostelic simulation　变构仿真
allostelic system　变构系统
allostelic system simulation　变构系统仿真
allowable value　允许值
alpha test　α 测试；透明度测试
alpha testing　透明度测试；α 测试
alpha value　α 值
alphabetical input　字母输入
alphanumerical input　字母数字输入
ALSP compatible　聚合级仿真协议兼容的
ALSP compliant　聚合级仿真协议相容的
ALSP-compliant simulation system　符合聚合级仿真协议的仿真系统
ALSP data　聚合级仿真协议数据

alternate　轮流的；交替的
alternate (*v.*)　轮流；交替
alternate key　替换键
alternate model　替代模型
alternate reality　平行实境
alternate reality game　平行实境游戏
alternating state　交替状态
alternative　比较；备选方案
alternative event　可选事件
alternative input　可供选择的输入
alternative model　可供选择的模型
alternative reality　另一种现实
alternative reality game　平行实境游戏
alternative simulation　可替代仿真
alternative simulation model　替代仿真模型
alternative solution　备选的解决方案
altricial system　晚成系统
altruism　无私，利他主义
ambiguity　含糊，不明确
ambiguity error　模糊误差
ambiguous　含糊的，不明确的
ambiguous data　多义数据，含义不明确的数据
ambiguous input　不明确的输入
ambiguous model　模糊模型
ambiguous representation　模糊表示（法）
ambiguous rule　模糊规则
amount of detail　细节量
analog　模拟；类比
analog computer　模拟计算机
analog-computer simulation　模拟计算机仿真
analog input　模拟输入

analog model 模拟模型
analog multilevel simulation 模拟多水平仿真
analog property 类比特性
analog simulation 模拟仿真
analog simulation software 模拟仿真软件
analog simulation study 模拟仿真研究
analog-to-digital converter 模拟-数字转换器
analogical 类似的；类推的
analogical model 类比模型
analogical understanding 类比理解
analogous 模拟的
analogy 类似，相似；类推；类比
analogy model 相似模型；类比模型
analysis 分析，分解；解析
analysis error 分析误差
analysis methodology 分析方法学
analysis of simulation result 仿真结果分析
analysis of variance 方差分析
analyst 分析者
analytic 分析的；解析的
analytic algorithm 解析算法
analytic complexity 分析复杂性
analytic model 解析模型
analytic simulation 分析仿真
analytic simulation technique 分析仿真技术
analytic trajectory 分析的（解析的）轨迹
analytical cognitive model 分析认知模型
analytical data compression 分析数据压缩
analytical method 分析方法
analytical model 分析模型
analytical modeling 分析建模；解析建模
analytical solution 解析的解决方案
analytical stability 解析的稳定性
analytically insolvable model 非解析解的模型
analytically solvable model 解析解的模型
analytically stable solution 解析稳定解
analyzer 分析器；检偏器
anapoiesis 再创造
anapoietic 再生
anatomical markup language 解剖学标记语言
ancestor 祖先；原型
ancestor model 先驱模型；原型模型
ancestor simulation 原型仿真；前人仿真
ancillary function 辅助作用
AND-input "与"输入
angular error 角度测量误差
animate (v.) 驱动；使有生气
animated 有生气的；活跃的
animation 动画
animation markup language 动画标识语言
animation model 动画模型
animation technology 动画制作技术
animation variable 动画变量
animator 动画片绘制者

annealing 退火
anomaly detection 异态检测
antecedent 前项；前提
antecedent part of a rule 规则的前件部分
antecedent variable 前因变量
antenna input 天线输入
anthropometrically correct model 人体测量校正模型
anti-interpretivism 反解释主义
anticipate (*v.*) 期望，盼望
anticipated 预先的
anticipated data 预期数据
anticipated emergence 预期出现
anticipated error 预期误差
anticipated event 预期事件
anticipated fact 预期的事实
anticipated input 期望输入
anticipated reality 预见的现实
anticipated situation 预期的形势
anticipation 预期；期望；早现
anticipation-based coupling 基于预测耦合
anticipation-based modeling 基于预期的建模
anticipation error 预期误差
anticipation mechanism 预期机制
anticipative 预期的
anticipative model 预期中的模型
anticipative system 预期系统
anticipatory 预期的，先行
anticipatory agent 预期代理
anticipatory analysis 预见性分析
anticipatory behavior 预期的行为
anticipatory mechanism 引导机制
anticipatory model 预想中的模型
anticipatory multisimulation 预期多仿真
anticipatory perceptual simulation 预期感知仿真
anticipatory relationship 先行关系
anticipatory simulation 预期仿真
anticipatory system 预期系统
anticipatory value 预期价值
antipositivism 非周期的
antisymmetric 反对称的
antithetic 异源；对偶
antithetic run 对照运行
antithetic simulation run 对偶仿真运行
antithetic simulation study 对偶式仿真研究
antithetic study 对偶研究
antithetic variable 对偶变量
antithetic variate 对偶变量
anthropogenic system 人为系统
aperiodic 非周期的
aperiodic system 非周期性系统
apparent reality 表面真实
appearance model 表观模型
append (*v.*) 追加
appended 附件的
applet 小应用程序
applicability 适用性；适应性
applicability check 适用性检查
applicability of an experimental frame to a model 实验框架对模型的适用性
applicability of backward integration 后向积分适用性
applicability of experimental frame 实

验框架的可用性
applicable 可应用的；可适用的
applicable experimental frame 适用的实验框架
application 应用；申请；要求；应用程序
application development framework 应用开发框架
application error 应用程序错误；应用误差
application layer 应用层
application-specific 特定于应用的
application-specific accreditation 特定于应用的确认
appraisal 评价；鉴定
appraisal value 估定价值
appraised value 估定价值；鉴定价值
apprehension 理解，了解
approach 进场；引道；接近；方法
approximate variable 近似变量
appropriate 适当的；相称的
appropriate simulation 适宜的仿真；合适的仿真
appropriateness 适合；相称；适当
approximate 近似的；大约的
approximate (v.) 近似；接近；大致估计
approximate DEVS morphism 近似离散事件系统规范同型
approximate morphism 近似同型
approximate reasoning model 近似推理模型
approximate simulation 近似仿真
approximate system 近似系统
approximate value 近似值

approximate zero-variance simulation 近似零方差仿真
approximated value 近似值
approximately correct 近似正确
approximation 逼近
approximation accuracy 逼近精度
approximation correctness 近似值正确性
approximation error 近似误差
approximation in modeling 渐进式建模
approximation order 逼近阶次
approximator 逼近器
apriori experiment 推理实验
arbitrary variable 任意变量
arc multiplicity 弧多样性
arcade game 街机游戏
archetypal model 原型模型
archetypally 原型地；典型地
archetype 原型
archetypical 原型的
archetypically 原型地
archieved model 实现的模型
architectural development 架构开发
architectural efficiency 架构效率
architectural style 构架风格
architecture 建筑学；架构；体系结构
architecture development 体系结构开发
architecture error 结构误差
architecture management 构建管理
argument 辐角；论证；自变量
argument passing 幅角传递；自变量传递
argument value 自变量值

argument variable　自变量
argumentation system　论证系统
arithmetic mistake　算法错误
array simulation　阵列仿真
arrival event　到达事件
arrival process　到达过程
art-directed simulation　以艺术为中心的仿真
artifact　人工制品；人工因素
artificial　人工的
artificial discontinuity　人为中断
artificial-immune system　人工免疫系统
artificial intelligence　人工智能
artificial life game　人工生命游戏
artificial neural network　人工神经网络
artificial reality　人工现实
artificial system　人工系统
artificial variable　人工变量
artificial world　虚拟世界；人工世界
artistic simulation　艺术仿真
as-fast-as-possible simulation　全速仿真
ascertainment error　断定误差
aspect　方位；方面
aspect modeling　外表建模
aspect of perception　感知面
aspect-oriented　面向方面的
aspect-oriented modeling　面向方面的建模
aspect-oriented perception　面向方面的感知
aspect-oriented transformation　面向方面的转变

assembled model　组合模型
assert (*v.*)　声明
assertion　诊断；断言
assertion checking　断言检查
assertional model　断言模型
assesment tool　评估工具
assess (*v.*)　评定；评价
assessed　评估
assessed data　评估数据
assessed value　评估价值
assessing　评价
assessment　评价；评估
assessment constraint　评估约束
assessment metric　评估测度；评价方法
assessment model　评估模型
assessment of a simulationist　由模拟主义者所做的评估
assessment of behavior generation technique　行为生成技术评估
assessment of data　数据评估
assessment of decision maker's alternatives　决策者选择评估
assessment of documentation　文档的评估
assessment of experimental condition　实验条件评价
assessment of experimental conditions　实验条件评估
assessment of experimental design　实验设计评估
assessment of modeler　建模器评估
assessment of modeling methodology　建模方法学评估
assessment of parameters of a model　模

型的参数评估
assessment of program correctness 程序正确性评估
assessment of program efficiency 程序效率评估
assessment of program reliability 程序可靠性评估
assessment of program robustness 程序鲁棒性的评估
assessment of programming technique 编程技术评估
assessment of pseudorandom number generator 伪随机数评估发生器
assessment of real-system data 实时系统数据评估
assessment of simulated data 仿真数据评估
assessment of simulation team 仿真团队评估
assessment of SL 仿真语言评估
assessment of software modeler 软件建模器评估
assessment of the applicability of an experimental frame to a model 实验框架对模型的适用性评估
assessment of the goal of the study 研究目标评估
assessment of the model 模型评估
assessment procedure 评估过程
assessor 鉴定器
asset 资产
asset value 资产价值
assigned value 分配值；预设值
assignment 指派；赋值
assimilate (v.) 同化（产）物

assimilation 同化作用
associated value 关联值
association 联想；关联
associative activation error 联合激活误差
associative control 关联控制
associative entity 关联实体
associative model 相关联的模型
associative Petri net 关联佩特里网
associative understanding 联想理解；关联理解
assumed 设想的；假设的
assumed goal 假定目标
assumed value 假定值
assumption 设想；假设
assumption error 假设误差
assumption function 假设函数
assurance 保证；担保
asymmetric 不对称的，非对称的
asymmetric information 不对称信息
asymmetric simulation 不对称仿真
asymmetrical distribution 不对称分布
asymmetrical function 非对称函数
asymmetrical wargame 非对称军事演习
asymmetry 不对称，非对称
asymptotic stability 渐近稳定性
asymptotic standard error 渐近标准误差
asymptotically stable system 渐近稳定的系统
asynchronized collaborative modeling 异步协调建模
asynchronized modeling 异步建模
asynchronous 异步的

asynchronous discrete-event simulation model 异步离散事件仿真模型
asynchronous input 异步输入
asynchronous simulation 异步仿真
asynchronous transfer mode 异步传输模式
asynchronous transmission 异步传输
atomic DEVS 原子离散事件系统规范
atomic DEVS model 原子型离散事件系统规范模型
atomic entity 原子实体
atomic model 原子模型
atomic model editor 原子模型编辑器
atomic model specification 原子模型规范
atomic multimodel 原子多模型
atomistic simulation 原子级模拟
attach (*v.*) 附着
attached 附属；附加；附着
attached menu 附属选单
attached variable 附属变量
attention 注意
attentive interaction 注意交互
attractor 吸引子
attribute 属性
attribute error 属性误差
attribute overloading 属性超载，属性过载
attribute ownership 属性所有权
attribute selection 属性选择
attribute similarity 属性相似度，属性相似性
attribute transformation 属性变换
attribute tree 属性树
attribute update 属性更新
attribute value 属性值
attributed 归因
attributed token 归属令牌
attribution error 归因误差
attributive entity 属性实体
audio input 音频输入
audio simulation 声音仿真
audit 审计
audit (*v.*) 检查；审计
audit policy 审计策略
audit trail 审计追踪
auditable data 审计数据
auditory 听觉的
auditory characteristic 听觉的特性
auditory feedback 听觉反馈
auditory interface 听觉接口
auditory monitoring 监听
auditory perception 听知觉
auditory sensation 听觉感知
augmented 增强；增广
augmented live simulation 增强实况仿真
augmented physical simulation 增强物理仿真
augmented reality 增强现实
augmented-reality game 增强现实游戏
augmented-reality simulation 增强现实仿真
augmented simulation 增强仿真
augmented virtuality 增强虚境
augmented vision 增强视觉
aural I/O 听觉输入/输出
auralization 可听化

authenticate (v.) 证明是真实的、可靠的或有效的；鉴别，鉴定
authenticated 已认证
authentication 证实；鉴别
authentication data 认证数据
authentication error 验证错误
authoring 授权的
authoring system 授权系统
authoritative 权威
authoritative data 权威数据
authoritative data source 权威数据源
authoritative model 权威模型
authority 批准；授权；许可
authorization testing 授权测试；核定测试
auto-modal 自动模态
autoepistemic model 自认知模型
autogenerated 自动生成的
automata 自动机
automata-based 基于自动机的
automata-based description 基于自动机的描述
automata-based technique 基于自动机的技术
automata model 自动机模型
automata network 自动机网络
automated analysis 自动分析
automated experiment 自动化实验
automated information system 自动化信息系统
automated model transformation 自动模型转换
automated parallelization of sequential simulation 序贯仿真的自动并行化
automated subject variable 自动被试变量
automated test tool 自动测试工具
automatic code generation 自动代码生成
automatic configuration 自动配置
automatic error correction 自动误差修正
automatic error detection 自动错误检测
automatic generation 自动生成
automatic parallelization 自动并行化
automatic program generation 自动程序生成
automatic semicode generation 自动半代码生成
automatic simulation model generation 自动仿真模型生成
automatic theorem analysis 自动定理分析
automatic update 自动更新
automatic variable 自动变量
automation 自动化
automatization process 自动化过程
automaton 自动机
autonomic system 自主系统；自治系统
autonomous 自主；自治的
autonomous agent 自治主体
autonomous behavior 自治的行为
autonomous model 自治模型
autonomous negotiation 自主协商
autonomous ordinary differential equation 自治常数微分方程
autonomous simulation mode 自主仿真模式

autonomous system 自治系统

autonomous time-invariant linear system 自治时不变线性系统

autonomous understanding 自主理解

autonomy 自主；自治

autonomy-oriented 面向自主性的

autonomy-oriented computing 自主型计算

autonomy-oriented modeling 面向自治的建模

autoplastic 自体移植的

autoplastic adaptation 自体适应

autoplastic-change 自体变化

autopoiesis 自创生；自我再制

autopoietic 自创生的；自我再生的

autopoietic model 自创生的模型

autopoietic simulation 自创生的仿真

autoregressive model 自回归模型

autoregressive modeling technique 自回归建模技术

autosimulation 自动仿真

autosimulative 自动仿真的

autostimulant 自体刺激的

autotelic game 自身具有目的的游戏

autotelic simulation 自身具有目的的仿真

autotelic system simulation 自身具有目的的系统仿真

auxiliary 辅助的

auxiliary assumption 辅助的假设

auxiliary parameter 辅助参数

auxiliary service 辅助服务

auxiliary value 辅助值；修正系数，校正系数

auxiliary variable 辅助变量

availability 可用性

average 平均的

average error 平均误差

average value 平均值

averaged value 平均值

aversive stimulus 厌恶刺激

avoidance 规避；预防

aware 知道的；意识到的

aware system 感知系统

awareness 意识

awareness-based coupling 基于感知的耦合

axiom 公理

axis 轴；轴线；体轴；中轴

azimuth error 方位误差

Ba

back-end 终端；后端

back-end analysis 后端分析

back-end interface 后端接口（界面）

back interpolation 向后插值

back-interpolation algorithm 后向插值算法

back-interpolation method 后向插值方法

back-interpolation technique 反向插值技术

back up (v.) 备份

back value 返回值

backcasting 后向估计

backcasting methodology 后向估计方法

backcasting model 后向模型

background 后台；背景

background execution 后台执行
background knowledge 背景知识
background task 后台任务
background variable 背景变量
backup 备份
backward 反向；反传；向后
backward Euler integration 向后欧拉积分
backward integration 向后积分
backward-reasoning model 反向推理模型
backward Runge-Kutta algorithm 后向龙格-库塔算法
backward simulation 反向仿真
backward validation 后向验证
backwards time causality 后向时间因果关系
balance error 平衡误差
balanced error 平衡误差
balanced input 平衡输入
band 波段；带环；带；下环；频带
band-structured 带结构化
band-structured matrix 带状结构矩阵
bandwidth 带宽
bargaining game 讨价还价博弈
base 基极；地基；底座，基座；库
base case simulation 基本案例仿真
base context 基础环境
base DEVS model 基础离散事件系统规范模型
base frame 基础框架
base model 基本模型
base object model 基本对象模型
base station random variable 基站随机变量

base value 基准值
base variable 基本变量
based 基于
baseline forcast 基线预测
baseline forecast 基线预测
baseline scenario 基线想定
baseline scenario simulation 基线想定仿真
baseline simulation 基线仿真
baseline simulation scenario 基线仿真想定
baseline solution 基准解决方案
basic 基础的
basic behavior 基础的行为
basic configuration 基础配置
basic coupling 基本耦合
basic input/output system 基本输入/输出系统
basic Petri net 基本佩特里网
basic simulator 基本仿真器
basic variable 基本变量
basis 基底；基元；底节
basis for model processing 模型处理基础
basis for modeling method 建模方法基础
batch input 批量输入
batch simulation 批量仿真
battlefield 战场
battlefield view 战场视图
battlefield visualization 战场可视化
Bayesian 贝叶斯
Bayesian Kriging model 贝叶斯克里格模型
Bayesian model 贝叶斯模型

Bayesian model merging 贝叶斯模型融合
Bayesian modeling 贝叶斯建模
Bayesian rule 贝叶斯规则
Bayesian spatio-temporal model 贝叶斯时空模型
be described 被描述的
bearing error 容忍误差
bearing value 承载值
behavior 行为
behavior abstraction 行为抽象
behavior analysis 行为分析
behavior complexity 行为复杂性
behavior composer 行为合成器
behavior concentration 行为集中
behavior configuration 行为配置
behavior credibility 行为可信度
behavior database 行为数据库
behavior discontinuity 行为不连续性
behavior display 行为显示
behavior documentation 行为文档
behavior emergence 行为涌现, 涌现行为
behavior evaluation 行为评价
behavior generation 行为生成
behavior generation parameter 行为产生参数
behavior generation technique 行为生成技术
behavior generator 行为发生器
behavior generator parameter 行为产生器参数
behavior graph 行为图
behavior management 行为管理
behavior morphism 行为同型
behavior-oriented 面向行为的
behavior processing 行为处理
behavior quality 行为质量
behavior reconfiguration 行为重置
behavior repository 行为库
behavior representation 行为表示(法)
behavior rule 行为规则
behavior space 行为空间
behavior specification 行为规范
behavior variable 行为变量
behavioral 行为的
behavioral abstraction 行为的抽象
behavioral accuracy 行为的准确性
behavioral anticipation 行为预期
behavioral assumption 行为的假设
behavioral characteristic 行为的特性
behavioral comparison 行为的比较
behavioral complexity 行为的复杂性
behavioral dependency 行为依赖
behavioral domain 行为论域
behavioral equivalence of systems 系统行为等效性
behavioral experiment 行为实验
behavioral factor 行为因素
behavioral hierarchy 行为层次结构
behavioral mapping 行为映射
behavioral metamodel 行为元模型
behavioral microsimulation model 行为的微观仿真模型
behavioral model 行为模型
behavioral model comparison 行为模型比较
behavioral model fidelity 行为模型的逼真度
behavioral model transformation 行为

模型转换　　　　　　　　　　　　　角色
behavioral modeling　行为建模
behavioral relation　行为关系
behavioral requirement　行为需求
behavioral risk factor　行为风险因素
behavioral script　行为脚本
behavioral self-organization　行为自组织
behavioral similarity　行为相似性
behavioral simulator　行为仿真器
behavioral substitutability　行为的可替代性
behaviorally adaptive simulation　行为适应性仿真
behaviorally anticipatory　行为预期
behaviorally anticipatory decision　行为预期决策
behaviorally anticipatory model　行为预期模型
behaviorally anticipatory multisimulation　行为预测多体仿真
behaviorally anticipatory simulation　行为预测仿真
behaviorally anticipatory system　行为预期系统
behaviorally equivalent systems　行为等价系统
behaviorally substitutable　行为上可替代的
belief-desire-intention language　信念-愿望-意图语言
belief-forming process　信念的形成过程
belief value　置信值
believable game character　可信的游戏角色
benchmark　基准；水准点；参照
benchmark (v.)　基准
benchmark model　基准模型
benchmark problem　基准问题
benchmarking　标杆
benchmarking analysis　基准分析
benefit of simulation　仿真效益
Bernoulli random variable　伯努利随机变量
Bernoulli variable　伯努利变量
best practice　最佳实践；最优方法
beta distribution　β分布
beta test　β测试
beta testing　β测试
beta variable　β变量
between-subject variable　组间变量
bi-dimensional topology　二维拓扑
bi-layer cellular automaton　双层元胞自动机
bi-variable　双变量
bi-variate relationship　双变量关系
bias　偏差；偏倚；偏压；偏置；偏移
bias error　偏差
bias-error of a measurement　测量偏差
biased data　有偏数据
biased error　有偏误差
biased estimator　有偏估值器
biased measurement　有偏测量
biased sample　有偏样本
bidirectional transformation　双向变化
bifocal model　双焦模型
bifocal modeling　双焦建模
bifurcated　分支
bifurcated methodology　分支方法学

bifurcation 分岔，分叉
bifurcation parameter 分歧参数
big data 大数据
big endian 大尾数法
big endian data format 大尾数法数据格式
big simulation 大型仿真
bilinear 双线性的
bilinear interpolation 双线性插值
bilinear system 双线性系统
bilinearization 双线性
bimodal 双峰的
binary logit 二类分对数
binary relation 二元关系
binary relationship model 二元关系模型
binary value 二进制数值
binary variable 二进制变量
binomial distribution 二项式分布
binomial random variable 二项式随机变量
bio-based 基于生物的
bio-input 生物输入
bio-inspired 生物激励；生物启发的
bio-inspired communication 生物启发的通信
bio-inspired computation 生物启发式计算
bio-inspired computing 生物启发式计算
bio-inspired method 仿生方法
bio-inspired model 仿生模型
bio-inspired modeling 仿生建模
bio-inspired simulation 仿生模拟
bio-inspired system 仿生系统
bio-inspired tool 仿生工具
bio-nano simulation 纳米生物仿真
biological modeler 生物学建模器
biological process 生物过程
biologically-inspired 受生物启发的
biologically-inspired algorithm 生物启发算法
biologically-inspired computation 生物启发式计算
biologically-inspired computing 生物启发式计算
biologically-inspired engineering 生物启发式工程
biologically-inspired simulation 生物启发式仿真
biologically-inspired system 生态激励系统
biomimetic simulation 仿生仿真
biosensor 生物传感器
biosimilar 生物仿制药；生物相似物
biosimilarity 生物相似性
biosimulation 生物仿真
bisimilar 互相似的
bisimilar system 互相似系统
bisimilarity 互相似性
bisimulation 互仿真
bisimulation equivalence 互仿真等价性
bisimulation equivalent 互仿真等价的
bit error 位错误；误比特
bivariate 双变量的
bivariate data 二元数据
black box 黑箱
black box approach 黑箱方法
black box approach in modeling 黑箱

建模方法

black box model　黑箱模型
black box testing　黑盒测试
black box validation　黑盒验证
black box verification　黑盒校核
blackboard model　黑板模型
blended algorithm　混合算法
blended learning simulation　混合学习仿真
blended simulation　混合仿真
block diagram　块图；方框图
block-lower-triangular form　块下三角形式
block model　块模型
block-oriented　面向模块的
block-oriented simulation　面向块的仿真
block-oriented simulation system　面向块的仿真系统
block-oriented SL　面向块的仿真语言
block-structured SL　块结构仿真语言
blocked design　块设计；区组设计
blurred boundary　界限模糊,边界模糊
board game　棋牌游戏
body animation markup language　人体动画标记语言
body of knowledge　知识体系
body of knowledge of model engineering　模型工程的知识体系
body of knowledge of modeling and simulation　建模与仿真知识体系
body of knowledge of simulation systems engineering　仿真系统工程的知识体系
body of knowledge specification　知识

体系规范

BOK (Body of Knowledge)　知识体系
BOK element　知识体系元素
bond graph　键合图
bond graph causalization　键合图因果关系
bond graph model　键合图模型
bond graph modeling　键合图建模
bond graph simulation　键合图仿真
Boolean model　布尔模型
Boolean value　布尔值
Boolean variable　布尔型变量
bottom-up construction　自下而上构建
bottom-up design　自下而上设计
bottom-up emergence　从底向上涌现
bottom-up reasoning　自底向上的推理
bottom-up testing　自下而上测试
Bouchaud-Mézard model　Bouchaud-Mézard 模型
bound　界限
bound validation　边界验证
bound variable　边界变量
boundary　边界
boundary analysis　边界分析
boundary condition　边界条件
boundary element method　边界元法
boundary symmetry condition　边界对称条件
boundary value　边界值
boundary value condition　边界值条件
boundary value problem　边界值问题
boundary value testing　边界值测试
bounded　有界的
bounded above　有上界的
bounded below　有下界的

bounded growth 有界增长
bounded input 有界输入
bounded-input bounded-output 有界输出有界输入
bounded-input bounded-output system 有界输入有界输出系统
bounded Petri net 有界佩特里网
bounded rationality 有界合理性
bounded time domain automaton 有界时域自动机
bounded variable 有界变量
brain-computer interface 脑机接口
brain-computer interface technology 脑机接口技术
brain-controlled 大脑控制
brain-controlled input 大脑控制输入
brain-inspired 大脑启发
brain-inspired computing 大脑启发式计算
brain-inspired modeling 大脑启发式建模
brain-machine interface 脑机接口
branch testing 分支测试
branched 支；分支
branched simulation 分支化的仿真
branching 分支
breadth first 广度优先
break (v.) 断裂；破译
breaking 分断；破碎；开断
brittle system 脆弱系统
broad-based 广泛的
broadcast 广播
broker 代理人
brokered model 代理模型
brokering 代理

Brownian model 布朗模型
browsing 浏览
buffer 缓冲
buffer value 缓冲值
buffered input 缓冲输入
buildable design 可塑设计
buildable design space 可塑设计空间
building 建筑物
built-in 内置的；固有的；嵌入的
built-in quality assurance 内置质量保证
built-in simulation 内置仿真
built-in simulator 嵌入仿真器
burst 猝发；突发信号
bus 总线
bus controller 总线控制器
business 公务；商务，商业；事务；企业
business game 商务博弈
business game management 商务博弈管理
business gaming 商务博弈
business model 商务模型
business requirements 业务需求，商务需求
business simulation game 商务仿真博弈
business simulation game management 商务仿真博弈管理
business training game 业务培训比赛
Byzantine behavior 拜占庭行为
Byzantine fault 拜占庭故障
Byzantine fault model 拜占庭故障模型
Byzantine model 拜占庭模型

Byzantine modeling 拜占庭建模

Cc

cache memory simulator 高速缓存模拟器
cache simulator 缓存模拟器
calculate (v.) 计算
calculated 计算出来的
calculated value 计算值
calculation 计算
calculation error 计算误差
calculation interval 计算时间间隔
calculation verification 计算校核
calibrate (v.) 校准
calibrated 校正，校准
calibrated data 校准数据
calibrated model 校准模型
calibrated simulation model 校准的仿真模型
calibrating 校准
calibration 定标；校正，校准
calibration data 校准数据
calibration error 校准误差
calibration model 校准模型
calling structure analysis 调用结构分析
calorific value 热值
cancel (v.) 取消
cancellation 注销
cancelled 取消的
candidate key 候选码；候选键
candidate model 候选模型
candidate submodel 候选子模型
candidate value 候选值

canonical description 规范描述
canonical description format 规范的描述格式
canonical form 正则形式；范式
canonical model 正则模型；典型模型
capability 能力
capability analysis 能力分析
capability-based 基于能力的
capability-based assessment 基于能力的评估
capability engineering-based experimentation 基于能力工程的实验
capability maturity model 能力成熟度模型
captologic simulation 计算机诱导性仿真
captology 计算机诱导性研究
capture 夺获；采集；俘获
capture error 捕获误差
cardinality of property 属性的基数
career 职业，职业生涯
Carleman linearization 卡莱曼线性化
Cartesian coordinate 笛卡尔坐标
cascade coupling 级联耦合
cascade coupling recipe 级联耦合方法
cascaded failure 级联失效
case-based simulation 基于案例的仿真
casual gamer 临时玩家
casual user 临时用户
catastrophic behavior 灾难性的行为
catastrophic behavior-discontinuity 灾变行为不连续性

catastrophic event 灾难性事件，灾变事件
catastrophic failure 灾难性故障，灾变失效
catastrophic SL 灾难性的仿真语言
categorical variable 分类变量
category 范畴；种类
Cauchy variable 柯西变量
causal 因果
causal assertion 因果断言
causal assignment 因果分配
causal bond graph 因果键合图
causal bond graph model 因果键合图模型
causal-descriptive model 因果描述模型
causal-descriptive modeling 因果描述建模
causal diagram 因果图
causal equation 因果方程
causal loop 因果循环
causal model 因果模型
causal modeling 因果建模
causal order 因果顺序
causal reasoning 因果推理
causal relation 因果关系
causal structure 因果结构
causal system 因果系统
causal traceability 因果追溯性
causal tracing 因果追溯
causality 因果关系
causality analysis 因果关系分析
causality bar 因果关系条
causality principle 因果原理
causalization 因果化

causalization of an equation system 方程系统的因果化处理
causalized 因果的
causalized equation system 因果方程系统
causally symmetric Bohm model 因果对称博姆模型
causation 因果关系
cause 原因
cause-and-effect 因果关系
cause-and-effect diagram 因果图
cause-and-effect language 因果语言
cause-effect 因果
cause-effect graphing 因果图表
cell 单元
cell-centered 以细胞为中心的
cell-centered modeling 以细胞为中心的建模
cell-DEVS 细胞离散事件系统规范
cell markup language 单元标记语言
cell model 细胞模型
cellular 蜂窝状的；细胞的；元胞的
cellular automata 元胞自动机
cellular automaton 细胞自动机
cellular automaton formalism 元胞自动机形式
cellular automaton model 元胞自动机模型
cellular automaton simulation 元胞自动机仿真
cellular DEVS model 元胞离散事件系统规范模型
cellular model 元胞模型
cellular SL 元胞仿真语言
cellular space model 元胞空间模型

censored dependent variable 受限因变量
center 中心
center-based simulation 集中式仿真
centered 居中的
central difference approximation 中心差分近似法
central difference formula 中心差分公式
central difference scheme 中心差分机制
central limit theorem 中心极限定理
central station 地面数据处理中心；中心站
centralized coordination 集中式协作
centralized scheduling mechanism 集中调度机制
centralized self-organizing system 集中式自组织系统
centralized synchronous approach 集中同步的方法
centralized system 集中式系统
centric 中心的
certain 确定的
certain coupling 一定的耦合
certain information 确定的信息
certainty 确定
certainty value 确定值
certifiable 可确认的
certificate 证书
certificate ($v.$) 认可
certificate of compliancy 相容性证明；合格证书
certification 证明；认证
certification authority 权威认证
certification commission 证书授权
certification data 认证数据
certification method 认证方法
certification methodology 认证方法学
certification process 证明(鉴定)过程
certified 被认证的；被证明的
certified data 被认证的数据
certified M&S professional 被认证的建模仿真专家
certified model 被认证的模型
certify ($v.$) 证明
ceteris paribus 在其他条件相同情况下
ceteris paribus assumption 在其他条件相同假设条件下
ceteris paribus law 在其他条件相同的规律下
ceteris paribus model 其他条件相同的模型
chain 链
chain of events 事件链
chain reaction system 链反应系统
challenge 挑战
challenged assumption 挑战性假设
chance 机会
chance variable 随机变量
change 变化
changeable model 可变模型
changing 变化，改变
channel 通道
chaos theory 混沌理论
chaotic 混沌的
chaotic dynamic system 混沌动力系统
chaotic dynamics 混沌动力学

chaotic error 混沌误差
chaotic model 混沌模型
chaotic simulation 混沌仿真
chaotic state 混沌状态
chaotic system 混沌系统
character 字符；特性
characteristic 特性
characteristic value 特征值
characterization 表征；特性
characterize (v.) 赋予……特色；表征
chart 图表
check 检查
check (v.) 检查
check algorithm 检查算法
checked 检查过的
checker 检验器
checking 检验；校验
checkpoint 检查点
chemical communication 化学通信
chemical sensation 化学感知
Chi-Squared random variable 卡方随机变量
chip 片
chip-level soft error 芯片级软错误
circle 圆
circular causality 循环的因果关系
circular variable 循环变量
clarity 清晰度
class 类
class attribute 分类属性
class diagram 类图
class generation 类生成
class hierarchy 类层次结构
class-subclass relationship 类-子类关系

class variable 类变量
class word 类词
classical general equilibrium model 经典一般平衡模型
classical simulation 经典仿真
classification 分类
classification code 分类代码
classification error 分类误差
classification model 分类模型
classroom learning 课堂学习
clause 子句；从句
clerical error 记录误差
client-server model 客户机-服务器模型
clock 时钟
clock input 时钟输入
clock synchronization 时钟同步
clock variable 时钟变量
clocked variable 时钟变量
cloning of a federate 联邦成员复制
closed cluster 闭群
closed-form model 封闭模型
closed-form simulation 封闭式仿真
closed-form solution 封闭解
closed game 封闭型博弈
closed-loop 闭环
closed-loop connection 闭环连接
closed-loop real-time simulation 闭环实时仿真
closed-loop simulation 闭环仿真
closed-loop simulation stability 闭环仿真稳定性
closed-loop system 闭环系统
closed system 封闭系统
closure 闭包

closure of formalism 形式下的闭包
closure principle 封闭性原则
closure under composition 组合形式下的闭包
closure under coupling 耦合形式下的闭包
closure under coupling property 耦合特性下的闭包
cloud and Cyber-physical system 云和赛博物理系统
cloud-based 基于云的
cloud-based framework 基于云的框架
cloud-based modeling framework 基于云的建模框架
cloud-based simulation 基于云的仿真
cloud-based simulation framework 基于云的仿真框架
cloud computing 云计算
cloud computing environment 云计算环境
cloud gaming 云博弈
cloud grid 云网格
cloud-hosted simulation 云托管仿真
cloud simulation 云仿真
cloud simulation environment 云仿真环境
cloud simulation tool 云仿真工具
cluster 集群
cluster formation 集群编队
cluster model 集群模型
cluster sectioning 集群切片
cluster simulation 分簇仿真
clustering 聚类
clustering algorithm 聚类算法

clustering state 聚类状态
clutter reduction 杂波抑制
co-model 协同模型
co-modeling 协同建模
co-modeling methodology 协同建模方法学
co-simulation 联合仿真
co-simulation component 联合仿真组件
co-simulation technique 协同仿真技术
coadaptation 协同适应
coalitional game 合作博弈
coarse data 粗糙数据
coarse-grain granularity 粗粒度
coarse-grained component 粗粒度的组件
coarse time-scale variable 粗粒度时间变量
coarseness 粗糙
coarser-grained component 粗粒度的组件
code 代码
code examination 代码核查
code generation 代码生成
code-generation tool 代码生成工具
code maintenance 代码维护
code of best practice 最佳应用守则
code verification 代码验证
codesign 协同设计
coding phase 编码阶段
coefficient 系数
coefficient event 系数事件
coefficient matrix 系数矩阵
coenetic variable 共生变量

coerce (v.) 强制
coercible 可强制的
coercible simulation 可强制仿真
coercion 强制
coercion simulation 强制仿真
coercive 强制的
coercivity simulation 矫顽磁力仿真
coersing simulation 逼真的仿真
coevolution 协同进化
coevolving system 协同进化系统
cognition 认知
cognitive 认知的
cognitive agent 认知主体
cognitive architecture 认知结构
cognitive aspect 认知方面
cognitive bias 认知偏差
cognitive complexity 认知的复杂性
cognitive-complexity individual 认知复杂性个体
cognitive coupling 认知耦合
cognitive ergonomics 认知工程学
cognitive experiment 认知实验
cognitive informatics 认知信息论
cognitive map 认知映射
cognitive model 认知模型
cognitive modeling 认知（系统）建模
cognitive-modeling architecture 认知建模体系架构
cognitive-modeling strategy 认知建模策略
cognitive monitoring 认知监控
cognitive perception 认知知觉
cognitive process 认知过程
cognitive response 认知响应
cognitive schema 认知模式

cognitive simulation 认知仿真
cognitive simulator 认知仿真器
cognitive state 认知状态
cognitive strategy 认知策略
cognitive system 认知系统
cognitive variable 认知变量
cognitized model 认知模型
coherence 连贯性；相关性；一致性
coherence-driven 一致性驱动
coherence relation 一致性关系
coherent framework 连贯性框架
coherent property 连贯特性
cohering simulation 凝聚仿真
cohesion 内聚性
Cokriging 协同克里格
Cokriging simulation 协同克里格仿真
collaborated virtual and augmented reality 协同虚拟和增强现实
collaborating modeling 协作建模
collaboration 协同
collaboration diagram 协作图
collaboration game 合作博弈
collaboration infrastructure 协作基本结构
collaboration modeling 协作建模
collaborative 协同的
collaborative agent 协同智能体
collaborative component-based simulation 基于协同组件的仿真
collaborative DEVS simulation 协同离散事件系统规范仿真
collaborative distributed-simulation 协同分布式仿真
collaborative environment 协同环境

collaborative M&S 协同式建模与仿真
collaborative mode 协同模式
collaborative model 协同模型
collaborative modeling 协作型建模
collaborative planning 协同规划
collaborative process 协同过程
collaborative requirements modeling 协同需求建模
collaborative simulation 协同仿真
collaborative simulation environment 协同仿真环境
collaborative simulation grid 协同仿真网格
collaborative simulation system 协同仿真系统
collaborative simulation technology 协同仿真技术
collaborative system 协同系统
collaborative system-of-systems 协同体系
collaborative technique 协同技术
collaborative technology 协同技术
collaborative virtual simulation 协同虚拟仿真
collapsed state 折叠状态
collect (v.) 收集
collecting 收集
collection 收集
collective behavior 集体行为
collective control 集体控制
collector 收集器
collocated Cokriging simulation 并置式克里格仿真
collocated cosimulation 并置式协同仿真
collocated simulation 并置式仿真
collocated variable 并置变量
collocation method 并置法
colored Petri net 着色佩特里网
colored Petri net simulator 着色佩特里网仿真器
colored structure digraph 着色结构有向图
colored token 彩色的标记
combat game 战斗游戏
combat model 作战模型
combat modeling 作战建模
combinational state 组合状态
combinatorial complexity 组合复杂性
combinatorial condition 组合条件
combinatorial game theory 组合博弈论
combined appearance model 组合外观模型
combined continuous/discrete model 连续/离散混合
combined continuous/discrete simulation 连续/离散混合仿真
combined discrete-event continuous-change SL 组合化离散事件连续变化仿真语言
combined error 组合错误
combined exercise 合成演练
combined learning 联合学习
combined model 组合模型
combined modeling 协同建模
combined simulation 联合仿真
combined simulation language 联合仿真语言

combined simulation system 联合仿真系统
combined SL 组合仿真语言
combined system 组合系统
combined-system simulation 联合系统仿真
command 指令
command-driven input 命令驱动的输入
command error 指令误差
commercial off-the-shelf model library 商业的现成模型库
commercial off-the-shelf simulation library 商业的现成仿真库
commercial off-the-shelf simulation package 商业的现成仿真工具包
common assumption 普遍的假设
common coupling 公共耦合
common database 公共数据库
common federation functionality 公共联邦功能
common formalism 公共形式
common framework 通用框架
common mode input 共模输入
common payoff function 共同支付函数
common reference model 通用的参考模型
common sense model 常识性模型
common sense reasoning 常识性推理
common sense variable 常识性变量
common use 共用
common-use M&S 通用建模仿真
common-use simulation 通用仿真
communal simulation 通信仿真

communication 通信
communication efficiency 通信效率
communication error 通信误差
communication interval 通信间隔
communication network 通信网络
communication skill 通信技术
communication skill enhancing 通信技能增强
communication time 通信时间
communicative agent 通信智能体
communicative model 通信模型
communicativeness 通信性
community 社区
community of practice 实践社区
compansated system 补偿系统
comparative analysis 比较分析法
comparison 比较
comparison of behavior 行为比较
comparison of structure 结构的对比
comparison testing 比较测试；对照测试
compartmental model 分段模型
compatibility 兼容性
compatibility determination 兼容性测定
compatibility model 兼容性模型
compatibility standard 兼容性标准
compatible 兼容的
compatible formalism 兼容形式
compatible interface 兼容的接口
compatible specification 兼容性规范
compensated error 补偿误差
compensating error 补偿误差
compensatory model 补偿模型
competence state 补偿状态

competition 竞争
competition game 竞争博弈
competition simulation 竞争仿真
competitive behavior 竞技行为，竞争性行为
competitive game 竞技游戏
competitive learning model 竞争学习模型
competitive model 竞争模型
competitive relationship 竞争关系，竞争性关系
competitive simulation 竞争模拟，竞争性仿真
competitive simulation game 竞争模拟博弈
competitive system 竞争系统
compilable code 可编译代码
compilation error 编译错误
compile (v.) 编译
compiled model 编译模型
compiler 编译器
complementary variable 互补变量
complete 完全
complete graph 完全图
complete information 完备信息
complete model 完备的模型
complete nodeset 完备的节点集
complete simulation 完全仿真
complete synchronization 完全同步
completely false 完全错误的
completely randomized design 完全随机化设计
completely true 完全正确
completeness 完整性
completeness check 完整性检查

complex 复杂的
complex adaptive system 复杂自适应系统
complex adaptive system modeling 复杂自适应系统建模
complex behavior 复杂的行为
complex data 复杂数据
complex event 复杂事件
complex evolutionary system 复杂进化系统
complex model 复杂模型
complex phenomenon 复杂现象
complex scenario 复杂情景
complex system 复杂系统
complex system modeling 复杂系统建模
complex-valued variable 复杂数值变量
complex variable 复变数，复变量
complexity 复杂性
complexity encapsulation 复合体封装
complexity measure 复杂性测量
complexity paradigm 复杂范式
complexity theory 复杂性理论
compliance 柔顺；合格；符合
compliance certificate 合格证书
compliance determination 合格测定
compliance test 合格测试；验证试验
compliance testing 合格测试；验证试验
compliance with standardization 符合标准
compliance with standards 符合标准
compliancy 相容性
compliant 符合的；相容的

complicated 复杂的
complicated model 复杂模型
complicated scenario 复杂情景
complicated system 复杂系统
component 组件
component allocation 组件分配
component-based 基于组件的
component-based approach 基于组件的方法
component-based architecture 基于组件的构架
component-based collaborated simulation 基于组件的协同仿真
component-based development 基于组件的开发
component-based distributed simulation 基于组件的分布式仿真
component-based modeling 基于组件的建模
component-based modeling paradigm 基于组件的建模范式
component-based paradigm 基于组件的范式
component-based simulation 基于组件的仿真
component-based simulation paradigm 基于组件的仿真范式
component-based simulation system 基于组件的仿真系统
component-based testing 基于组件的测试
component certification 组件认证
component class 组件类
component diagram 组件图
component encapsulation 组件封装
component failure 组件失效
component integration standard 组件集成标准
component interaction 组件交互
component level 元件级
component-level misbehavior 组件级别的不当行为
component-level performance 组件级性能
component-level property 组件级属性
component library 元件库
component model 组件模型
component model repository 组件模型库
component model template 组件模型模板
component object model 组件对象模型
component object model interface 组件化对象模型接口
component of a model 模型组件
component-reuse methodology 可重用组件方法论
component simulation 组件仿真
component system 组件系统
component type 组件类型
componential 成分的
componential analysis 成分分析
componential model 组件的模型
componentized 组件化
componentized model 组件化模型
comportmental model 行为模型
composability 可组合性
composability constraint 可组合约束
composability level 可组合水平

composability methodology 可组合性方法学
composability technology 可组合性技术
composability theory 可组合性理论
composable 可组合的
composable environment 可组合环境
composable federation 可组合联邦
composable framework 可组合框架
composable model 可组合的模型
composable scenario 可组合想定
composable simulation 可组合式仿真
composable synthetic environment 可组合合成环境
composant 组件
compose (v.) 组成
composer 合成器
composite 复合；合成
composite activity 复合活动
composite attribute 组合属性
composite behavior 复合行为
composite error 复合误差
composite message 复合消息
composite model 复合模型
composite-model execution 复合模型运行
composite modeling 复合建模
composite simulation 复合仿真
composite video input 复合视频输入
composition 复合
composition axiom 组合公理
composition complexity 组和复杂性
composition instance 构造实例
composition language 复合语言
composition service 组合服务
composition technique 组合技术
compositional model 组合模型
compositional modeling 合成建模
compound game 复合博弈
comprehensibility 可理解性
comprehensible input 可理解的输入
comprehensible model 可理解的模型
comprehension 理解
comprehensive 综合的
comprehensive core body of knowledge 综合核心知识体系
comprehensive evaluation 综合评价
comprehensive similarity theory 综合相似性理论
comprehensive simulation system 综合仿真系统
comprehensive system 综合系统
compress (v.) 压缩
compressed 被压缩的
compressed time 压缩的时间
compressed time-series 压缩时间序列
compressed-time simulation 压缩实时仿真
compression 压缩
compression method 压缩方法
computability 可计算性
computable 可计算的
computable model 可计算模型
computable transition 可计算变迁
computation 计算
computation-intensive 计算密集的
computation method 计算方法
computation time 计算时间
computational 计算的
computational activity 计算活动

computational agent　计算智能体
computational causality　计算的因果关系
computational closure　计算闭包
computational complexity　计算的复杂性
computational discovery　计算发现
computational emergence　计算涌现
computational error　计算误差
computational experiment　计算实验
computational experimentation　计算实验
computational federation　计算联邦
computational geometry　计算几何学
computational instability　计算不稳定性
computational intelligence　计算智能
computational interoperability　计算互操作性
computational logic　计算逻辑
computational method　计算方法
computational model　计算模型
computational modeling　计算建模
computational process　计算过程
computational replicability　计算可复制性
computational resource　计算资源
computational result　计算结果
computational simulation　计算仿真
computational stability　计算稳定性
computational system　计算系统
computationally feasible model　计算上可行的模型
computationally imposed　计算强加的
computationally imposed simplification　计算强加的简化
computed situation　情景计算
computed value　计算数值
computer　计算机
computer-aided　计算机辅助的
computer-aided design　计算机辅助设计
computer-aided experiment　计算机辅助实验
computer-aided experimentation　计算机辅助实验
computer-aided model verification　计算机辅助的模型校核
computer-aided modeling　计算机辅助的建模
computer-aided simulation　计算机辅助仿真
computer-aided training　计算机辅助训练
computer-aided VV&A　计算机辅助的校核、验证与确认
computer-assisted　计算机辅助的
computer-assisted game　计算机辅助博弈
computer-assisted methodology　计算机辅助方法学
computer-assisted model　计算机辅助的模型
computer-assisted simulation　计算机辅助的仿真
computer-assisted simulation proof　计算机辅助仿真证明
computer-assisted understanding　计算机辅助理解
computer-assisted use　计算机辅助用

途

computer-based 基于计算机的

computer-based experiment 基于计算机的实验

computer-based game 基于计算机的博弈

computer-based simulation 基于计算机的仿真

computer-based training 基于计算机的培训

computer-centric model 以计算机为中性的模型

computer-driven simulacre 计算机驱动的仿真

computer error 计算机误差

computer game 计算机游戏

computer game control 计算机游戏控制

computer game design 计算机游戏设计

computer game development 计算机游戏开发

computer gaming 电脑游戏

computer-generated 计算机生成

computer-generated imagery 计算机生成的影像

computer graphics 计算机图形学

computer hardware 计算机硬件

computer-interpretable computational model 计算机可解释的计算模型

computer-interpretable model 计算机可解释的模型

computer-mediated simulation 计算机介导仿真

computer model 计算机模型

computer network 计算机网络

computer-network simulation 计算机网络仿真

computer processable 计算机可处理的

computer-processable model 计算机可处理的模型

computer processing speed 计算机处理速率

computer-readable model 计算机可读的模型

computer resource 计算机资源

computer simulation 计算机仿真

computer simulation game 计算机仿真博弈

computer software 计算机软件

computer war game 计算机战争游戏

computerization 计算机化

computerization error 计算机化误差

computerized applicability check 电脑适用性检查

computerized completeness check 电脑完整性检查

computerized consistency check 电脑一致性检查

computerized documentation 计算机化文件

computerized game 计算机游戏

computerized management game 计算机管理对策

computerized model 计算机化的模型

computerized model-verification 计算机处理的模型校核

computerized simulation 计算机化的仿真

computing　计算
conceived reality　设想的现实
concentration　集中
concept　概念
concept identification　概念识别
concept validation　概念验证
conceptual　概念的
conceptual abstraction　概念抽象
conceptual agent model　概念智能体模型
conceptual analysis　概念分析
conceptual attribute　概念属性
conceptual causal diagram　概念因果图
conceptual-change　概念的变化
conceptual context　概念上下文
conceptual data model　概念数据模型
conceptual description　概念描述
conceptual design　概念设计
conceptual domain　概念域
conceptual error　概念误差
conceptual experiment　概念实验
conceptual experimental frame　概念实验框架
conceptual foundation　概念基础
conceptual framework　概念架构
conceptual interoperability　概念互操作性
conceptual interoperability level　概念互操作水平
conceptual interoperability model　概念互操作模型
conceptual metamodel　概念元模型
conceptual model　概念模型
conceptual model attribute　概念模型属性
conceptual model development　概念模型开发
conceptual model of the mission space　任务空间概念模型
conceptual model quality　概念模型品质
conceptual model validation　概念模型验证
conceptual model verification　概念模型校核
conceptual modeling　概念建模
conceptual modeling formalism　概念建模形式
conceptual modeling framework　概念建模框架
conceptual modeling language　概念建模语言
conceptual modeling technique　概念建模技术
conceptual representation　概念表示(法)
conceptual requirements　概念需求
conceptual reusable simulation model　概念可重用仿真模型
conceptual scenario　概念想定
conceptual scenario metamodel　概念想定元模型
conceptual schema　概念模式
conceptual simulation　概念仿真
conceptual simulation model　概念仿真模型
conceptual simulator　概念仿真器
conceptual system　概念系统
conceptual validity　概念正确性

conceptualization 概念化
concern-oriented 面向关系的
concern-oriented modeling 面向关系的建模
concerted model 协同建模
conclusion part of a rule 规则的结论部分
concomitant variable 伴随变量
concrete model 具体的模型
concrete primitive 具体的原语
concrete value 具体值
concurrent 共点；同时发生的
concurrent engineering 并行工程
concurrent-logic variable 并行逻辑变量
concurrent model 并行模型
concurrent process analysis 并行处理分析
concurrent processing 并行处理
concurrent simulation 并发仿真
concurrent simulator 并发仿真器
concurrent system 并发系统
concurrent task 并发任务
condensed-time simulation 简明的时间模拟
condition 条件
condition testing 条件测试
condition variable 条件变量
conditional 有条件的
conditional event 条件事件
conditional logit 有受约束的分对数
conditional model 条件模型
conditional Monte Carlo simulation 条件蒙特卡罗仿真
conditional sentence 条件语句

conditional simulation 条件仿真
conditional value 条件值
conditional variable 条件变量
conditioned stimulus 条件刺激
confederation 联盟
confidence 置信度
confidence assessment 信任（置信度）评估
confidence interval 置信区间
confidence interval technique 置信区间技术
confidence limit 置信限制
confidence measure 置信测度
confidence region 置信区域
confidence value 置信度评价
configurability 可配置性
configurable model 可配置的模型
configuration 配置
configuration error 配置误差
configuration management 配置管理
confirmation 认证
confirmation error 认证误差
conflict management 冲突管理
conflicting rule 冲突规则
confluent DEVS transition-function 合流离散事件系统规范的转换函数
confluent transition-function 合流转换函数
conformance 一致性
conformity 一致
confound 混淆
confounder 干扰因子
confounding factor 混杂因素
confounding variable 混杂变量
confusion 错乱性

congruent 同余的；一致的
congruent behavior 一致的行为
conjectural model 推测的模型
conjoint simulation 联合仿真
conjugate gradient algorithm 共轭梯度算法
conjunctive coupling 关联耦合
conjunctive coupling recipe 关联耦合配方
conjunctive model 关联模型
connected experience 关联经验
connection 连接
connection error 连接误差
connectionist model 连接模型
connectivity 连通性
connectivity of operations perspective 操作连通性的角度
connector 连接器
conscious experiment 意识实验
consensus reality 一致性现实
consequence 结论
consequent 后项；结果
conservation of energy 能量守恒
conservation of mass 质量守恒
conservation of momentum 动量守恒
conservation principle 守恒原理
conservative assumption 保守的假设
conservative event simulation 保守事件仿真
conservative method 保守方法
conservative parallel simulation 保守并行仿真
conservative simulation 保守仿真
conservative synchronization protocol 保守同步协议
conservative system 守恒系统
conservative time management 保守时间管理
considered system 被考虑系统
consistency 一致性
consistency check 一致性检查
consistency checker 一致性检验器
consistency checking 一致性检查
consistency condition 一致性条件；相容性条件
consistency criterion 一致性准则
consistency error 一致性误差
consistency management 一致性管理
consistency of coupling 一致性耦合
consistency of goal components 目标分量的一致性
consistent 一致的
consistent data 一致性数据
consistent discretization scheme 一致离散化方案
consistent emergent behavior 一致紧急行为
consistent initial condition 一致初始条件
consistent model 一致性的模型
consistent modeling 一致性建模
console game 控制台游戏；主机游戏
constant 常数
constant input 常量输入
constant-sum trust game 常和信任博弈
constant value 常值，恒定值
constant variable 定常变量
constituent system 合成系统
constrained 约束的

constrained active appearance model 约束主动外观模型
constrained conditional model 约束条件模型
constrained factor model 约束因子模型
constrained model 约束模型
constrained optimization model 约束优化模型
constrained simulation 约束仿真
constrained variable 限定变量，约束变量
constraint 约束
constraint analysis 约束分析
constraint decision defuzzification 约束决策逆模糊化
constraint-driven 约束驱动
constraint-driven multimodel 约束驱动的多模型
constraint equation 约束方程
constraint error 约束误差
constraint model 约束模型
constraint modeling 约束建模
constraint system 约束系统
construct (v.) 构成；建立
construction 施工；构造
constructive emotional behavior 构造性情绪行为
constructive entity 构造性实体
constructive model 构造性模型
constructive simulation 构造性仿真
constructive simulation integration 构造仿真集成
constructive training 建构性训练
constructive training environment 构造性训练环境
constructive training simulation 构造性学习仿真
constructive training system 有帮助的训练系统
constructivism 构成主义
constructivist 构成主义者
constructivist methodology 构成主义方法论
constructivist paradigm 构成主义范式
constructivist theory 构成主义理论
consultant 咨询者
consumer 用户；消费者
consumer data 消费者数据
contact lens display 接触镜显示
container model 容器模型
containment 包容性；安全壳
content 内容
content accuracy 内容正确性
content aggregation model 内容聚合模型
content coupling 内容耦合
content description layer 内容描述层
contest game 竞技游戏
context 情境（上下文）
context aggregation 情境聚合
context analysis 情境分析
context-aware 情境感知的
context-aware input 情境感知输入
context-aware system 情境感知系统
context awareness 情境感知
context compatibility 情境兼容性
context comprehension 情境理解
context design 情境设计
context formalization 情境形式化

context-free model　情境无关模型
context-free simulation　情境无关仿真
context identification　情境识别
context inference　情境推理
context management　情境管理
context model　情境模型
context modeling tool　情境建模工具
context representation　情境表达
context-sensitive　上下文相关的
context-sensitive application　上下文相关的应用
context-sensitive help　上下文相关的帮助
context-sensitive input　上下文相关的输入
context-sensitive system　上下文相关的系统
context-unaware input　上下文未知的输入
context variable　上下文变量
context-dependent model　上下文相关的模型
contextual　上下文的
contextual analysis　情境分析
contextual compatibility　上下文兼容性
contextual concept　语境概念
contextual dependency　上下文相关
contextual help　上下文帮助
contextual understanding　语境理解
contextual validation　上下文验证
contextualized emergent behavior　情境化突现行为
contingency　偶然性；应急
contingency model　意外事件模型
contingent　偶然的（事件）
contingent condition　不确定条件
contingent event　偶然事件
continual planning　持续规划
continuation condition　延拓条件
continuity　连续性
continuity test　连续性测试
continuous　连续的
continuous activity　连续活动
continuous adaptation　连续适应性
continuous attribute　连续属性
continuous-change simulation　连续变化仿真
continuous-change variable　连续变化的变量
continuous data　连续数据
continuous dependent variable　连续相关变量
continuous/discrete SL　连续/离散仿真语言（SL）
continuous distribution　连续分布
continuous game　连续对策
continuous global optimization　连续全局最优化
continuous kernel game　连续核心对策
continuous Markov chain　连续马尔可夫链
continuous model　连续模型
continuous modeling　连续建模
continuous multimodel　连续多模型
continuous ordinal variable　连续顺序变量
continuous phenomenon　连续现象
continuous problem　连续问题

continuous random variable 连续随机变量
continuous run 连续运行
continuous segment 连续片段
continuous simulation 连续仿真
continuous-simulation language 连续仿真语言
continuous SL 连续仿真语言
continuous space 连续空间
continuous-space continuous-time model 空间连续时间连续模型
continuous-space discrete-time model 空间连续时间离散模型
continuous-space model 空间连续模型
continuous-space system 空间连续系统
continuous-state model 连续状态模型
continuous system 连续系统
continuous-system modeler 连续系统建模器
continuous-system simulation 连续系统仿真
continuous time 持续时间
continuous-time continuous simulation 连续时间连续仿真
continuous-time linear system 连续时间线性系统
continuous-time model 连续时间模型
continuous-time simulation 连续时间仿真
continuous-time SL 连续时间仿真语言
continuous-time state-space model 连续时间状态空间模型
continuous-time system 时间连续系统
continuous-time system SL 连续时间系统仿真语言
continuous-time variable 连续时间变量
continuous trajectory 连续轨迹
continuous-valued variable 连续赋值变量
continuous variable 连续变量
continuous-variable model 连续变量模型
continuous-variable simulation optimization 连续变量仿真优化
continuous-change 连续变化的
continuous-change model 连续变化模型
continuously increasing variable 连续增长变量
contour plot 等高线图；等值线图
contract management 合同管理
contractivity 收缩性
contradiction 矛盾
contradiction of components of a goal 目标分量的矛盾
contradiction of goal components 目标分量的矛盾性
contradictory rule 矛盾规则
control 控制
control ($v.$) 控制
control accuracy 控制精度
control algorithm 控制算法
control analysis 控制分析
control coupling 控制耦合
control error 控制误差

control flow analysis 控制流分析
control model 控制模型
control segment 控制片段
control state 控制状态
control station 控制站
control system 控制系统
control-theoretic approach 控制论方法
control variable 控制变量
controllability 可控性
controllable 可控的
controllable assumption 可控的假设
controllable component 可控的组件
controllable factor 可控因素
controllable input variable 可控制的输入变量
controllable plant 可控装置
controllable system 可控系统
controllable variable 可控制的变量
controlled 受控的
controlled experiment 受控实验
controlled experimentation 受控实验
controlled Petri net 受控佩特里网
controlled state 受控状态
controlled system 受控系统
controlled variable 被控变量，受控变量
controller 控制器
convection-diffusion equation 对流扩散方程
convention 常规
conventional coupling 常规耦合
conventional dependency 常规相关性
conventional experience 常规经验
conventional experimentation 常规实验
conventional game 常规游戏
conventional input 常规输入
conventional model 常规模型
conventional paradigm 常规范式
conventional simulation 常规仿真
convergence error 收敛误差
convergence model 集中式模型
convergence order 收敛阶次
convergence simulation 收敛仿真
convergent boundary simulation 收敛边界仿真
convergent simulation 收敛的仿真
convergent validity 收敛的有效性
converging simulation 正在收敛的仿真
conversion 转换
conversion value 转换值
converted sensory data 被转换感知数据
converted value 被转换值
converter 转换器
convex game 凸对策；凸博弈
convex Kriging 凸克里格法
cooperate (v.) 协作
cooperation 协作
cooperation simulation 协作仿真
cooperative behavior 协作行为
cooperative development 合作开发
cooperative game 合作博弈
cooperative game theory 合作博弈理论
cooperative learning simulation 协作学习仿真
cooperative relationship 协作关系

cooperative simulation 协同仿真
cooperative simulation effort 协同仿真的尝试
cooperative simulation game 协同仿真对策
cooperative system 协作系统
coopetition 合作竞争
coopetition game 合作竞争对策
coopetition simulation 合作竞争仿真
coopetitive 合作的，协同的
coopetitive behavior 合作性行为，协同性行为
coopetitive relationship 合作性关系，协同性关系
coopetitive simulation 协同仿真，合作性仿真
coopetitive simulation game 协同仿真博弈
coordinate 坐标
coordinate transformation 坐标变换，坐标转换
coordinated 协调的
coordinated time advancement 协调的时间推进
coordinated universal time 协调世界时
coordination 协调
coordination game 协调博弈
coordination language 协调语言
coordination model 协调模型
coordination system 协调系统
coordination variable 协调变量
coordinator 协调者
coprocessor 协处理器
copying error 仿形误差

core body of knowledge 核心知识体系
core competency game 核心竞争力博弈
core ontology 核心本体论
core topic 核心主题
correct 正确的
correct (v.) 修正
correct assumption 修正假设
correct data 正确的数据
correct model 修正模型
correct output 正确输出
correct prediction 修正预测
correct result 修正结果
correct self-model 修正自模型
correctability 可修正性
correctable 可修正的
corrected error 被修正的误差
correcting 纠正
correction 校正
corrective maintenance 修正性维护
corrective model maintenance 修正性模型维护
correctness 正确性
correctness analysis 正确性分析
correctness determination 正确性测定
correctness proof 正确性证明
correctness verification 正确性校核
corrector 校正子
correlated error 相关误差
cosimulation 联合仿真
cosimulation component 联合仿真组件
cosimulation technique 联合仿真技术
cost 成本；费用；代价

cost analysis for simulation-based acquisition 基于仿真采办的成本分析

cost function 代价函数

cost model 成本模型

cost of simulation 仿真成本

cost requirements 成本需求

COTS (Commercial Off-The-Shelf) software 商用成品软件

counterfactual simulation 反事实的仿真

counterfeit 伪造

counterintuitive behavior 直觉行为

counterintuitive result 违反直觉的结果

couple (v.) 耦合；连接

coupled 耦合的

coupled algorithm 耦合算法

coupled continuous model 耦合的连续模型

coupled DEVS 耦合离散事件系统规范

coupled DEVS behavior 耦合离散事件系统规范行为

coupled DEVS model 耦合离散事件系统规范模型

coupled discrete model 耦合离散模型

coupled human and material system 耦合人与材料系统

coupled linear model 耦合线性模型

coupled linear/nonlinear model 耦合线性/非线性模型

coupled model 耦合的模型

coupled-model editor 耦合模型编辑器

coupled multiformalism 耦合多种形式的

coupled multimodel 耦合多模型

coupled network 耦合网络

coupled nonlinear-model 耦合非线性模型

coupled ODE 耦合的常微分方程

coupled partial differential equation 耦合偏微分方程

coupled PDE 耦合型偏微分方程

coupled program 耦合程序

coupled simulation 耦合仿真

coupled simulation software 耦合仿真软件

coupled successor simulation study 耦合后继仿真研究

coupled system 耦合系统

coupled system model 耦合系统模型

coupled variable-component model 耦合可变成分模型

coupled variable-structure model 耦合可变结构模型

coupler 耦合器

coupling 耦合（性）

coupling-based 基于耦合的

coupling-based technique 基于耦合的方法

coupling closure 耦合闭包

coupling constraint 耦合约束

coupling function 耦合函数

coupling matrix 耦合矩阵

coupling recipe 耦合方法

coupling scheme 耦合机制

coupling strength 耦合强度

coupling topology 耦合拓扑

coupling with variable connection 可变联接的耦合
course 课程
course error 行为误差
course-grain approach 行为粒度方法
course-grain model 行为粒度模型
covariable 可协变的
covariate 协变量
covariate dynamics 协变量动力学
coverability tree 可覆盖树
created 创造的
created model 创造的模型
created reality 创造的实体
creation error 创造误差
creative aspect 创造性的方面
creative aspects of modeling 建模的创造性方面
creative system 创造性系统
credibility 可信性
credibility determination 可信性测定
credibility metric 可信性度量
credible 可信的
credible input 可信的输入
credible model 可信的模型
credible simulation 可信仿真
credible solution 可信解
creep 爬行
crisis management game 危机管理对策
crisp 危机
crisp parameter 危机参数
crisp rule 危机规则
crisp rule-base 危机规则的
crisp value 危机值
crisp variable 危机变量
crispness 危机性
criteria 标准，准则
criteria development 标准开发
criteria vector 标准矢量
criterion 标准；判据
criterion for acceptability 可接受性准则
criterion variable 标准变量
critical assumption 临界假设
critical component 关键组成
critical-event simulation 关键事件仿真
critical multiplism 临界乘数
critical state-transformation 临界状态变换
critical state-transition 临界状态转换
critical value 临界值
critical variable 临界变量
critical-event 关键事件
critical-event game 关键事件对策
cross-entropy method 互熵方法
cross functional integration 交互能力一体化
cross impact analysis 交互影响分析
cross model 十字形模型
cross model comparison 交叉模型比较
cross model validation 交叉模型验证
cross model validity 交叉模型有效性
cross-network communication 交叉网络通信
cross validation 交叉核实
cross-validation procedure 交叉验证过程
cross validity 交叉有效性

crossing 相交，交叉
crowd behavior 群体行为
crowd simulation 群体仿真
crowdsourced behavior 众包行为
crowdsourced data 众包数据
crowdsourced data analysis 众包数据分析
crowdsourced data collection 众包数据收集
crowdsourced data entry 众包数据输入
crude model 原始的模型
cubic interpolation 三次插值
cubic spline interpolation 三次样条插值
cultural bias error 文化差异
cultural complexity 文化复杂性
cultural environment 文化环境
cultural feature 文化特征
cultural features data 文化特征数据
cultural filter 文化过滤
cultural perception error 文化感知误差
cultural value 文化值
cumulative distribution function 累积分布函数
cumulative error 累积误差
current 当前的
current data 当前数据
current event 当前事件
current model 当前模型
current state 当前状态
current time 当前时间
current value 当前值

curriculum 课程
curve 曲线
customer 客户
customer acceptance 客户接受度
customizability 可定制性
customizable simulation 可定制化仿真
customizable system 可定制的系统
customization 用户化，客户化
customization option 自定义选项
customized simulation 定制仿真
Cyber-physical state 赛博物理状态，信息物理状态
Cyber-physical system 赛博物理系统，信息物理系统
Cyber-physical system simulation 赛博物理系统仿真，信息物理系统仿真
Cyberagent 赛博代理
cybernetics 控制论
Cybersecurity 赛博安全
Cyberspace 赛博空间
Cybersystem 赛博系统
cycle 周期
cycle-based 基于周期的
cycle-based simulation 基于周期的仿真
cyclic algorithm 循环算法
cyclic behavior 周期的行为
cyclic event 周期事件
cyclic metamorphic multimodel 循环突变多模型
cyclic multimodel 循环多模型
cyclical time-scale variable 周期时间尺度变量

Dd

damping 阻尼
damping error 阻尼误差
damping error sign 阻尼误差符号
damping matrix 阻尼矩阵
damping plot 阻尼图
dangerous event 危险事件
data 数据
data acceptability 数据合格；数据可接受性
data accuracy 数据精确性
data acquisition 数据采集
data administration 数据管理
data administrator 数据管理员
data aggregation 数据聚集
data analysis 数据分析
data-analysis technique 数据分析技术
data analytics 数据分析法
data animation 动画数据
data appropriateness 数据适当性
data architecture 数据架构
data assessment 数据评估
data association 数据关联
data attribute 数据属性
data authentication 数据认证
data availability 数据可用性
data-based 基于数据的
data-based model 基于数据的模型
data-based modeling 基于数据的建模
data calibration 数据校准
data center 数据中心
data center layer 数据中心层
data center transformation 数据中心变换

data centric 数据中心的
data-centric approach 以数据为中心的方法
data-centric computing 以数据为中心的计算
data certification 数据认证
data characteristic 数据特性
data cleansing 数据清理
data collection 数据收集
data collection assessment 数据采集评估
data collection phase 数据收集阶段
data collector 数据收集器
data completeness 数据完整性
data compression 数据压缩
data consistency 数据一致性
data consumed by M&S 建模与仿真使用的数据
data consumer 数据消费者
data correctness 数据正确性
data cost 数据成本
data coupling 数据耦合
data currency 数据流通
data customer 客户数据
data density 数据密度
data dependency analysis 数据相关性分析
data dictionary 数据词典
data dictionary system 数据词典系统
data-directed 数据导向的
data-directed inference 数据导向的推论
data documenting 数据建档
data dredging 数据挖掘
data-driven 数据驱动的

data-driven decision 数据驱动的决策
data-driven error 数据驱动误差
data-driven intelligence 数据驱动智能
data-driven method 数据驱动方法
data-driven methodology 数据驱动方法论
data-driven modeling 数据驱动的建模
data-driven process 数据驱动过程
data-driven reasoning 数据驱动推理
data-driven simulation 数据驱动仿真
data-driven system 数据驱动系统
data element 数据元素
data element standardization 数据单元标准化
data encryption 数据编码；数据加密
data entity 数据实体
data entry 数据入口
data-entry error 数据入口误差
data error 数据误差
data exchange standard 数据交换标准
data expertise 数据专家知识
data extraction 数据提取
data farming 数据耕耘
data fidelity 数据逼真度
data filtering 数据过滤
data-filtering mechanism 数据滤波机理
data fishing 数据捕捞
data flow 数据流
data flow analysis 数据流分析
data flow diagram 数据流图
data flow machine 数据流机
data flow modeling 数据流建模
data flow testing 数据流测试
data flow variable 数据流变量
data forensics 数据取证
data format 数据格式
data format converter 数据格式转换
data fusion 数据融合
data generator 数据发生器
data governance 数据治理
data grid 数据栅格
data hiding 数据隐匿
data input 数据输入
data instrumentation 数据仪器
data integrity 数据完整性
data integrity requirements 数据完整性需求
data-intensive 数据密集的
data-intensive computing 数据密集型计算
data-intensive simulation 数据密集型仿真
data-intensive system 数据密集型系统
data interchange 数据交换
data interchange standard 数据交换标准
data interface testing 数据接口测试
data issue 数据发布
data layer 数据层
data layout 数据规划
data logger 数据记录器
data maintenance 数据维护
data management 数据管理
data meaning 数据含义
data mediation 数据调解
data metamodel 数据元模型

data mining 数据挖掘
data mining language 数据挖掘语言
data model 数据模型
data model transformation 数据模型变换
data modeling 数据建模
data modeling method 数据建模方法
data organization 数据组织
data parallel execution 数据并行执行
data point 数据点
data preparation tool 数据准备工具
data produced by M&S 由建模与仿真产生的数据
data producer 数据生成器
data-producer certification 数据生成器认证
data proponent 数据支持者
data quality 数据质量
data quality assurance 数据治理保证
data relevance 数据相关性
data repository 数据仓库
data representation 数据描述
data representation model 模型数据描述
data requirements 数据需求
data resolution 数据分辨率
data resource 数据资源
data scaling 数据缩放
data scientist 数据科学家
data security 数据安全
data semantics 数据语义学
data sensor 数据传感器
data set 数据集合
data sharing 数据分享
data smoothing method 数据滤波方法
data snooping 数据探测
data source 数据源
data sourcing 数据采购
data space 数据空间
data standard 数据标准
data standardization 数据标准化
data steward 数据管理员
data stream 数据流
data structure 数据结构
data-structured coupling 数据结构耦合
data structuring 数据结构化
data synchronization 数据同步
data template 数据模板
data transfer 数据传送
data transformation 数据转换
data-transformation technique 数据转换技术
data unit 数据单位
data update 数据更新
data user 数据用户
data V&V 数据校核与验证
data validation 数据确认
data validity 数据有效性
data value 数据值
data verification 数据校核
data verification and validation 数据校核与验证
data verification, validation and certification 数据校核、验证与认证
data visualization 数据可视化
data VV&A 数据校核、验证与确认
data VV&C 数据校核、验证与认证
data warehouse 数据仓库
database 数据库

database administration　数据库管理
database administrator　数据库管理员
database design　数据库设计
database directory　数据库目录
database error　数据库错误
database management system　数据库管理系统
database system　数据库系统
data-representation technique　数据表达技术
dataset　数据集
deactivation　钝化；失活
dead code　死代码
dead-end rule　终止规则
dead reckoning　航位推算法
dead variable　死变量
deadlock　死锁
deadlock avoidance　死锁避免
deadlock detection　死锁探测
deadlock state　死锁状态
debugging　调试
debugging phase　调试阶段
debugging run　调试运行
decentralized coordination　分布式协调
decentralized model　分布式模型
decentralized self-organizing system　分布式自组织系统
decentralized system　分布式系统
decidable　可决定的
decimal value　十进制数值
decision　决定；决策
decision criterion　决策准则
decision error　决策误差
decision maker　决策者

decision making　决策
decision making skill enhancing　决策技能增强
decision model　决策模型
decision simulation　决策仿真
decision skill　决策技术
decision support　决策支持
decision-tree model　决策树模型
decision variable　决策变量
decision variable space　决策变量空间
declaration　声明
declaration of a variable　变量声明
declarative　陈述性，说明性
declarative agent communication language　说明性智能体通信语言
declarative agent coordination language　说明性智能体协调语言
declarative agent language　说明性智能体语言
declarative agent technology　陈述性智能体技术
declarative approach　陈述性方法
declarative knowledge　陈述性知识
declarative language　说明性语言
declarative model-transformation language　说明性模型变换语言
declarative modeling　说明性建模
declarative part of a model　模型的说明性部分
declarative programming　说明性程序设计
declarative programming paradigm　说明性程序设计范式
declarative representation　说明性表示
declarative SL　说明性仿真语言

declarative statement　说明性语句
declarative technology　陈述性技术
declare (v.)　声明
declared random variable　声明的随机变量
declared variable　声明的变量
decomposability　可分解性
decomposable　可分解的
decomposable activity　可分解的活动
decomposable game　可分解对策
decompose (v.)　分解
decomposed activity　分解活动
decomposed trajectory　分解轨迹
decomposition　分解
decomposition method　分解方法
decomposition model　分解模型
decomposition tree　分解树
decompositional model　分解的模型
decouple (v.)　解耦
decoupled　解耦的
decoupled model　解耦模型
decoupling　解耦
dedicated model　专用模型
deduce (v.)　推导
deducible emergence　可推断的出现
deduct (v.)　演绎；推断，推论
deduction　演绎；推论
deduction model　推演模型
deductive　演绎的
deductive coherence　演绎一致性
deductive error　演绎误差
deductive logical fallacy　演绎逻辑上的谬论
deductive method　演绎方法
deductive model　演绎的模型

deductive modeling　演绎法建模
deductive reasoning　演绎推理
deep model　深入的模型
deep understanding　深入理解
default　默认
default parameter　默认参数
default value　默认值
defect　缺陷
defect density metric　缺陷密度度量
defective　有缺陷的
defined allowable value　定义的允许值
defined variable　定义的变量
definite　明确的
definition　定义
definition error　定义误差
deflection error　偏转误差
deformable model　可变形的模型
defuzzificate (v.)　去模糊
defuzzificated　去模糊的
defuzzificated value　去模糊化值
defuzzification　去模糊化
defuzzification method　去模糊化方法
defuzzifier　去模糊器
degeneracy test　简并测试
degenerate distribution　退化分布
degree　程度
degree of coupling　耦合程度
degree of falsity　虚假程度
degree of freedom　自由度
degree of membership　隶属度
degree of model validity　模型有效性程度
degree of truth　真实程度
deictic　直证的

deictic device　指示装置
delay　延时；滞后
delay-differential equation　滞后微分方程
delayed transition　延时转换
deletion error　删除错误
deliberately distorted reality　故意歪曲事实
deliberative agent　审议主体
deliberative coherence　审议连贯
deliberative system　审议制度
delimiterless input　无分隔符输入
delivered model　交付模型
delivery　交付；投放；传递；发送
Delphi technique　德尔菲技术
demon　"幽灵"
demon-controlled simulation　"幽灵"控制的仿真
demonstration　示范
denotational model　指示模型
dense connection　密集连接
dense output　密集输出
density　密度
density estimation　密度估计
depend (v.)　依赖
dependability　可靠性
dependability evaluation　可靠性评价
dependable　可靠的
dependable system　可信系统
dependence　依赖
dependency　依存；相关
dependency closure　相关闭包
dependent　相关的；依赖的
dependent assumption　依靠的假设
dependent model　从属模型

dependent variable　因变量
deployed model　部署的模型
deployment　部署
deployment diagram　部署图
deployment model　部署模型
depth first　深度优先
depth of perception　感知深度
derivability　可导出、可微分的性质或状态
derivability of a variable　变量的可导性
derivability variable　可微变量
derivable　可诱导的；可推论的
derivable experimental frame　可推论出的实验框架
derivative　推导；派生；导数
derivative assumption　衍生假设
derivative discontinuity　导数的不连续性
derivative section　衍生部分
derived　衍生的
derived behavior　导出的行为
DES (Discrete Event Simulation)　离散事件仿真
descendant model　后继模型
describe (v.)　描述；说明
described　描述的
description　描述；说明
description error　描述误差
description technique　描述技术
descriptive　描述的，叙述的
descriptive approach　描述的方法
descriptive assessment　描述的评估
descriptive complexity　描述的复杂性
descriptive data　描述的数据

descriptive decision 描述性决策
descriptive format 描述格式
descriptive knowledge 描述性知识
descriptive language 描述语言
descriptive model 描述的模型
descriptive model analysis 描述性模型分析
descriptive modeling 描述性建模
descriptive ontology 描述本体
descriptive simplicity 描述性简化
descriptive simulation 描述性仿真
descriptive understanding 描述性理解
descriptive variable 描述性变量
descriptor 描述符
deselected action 取消的活动
design 设计
design acceptability 设计可接受性
design conformance 设计一致性
design criterion 设计准则
design document 设计文档
design documentation 设计文件
design error 设计误差
design for change 设计变更
design guideline 设计准则
design language 设计语言
design methodology 设计方法学
design of experiment 实验设计；试验设计
design of simulation experiment 仿真实验设计
design pattern 设计模式
design phase 设计阶段
design requirement 设计需求
design selection 设计选型
design simulator 设计仿真器
design space 设计空间
design specification 设计规范
design validation 设计验证
design variable 设计变量
design verification 设计校核
designed experiment 设计实验
designed experimentation 设计实验
designed for change 设计变更
designed model 设计的模型
designed system of interest 设计的收益系统
desirability 期望
desirable feature 需要的特性
desirable system 预期系统
desired behavior 期望的行为
desired performance 预期性能
desired value 预期值
desk checking 桌面检查
DESS-DEVS 稳态双回波离散事件系统规范
destructive thought experiment 破坏性思维实验
detached eddy simulation 分离涡仿真
detail 细节，详情
detail complexity 细节复杂性
detailed data 详细的数据
detailed model 详细的模型
detailed modeling 详细的建模
detailed system 详细的系统
detailed system modeling 详细的系统建模
detailedness 详细
detectable 可探测
detectable failure 可检测到的错误
detected error 侦查误差

detected event 发现事件
detecting 探测
detection 探测
detection efficiency 检测效率
detection error 探测误差
detector 探测器
determination 决心；测定
determined 决定了的
determinism 决定论
deterministic 确定性的
deterministic algorithm 确定性算法
deterministic assumption 确定性假设
deterministic context-free L-system 确定性上下文无关的林氏系统
deterministic DEVS 确定性离散事件系统规范
deterministic distribution 确定性分布
deterministic experiment 确定性实验
deterministic function 确定性函数
deterministic model 确定性模型
deterministic optimization 确定性优化
deterministic simulation 确定性仿真
deterministic system 确定性系统
deterministic timed transition 确定性时间变迁
deterministic transition 确定性变迁
deterministic variable 决定性的变量
develop (v.) 开发
developer 开发者
development 发展；开发
development life cycle 开发生命周期
development methodology 开发方法学
development model 开发模型
development paradigm 开发范式
development process 开发过程
development requirements 发展需求
development risk 开发风险
development step 开发步骤
developmental model 开发的模型
device 设备
device error 设备误差
device latency 设备延迟
DEVS (Discrete Event System Specification) 离散事件系统规范
DEVS-based 基于离散事件系统规范的
DEVS-based tool 基于离散事件系统规范的工具
DEVS coordinator 离散事件系统规范协调器
DEVS coupling 离散事件系统规范耦合
DEVS dynamic structure 离散事件系统规范动态结构
DEVS experimental frame realization 离散事件系统规范实验框架的实现
DEVS external transition function 离散事件系统规范的外部转换函数
DEVS formalism 离散事件系统规范形式
DEVS generator 离散事件系统规范发生器
DEVS global state transition function 离散事件系统规范的全局状态转换函数
DEVS hierarchical form 离散事件系统规范的分层形式
DEVS internal transition function 离散

事件系统规范的内部转换函数
DEVS model 离散事件系统规范模型
DEVS parameter morphism 离散事件系统规范参数同型
DEVS pathways 离散事件系统规范路径
DEVS root-coordinator 离散事件系统规范跟协调器
DEVS simulation 离散事件系统规范仿真
DEVS simulation protocol 离散事件系统规范仿真协议
DEVS simulator 离散事件系统规范仿真器
DEVS system entity structure 离散事件系统规范系统实体结构
DEVS system morphism 离散事件系统规范系统同型
DEVS time advance function 离散事件系统规范的时间推进函数
DEVS transducer 离散事件系统规范转换器
diagnosis 诊断
diagnostic error 诊断误差
diagnostic model 诊断模型
diagnostic system 诊断系统
diagonal matrix 对角矩阵
diagram 图
dialectical system 辨证的系统
dichotomous variable 两分法变量
dictionary 词典，字典
diffential algebraic problem 微分代数问题
difference 差分
difference equation 差分方程

difference equation model 差分方程模型
difference equation simulation 差分方程仿真
differencing 差分
differential algebraic equation 微分代数方程
differential algebraic equation solver 微分代数方程求解器
differential analyzer 微分分析器
differential equation 微分方程
differential equation model 微分方程模型
differential equation solver 微分方程求解器
differential game 微分对策
differentiation 微分法
differentiation operator 微分算子
diffusive behavior 扩散的行为
digital 数字
digital analog simulation 数字模拟仿真
digital analog SL 数字模拟仿真语言
digital certificate 数字证书
digital computer 数字计算机
digital computer simulation 数字计算机仿真
digital continuous-system SL 数字连续系统仿真语言
digital data 数字化数据
digital differential analyzer 数字微分分析仪
digital discrete-system SL 数字离散系统仿真语言
digital experiment 数字实验

digital footprint 数字足迹
digital game 数字博弈
digital hardware 数字硬件
digital human modeling 数字人建模
digital input 数字输入
digital media 数字媒体
digital online SL 数字在线仿真语言
digital property 数字特性
digital quantum simulation 数字量子仿真
digital realization 数字实现
digital simulation 数字仿真
digital simulation program 数字仿真程序
digital simulation software 数字仿真软件
digital simulation study 数字仿真研究
digital SL 数字仿真语言
digital source SL 数字源仿真语言
digital-to-analog converter 数字-模拟转换器
digital trace 数字追踪
digital twin 数字双胞胎,数字化双胞胎,数字孪生
digitization error 数字化误差
digraph 有向图
digraph model 有向图模型
dilemma 困境
dimension 维度
dimension analysis 维度分析
dimensional analysis 层次分析
dimensional data modeling 空间数据建模
dimensionless 无量纲的
dimorphic 二态的

dimorphic behavior 二态的行为
direct communication 直接的通信
direct coordination 直接协调
direct coupling 直接耦合
direct input 直接输入
direct input device 直接输入设备
direct interaction 直接交互
direct manipulation interface 直接操纵接口
direct measurement technique 直接量测技术
direct multibody system dynamics 直接多体系统动力学
direct numerical simulation 直接数字仿真
direct problem 直接的问题
direct simulation 直接仿真
direct solution method 直接解决方法
directed 导向的
directed graph 有向图
directed system-of-systems 有向的系统的系统
direction transformation 方向变换
directly coupled 方向耦合
directory 目录
DIS (Distributed Interactive Simulation) 分布式交互仿真
DIS compatible 分布交互仿真兼容的
DIS compliant 符合分布交互仿真的
DIS-compliant simulation system 符合分布交互仿真协议的仿真系统
DIS protocol data 分布交互仿真协议数据
disaggregate 解聚
disaggregate (v.) 解聚

disaggregated 解聚
disaggregated model 分解的模型
disaggregation 解聚
discernible 可辨别的
discipline 学科；规律
disconfirmation 失验
disconnected simulation 非连接仿真
discontinuity 中断，不连续
discontinuity detection 中断探测
discontinuity handling 非连续处理
discontinuity-handling algorithm 中断处理算法
discontinuity localization 间断性定位
discontinuity-processing algorithm 间断性处理算法
discontinuous 不连续的
discontinuous-change model 非连续变化模型
discontinuous-change variable 间断变化的变量
discontinuous function 不连续函数
discontinuous hyperbolic PDE 不连续双曲型偏微分方程
discontinuous model 非连续的模型
discontinuous simulation 非连续的仿真
discontinuous state-variable 不连续状态变量
discontinuous system 非连续系统
discontinuous variable 不连续变量
discovered error 被发现的错误
discovery 发现
discovery learning 发现式学习
discrepancy 差异性
discrete 离散的
discrete-arithmetic-based simulation 基于离散算法的仿真
discrete-arithmetic-based 基于离散算法的
discrete attribute 离散属性
discrete-change 离散变化
discrete-change model 离散变化模型
discrete-change simulation 离散变化仿真
discrete-change variable 离散变化的变量
discrete-control variable 离散控制的变量
discrete damping 离散阻尼
discrete data 离散数据
discrete distribution 离散分布
discrete-event 离散事件
discrete-event abstraction 离散事件抽象
discrete-event-based multimodeling methodology 基于离散事件的多重建模方法学
discrete-event formalism 离散事件形式
discrete-event line simulation 离散事件线仿真
discrete-event list 离散事件列表
discrete-event model 离散事件模型
discrete-event modeling 离散事件建模
discrete-event simulation 离散事件仿真
discrete-event system 离散事件系统
discrete-event system specification 离散事件系统规范

discrete-event system specification model　离散事件系统规范模型
discrete Fourier transform　离散傅里叶变换
discrete Markov process　离散马尔可夫过程
discrete model　离散模型
discrete multimodel　离散多模型
discrete net　离散的网
discrete random variable　离散随机变量
discrete simulation　离散仿真
discrete-simulation environment　离散仿真环境
discrete-simulation language　离散仿真语言
discrete SL　离散仿真语言
discrete space　离散空间
discrete-space continuous-time model　空间离散时间连续模型
discrete-space discrete-time model　空间离散时间离散模型
discrete-space model　离散空间模型
discrete-state equation　离散状态方程
discrete-state model　离散状态模型
discrete-state system　离散状态系统
discrete-state variable　离散状态变量
discrete system　离散系统
discrete-system modeler　离散系统建模器
discrete-system simulation　离散系统仿真
discrete time　离散时间
discrete-time continuous simulation　离散时间的连续仿真
discrete-time controller　离散时间控制器
discrete-time difference equation model　离散时间差分方程模型
discrete-time flow mechanism　离散时间的流机制
discrete-time linear system　离散时间线性系统
discrete-time method　时间离散方法
discrete-time model　离散时间模型
discrete-time simulation　离散时间仿真
discrete-time simulation system　离散时间仿真系统
discrete-time state-space model　离散时间状态空间模型
discrete-time system　离散时间系统
discrete-time systems theory　离散时间系统理论
discrete-time variable　离散时间的变量
discrete trajectory　离散轨迹
discrete-valued variable　离散值的变量
discrete variable　离散变量
discrete-variable simulation optimization　离散变量仿真优化
discretization　离散化
discretization error　离散化误差
discretization method　离散化方法
discretization point　离散点
discretization scheme　离散化机制
discretize (v.)　离散
discretized　离散化的
discretized variable　离散化变量

discriminant variable　判别式变量
discrimination net model　辨识网模型
discriminative model　辨识模型
discriminative stimulus　辨识激励
discriminatory analysis　判别分析
disjoint　分离
disjointness　分离
disjunctive coupling　析取耦合
disjunctive model　非连贯的模型；分离的模型
disjunctively coupled　非连贯的耦合
disk error　磁盘错误
disk-seek error　磁盘查找错误
disk write error　磁盘写错误
displacement variable　位移变量
display　显示
display-based simulation　基于显示的仿真
display device　显示设备
disposal value　清算价值
disprove (v.)　反驳
dissimilar　相异的
dissimilar model　非相似的模型
dissimilar simulation　非相似仿真
dissimilarity　不同；相异点
dissimilarity criterion　相异判据
dissimilate (v.)　异化
dissimilated　被异化的
dissimilating　异化
dissimilation　异化
dissimilative　异化的
dissimilatory　异化的
dissimilitude　相异
dissimulate (v.)　掩饰
dissimulated　掩饰

dissimulating　掩饰
dissimulation　掩饰
dissimulative　掩饰的
dissimulator　掩饰器
distance transformation　距离变换
distorted model　扭曲的模型
distorted reality　扭曲的现实
distracting input　分散的输入
distractive technology　扰动技术
distribute (v.)　分布；分发
distributed　分布式
distributed agent simulation　分布式代理仿真
distributed algorithm　分布式算法
distributed and parallel computing　分布式和并行计算
distributed application　分布式应用程序
distributed asymmetric simulation　分布式非对称仿真
distributed collaboration　分布式协同
distributed computational model　分布式计算模型
distributed computing　分布式计算
distributed DEVS simulation　分布式离散事件系统规范仿真
distributed event-based system　基于事件的分布式系统
distributed event-driven simulation　分布式事件驱动仿真
distributed execution　分布式执行
distributed exercise　分布式演练
distributed experimental frame　分布式实验框架
distributed federate　分布式联邦成员

distributed federation 分布式联邦

distributed heterogeneous simulation 分布式异构仿真

distributed high-performance computing 分布式高性能计算

distributed interactive simulation 分布式交互仿真

distributed lazy simulation 分布式懒惰仿真

distributed M&S 分布式建模与仿真

distributed mission training 分布式任务训练

distributed model 分布式模型

distributed model checking 分布式模型检查

distributed model development 分布式模型开发

distributed model-repository 分布式模型仓库

distributed-parameter model 分布式参数模型

distributed-parameter system 分布参数系统

distributed-parameter system simulation 分布参数系统仿真

distributed problem solving 分布式问题求解

distributed processor 分布式处理器

distributed real-time simulation 分布式实时仿真

distributed scheduling mechanism 分布式调度机制

distributed simulation 分布式仿真

distributed simulation architecture 分布式仿真体系结构

distributed simulation environment 分布式仿真环境

distributed simulation framework 分布式仿真架构

distributed simulation protocol 分布式仿真协议

distributed simulation technology 分布式仿真技术

distributed simulator 分布式仿真器

distributed system 分布式系统

distributed system design 分布式系统设计

distributed-system SL 分布式系统仿真语言

distributed systems architecture specification 分布式系统体系结构规范

distributed training 分布式训练

distributed virtual environment 分布式虚拟环境

distributed web-based simulation 基于万维网的分布式仿真

distribution 分布状态；分发

distribution function 分布函数

divided attention 分散注意

DNA-based 基于DNA的

DNA-based coupling 基于DNA的耦合

DNA-based model 基于DNA的模型

DNA-based modeling 基于DNA的建模

DNA-based simulation 基于DNA的仿真

DNA computing DNA计算

DNA machine DNA机

docking 装载
document 文档
document (v.) 评注
document centric 文档中心的
document-centric approach 以文档为中心的方法
document repository 文件库
documentation 文件编制
documentation checking 文档检查
documentation framework 文档框架
documentation of result 结果文档
documentation of VV&T certification 校核、验证与测试证书文档
documentation standard 文档标准
documentation template 文档模板
documentation tree 文档树
documenting 建档
domain 论域；定义域；区域；域
domain analysis 领域分析
domain application 域名注册；领域应用
domain boundary 领域边界
domain dependent 域相关
domain expert 领域专家
domain independent 独立域
domain knowledge 领域知识
domain model 领域模型
domain of analytical stability 解析域稳定性
domain of intended application 拟申请的域名注册
domain ontology 领域本体
domain-oriented 面向领域的
domain-oriented sensitivity analysis 面向领域的灵敏度分析
domain requirements 领域需求
domain-specific 特定领域
domain-specific agent 特定领域智能体；特定领域代理
domain-specific interface 特定域接口
domain-specific knowledge 特定领域知识
domain-specific language 指定域语言
domain-specific library 特定领域库
domain-specific model 特定领域模型
domain-specific modeling 特定领域建模
domain-specific modeling approach 指定域建模方法
domain-specific modeling language 指定域建模语言
domain-specific simulation language 指定域仿真语言
domain variable 域变量
dominance model 主导模型
dominant dynamics 主导动力学
dominant pole 主导极点
dominant variable 主导变量
dormant aspect 休眠的方面
dormant aspect of model 模型的休眠方面
dormant model 休眠模型
dormant state 休眠状态
dos & don'ts in M&S 建模与仿真中的做与不做
double pole 双极
downward causation 向下因果关系
drag-and-drop interaction 拖放式交互
driven 驱动
driver 驱动器

driver error 驱动误差
driver training 驾驶员训练
driving game 赛车游戏
driving simulator 驾驶仿真器
driving variable 操纵变量
drop-down menu 下拉式菜单
dry laboratory 干燥实验室
DS-DEVS 动态结构-离散事件系统规范
dual use technology 两用技术
dual variable 对偶变量
dubious assumption 不确定的假设
Dublin core metadata template 都柏林核心元数据模板
Duhem problem Duhem 问题
Duhem-Quine problem Duhem-Quine 问题
Duhem-Quine thesis Duhem-Quine 命题
dummy variable 虚构的变量
dumping error 转存错误
dumping error function 转存错误函数
duplicate output 重复输出
duplication 复制
dyn-DEVS 动态离散事件系统规范
dynamic 动态
dynamic adaptation 动态自适应
dynamic assumption 动态的假设
dynamic behavior 动态的行为
dynamic check 动态的检查
dynamic complexity 动态复杂性
dynamic composability 动态的组合性
dynamic composition 动态的组合
dynamic coupling 动态耦合
dynamic data 动态数据
dynamic data-driven application system 动态数据驱动的应用系统
dynamic data-driven simulation 动态数据驱动仿真
dynamic data-structure system 动态数据结构系统
dynamic data-update 动态数据更新
dynamic decoupling 动态解耦
dynamic DEVS 动态离散事件系统规范
dynamic documentation 动态文件
dynamic environment 动态环境
dynamic error 动力学误差
dynamic error-analysis 动态错误分析
dynamic event sequence 动态事件序列
dynamic federate-composition 动态联邦组成
dynamic federate-coupling 动态联邦耦合
dynamic-fidelity modeling 动态保真度建模
dynamic game 动态博弈
dynamic hypergame 动态超对策
dynamic interaction 动态交互
dynamic interoperability 动态互操作性
dynamic-interoperability level 动态互操作水平
dynamic knowledge 动态知识
dynamic lighting 动态光照
dynamic microsimulation-model 动态微仿真模型
dynamic mode-update mechanism 动态模式更新机制

dynamic model	动态模型
dynamic model composability	动态模型组合性
dynamic model-composition	动态模型组合
dynamic model-coupling	动态模型耦合
dynamic model discovery	动态模型发现
dynamic model-documentation	动态模型文件
dynamic model location	动态模型位置
dynamic model structure	动态模型结构
dynamic model switching	动态模型转换
dynamic model update	动态模型更新
dynamic model updating	动态模型更新
dynamic modeling	动态建模
dynamic modeling formalism	动态模型形式
dynamic natural environment	动态自然环境
dynamic network	动态网络
dynamic ontology	动态本体
dynamic ontology sharing	动态本体共享
dynamic personality filter	动态个性过滤器
dynamic program check	动态程序检查
dynamic reality	动态实体
dynamic relationship	动态关系
dynamic scenario	动态脚本
dynamic similarity	动态相似性
dynamic simulation	动态仿真
dynamic simulation-composition	动态仿真组合
dynamic simulation-linking	动态仿真连接
dynamic simulation-service	动态模拟服务
dynamic simulation-update	动态仿真更新
dynamic simulation-updating	动态仿真更新
dynamic software-architecture	动态软件结构
dynamic structural-model	动态结构化模型
dynamic structure	动态结构
dynamic-structure automaton	动态结构自动机
dynamic structure DEVS	动态结构离散事件系统规范
dynamic-structure model	动态结构模型
dynamic-structure multimodel	动态结构多模型
dynamic-structure network	动态结构网络
dynamic submodel replacement	动态子模型替换
dynamic switching	动态转换
dynamic system	动态系统
dynamic system simulation	动态系统仿真
dynamic technique	动态方法，动态

技术
dynamic update 动态更新
dynamic updating 动态更新
dynamic validity 动态有效性
dynamic VV&T technique 动态校核、验证与测试技术
dynamically bound 动态划界
dynamically-bound variable 动态划界的变量
dynamically changing 动态地改变
dynamically composable 动态地可组合的
dynamically-composable federation 动态可组合联邦
dynamically-composable simulation 动态可组合仿真
dynamically coupled 动态耦合的
dynamically-coupled model 动态耦合模型
dynamically-coupled multimodel 动态耦合多模型
dynamically-scoped variable 动态作用域变量
dynamics 动力学

Ee

e-glass display 电子玻璃显示
e-glass visualization 电子玻璃可视化
e-lens display 电子透镜显示
early application 早期应用
econometric model 计量经济模型
economic impact of simulation 仿真经济影响
economics of simulation 仿真经济

eddy simulation 涡流仿真
edge computing 边缘计算
edge simulation 边缘仿真
editor 编辑器，编辑程序
educate (v.) 教育
education 教育
education technique 教育方法；训练技术
educational 教育的
educational game 教育游戏
educational simulation game 寓教于乐的仿真游戏
educative 教育的
edutainment 教育的
effect 效应；结果
effective decision 有效决策
effectiveness 能行性；有效性
effectiveness metric 有效性度量
effectiveness of mutation 变异有效性
effects-based modeling 基于效果的建模
effects of simulation exposure 模拟曝光效应
efficiency 效率
effort 力；成就
effort variable 可变作用力
eigenvalue 特征值
eigenvalue matrix 矩阵特征值
eigenvector 特征向量
elaborated variable 详尽说明的变量
elaboration 精心制作
elapsed time 已逝时间
electronic contact lens display 电子隐形眼镜显示
electronic game 电子游戏

electronic game industry 电子游戏产业
electronic gaming and simulation 电子游戏与仿真
electronic transcription error 电子转录错误
element 元素
element alignment 元素对齐
element grouping 元素分组
elimination 消除；消元法
elliptic PDE 椭圆型偏微分方程
embedded 嵌入式
embedded agent 嵌入式智能体
embedded interaction 嵌入式交互
embedded model 嵌入式模型
embedded simulation 嵌入式仿真
embedded system 嵌入式系统
embedded training 嵌入式训练
embedded training system 嵌入式训练系统
embedded value 嵌入式数值
embedding 嵌入
emerge (v.) 出现
emerged 出现的
emerged feature 出现的特征
emergence 突现；涌现
emergence complexity 涌现复杂性
emergence condition 涌现条件
emergence modeling 涌现建模
emergence of a property 特性的涌现
emergence prediction 涌现预测
emergence simulation 涌现仿真
emergence simulation approach 涌现仿真方法
emergency egress simulation 紧急出口仿真
emergent 涌现的，紧急的
emergent behavior 突现行为；应急行为
emergent behavior repository 突现行为库
emergent characteristic 涌现特征
emergent cluster 紧急集群
emergent complexity 突发复杂性
emergent condition 紧急状况
emergent dynamics 涌现动力学
emergent entity 突现的实体
emergent feature 涌现特征
emergent function 紧急功能
emergent functionality 紧急功能性
emergent game 应急对策
emergent gaming platform 应急对策平台
emergent input 紧急输入
emergent macro-behavior 紧急性宏观行为
emergent misbehavior 突发性不当行为
emergent output 紧急输出
emergent output function 紧急输出操作
emergent paradigm 应急范式
emergent phenomenon 涌现现象
emergent platform 应急平台
emergent problem 涌现的问题
emergent property 涌现的特性
emergent self-model 应急自模型
emergent simulation 应急仿真
emergent simulation platform 应急仿真平台

emergent state 涌现状态；紧急状态
emergent structure 应急结构
emergent substance 紧急物资
emergent system characteristic 应急系统特征
emergent system property 应急系统特性
emergent transition 紧急转换
emergent transition function 紧急转换函数
emergentism 突现论；涌现主义
emergentist 突现论者
emerging 新兴的
emerging behavior 新兴的行为
emerging condition 发生的条件
emerging context 发生的情境
emerging feature 新兴的特征
emerging paradigm 新兴的范式
emerging phenomenon 涌现现象
emerging practice 新兴的实践
emerging problem 涌现的问题
emerging property 涌现特性
emerging simulation system 涌现仿真系统
emerging successor model 新继任者模型
emerging system 涌现系统
emerging technology 新兴的技术
emerging trend 新趋势
emit (v.) 发射
emitted 发出的
emitter 发射极
emitting 发出
emotion 情绪
emotion-induced 情感诱导

emotion markup language 情感标记语言
emotion model 情感模型
emotional behavior 情绪的行为
emotional coupling 情绪耦合
emotional experiment 情绪实验
emotional expression 情绪表现
emotional filter 情感渗透
emotional input 情绪化输入
emotional interaction 情感交互
emotional interface 情感界面
emotional model 情感模型
emotional modeling 情感建模
emotional parameter 情感参数
emotional response 情绪响应
emotional state 情感状态
emotional trigger 情感触发
emotionally coupled 情绪上耦合的
emotionally-intelligent interactive system 情感智能交互系统
emotionally intelligent interface 情感智能界面
emotive response 感情响应
empiric 经验主义的
empirical 经验的
empirical analysis 经验分析
empirical confirmation 经验的认证
empirical consequence 经验成果
empirical data 经验数据
empirical dependability 经验依赖
empirical distribution 经验分布
empirical evaluation 经验评价
empirical experiment 经验试验
empirical knowledge 经验知识
empirical measure 经验测量

empirical model 经验模型
empirical model design 经验模型设计
empirical model development 经验模型开发
empirical model validation 经验模型验证
empirical model validity 经验模型有效性
empirical modeling 经验建模
empirical modeling algorithm 经验建模算法
empirical modeling application 经验建模应用
empirical modeling language 经验建模语言
empirical modeling principle 经验建模原理
empirical modeling tool 经验建模工具
empirical source 经验源
empirical study 经验性研究；实证研究
empirical validity 实证效度
empirically derived 经验推导的
empirically-derived analytical model 基于经验推导的解析模型
empirically-derived model 基于经验推导的模型
empiricism 经验论
emulate (v.) 模拟
emulated 仿真的
emulated behavior 仿真行为
emulated instrument 仿真仪器
emulation 仿真（计算机）
emulation methodology 模拟方法论
emulation model 仿真模型
emulation program 模拟程序
emulative 模拟的
emulator 模拟器
enabled 使能的
enabled instance 启用实例
enabled rule 激活的规则
enabler 使能器
enabling 使能
enabling rule 启动规则
enabling simulation 使能仿真
enabling-simulation interoperability 使能仿真互操作性
enabling technology 使能技术
encapsulated 封装的
encapsulated complexity 封装的复杂性
encapsulation 封装
enclave model 飞地模型
enclave model set-up 飞地模型建立
encoding error 编码错误
encrypt (v.) 加密
encrypted 加密的
encryption 加密
encryption data 加密数据
encryption error 加密错误
end-firing phase 发射末端
end-to-end model 端到端模型
end-user customization tool 终端用户定制工具
endian 尾段
ending state 结束状态
endo-model 内部模型
endogenous 内生的；内源性；内源的
endogenous assumption 内生假设

endogenous covariate 内生协变量
endogenous event 内生事件
endogenous input 内生输入
endogenous variable 内生变量
endomodel 内部模型
endomorph 内胚层
endomorphic 自同态的
endomorphic model 自同态模型
endomorphic simulation 自动态仿真
endomorphic simulation model 自同态仿真模型
endomorphic simulator 自同态仿真器
endomorphic system 自同态系统
endomorphism 自同态
endomorphy 内胚层体型
energy 能量
energy conservation law 能量守恒定律
energy flow 能量流
energy transducer 能量转换器
engagement training 参与培训；交战训练
engine 引擎
engineered emergence 工程突现
engineering 工程（学）
engineering design 工程设计
engineering requirements 工程需求
engineering simulation 工程仿真
engineering simulator 工程仿真器
enhance (v.) 增强
enhanced 增强的
enhanced reality 增强现实
enhancing 增强
enriched reality 丰富现实
entangled state 纠缠态

entanglement 纠缠
enterprise 企业
enterprise model 企业模型
enterprise simulation model 企业仿真模型
entertainment 娱乐
entertainment game 娱乐游戏
entertainment model 娱乐模型
entertainment simulation 娱乐仿真
entity 实体
entity behavior 实体行为
entity coordinate 实体坐标
entity flow 实体流
entity level 实体层次
entity-level simulation 实体级仿真
entity model 实体模型
entity perspective 实体视角
entity-relationship diagram 实体关系图
entity-relationship model 实体关系模型
entity structure 实体结构
entry 入口
environment 环境
environment error 环境误差
environment-mediated 环境调和的
environment-mediated self-adaptive system 环境调和的自适应系统
environment-mediated self-organizing system 环境调和的自组织系统
environment model 环境模型
environment representation 环境表示（法）
environment variable 环境变量
environmental data 环境数据

environmental effect 环境效应
environmental effect model 环境效应模型
environmental entity 环境实体
environmental event 环境事件
environmental feature 环境特征
environmental interaction 环境交互
environmental method 环保方法
environmental model 环境模型
environmental modeling 环境建模
environmental representation 环境表征
environmental simulation 环境仿真
environmental variable 环境变量
epistemic arrogance 认知傲慢
epistemic model 知识模型
epistemic opacity 认知的不透明度
epistemic trust 认知信任
epistemic uncertainty 认知不确定性
epistemological 认识论的
epistemological emergence 认识论的涌现
epistemological perspective 认识论视角
epistemological shift 认识论转变
epistemology 认识论
epistemology-based classification 基于认识论的分类
epistemology of M&S 建模与仿真的认识论
equation 方程
equation-based model 基于方程的模型
equation model 方程模型
equation-oriented simulation 面向方程的仿真
equation-oriented simulation system 面向方程的仿真系统
equilibrium 平衡
equilibrium condition 平衡条件
equilibrium model 平衡模型
equipment 设备，装置
equipment error 设备误差
equivalence 等价
equivalence morphism 等价同型
equivalence of systems 系统等价性
equivalence partitioning testing 等价划分测试
equivalence relation 等价关系
equivalencing 等价；使等价
equivalent 相等的；等价物
equivalent input 等效输入
equivalent model 等效模型
equivalent system 等效系统
equivocation 疑义度
ergodic 遍历的；遍历性
ergodic behavior 遍历行为
ergodic hypothesis 遍历假设
ergodic system 遍历系统
ergodic theory 遍历理论
ergodicity 遍历性
ergonomic 人机工程
ergonomics 人机工程学
Erlang random variable 厄兰随机变量
erratic error 不规则误差
erroneous 不正确的
erroneous behavior 错误的行为
erroneous concept 错误的概念
erroneous perception 错觉
erroneous reasoning 错误推理

erroneously 不正确地
error 误差
error accumulation 误差累积
error ambiguity 误差模糊
error analysis 错误分析；误差分析
error assessment 错误评估；误差评估
error band 误差带
error bound 误差界
error burst 错误猝发
error capture 错误俘获
error character 错误字符
error checking 错误检查
error checking code 误差校验代码；错误检查代码
error code 错误代码
error coefficient 误差系数
error concealment 错误隐藏
error condition 错误条件
error containment 容许误差
error control 误差控制
error-controlled simulation 误差控制仿真
error correcting 误差校正
error-correcting analysis 纠错分析
error correcting code 纠错代码
error correction 误差校正；错误纠正
error data 错误数据
error detecting code 错误检测代码
error detecting system 错误检测系统
error detection 错误探测
error-detection model 误差检测模型
error due to misunderstanding 由误解引起的错误
error elimination 误差消除
error estimation 误差估计
error flag 错误标志
error function 误差函数
error growth 误差增长
error handler 错误处理程序
error handling 错误处理
error indication 误差显示
error layer 误差层
error log 错误运行日志
error measurement 误差测量
error message 错误消息
error mining 误差挖掘
error model 误差模型
error multiplication 误码增殖
error of accepting wrong model 存伪错误模型
error of omission 遗漏误差
error of rejecting valid model 弃真错误模型
error of solving wrong problem 校正错误问题的误差
error-prediction model 误差预测模型
error prevention 差错防范
error probability 错误概率
error processing 错误处理
error ratio 错误率
error recovery 错误恢复
error reduction 误差减少
error spread 误差传播
error tolerance 误差容限
error type 错误类型
errorist 总是犯错却自认为正确的人
escapist simulation 逃避现实仿真
essential element 主要元素
essential variable 基本变量
essentially cascade coupling 本质级联

耦合
essentially cascade coupling recipe 基本级联式配方
estimate (v.) 估计
estimated 估计；推算
estimation 估计
estimation error 估计误差
estimation theory 估计理论
estimator 估计器
ethical 合规的；伦理的
ethical assessment 合乎规格的评估
ethical assessment of the goal 目标的合规评估
ethical assessment of the goal of the study 研究目标的合规评估
ethical assessment of the study 研究的合规评估
ethical assessment of value system 价值系统的合规评估
ethical error 伦理错误
ethical evaluation 伦理评价
ethical goal-assessment 伦理目标评价
ethical issue 伦理问题
ethical principle 伦理原则
ethical rationality 伦理合理性
ethical rule 伦理规则
ethical simulation 伦理仿真
ethics 伦理学
Euler angle 欧拉角
Euler integration 欧拉积分
Euler integration method 欧拉积分方法
Eulerian model 欧拉模型
evaluate (v.) 评价
evaluated input 评估输入

evaluated source of input 评估输入源
evaluated understanding 评价认识
evaluation 评价
evaluation data 评估数据
evaluation data set 评估数据集合
evaluation goal 评估目标
evaluation metric 评价指标
evaluation of input 输入评估
evaluation of source of input 输入源评估
evaluative model 评估模型
evaluative model analysis 评价模型分析
evaluative simulation 可评估的仿真
evaluative understanding 评价性理解
evaluator 计算器；求值程序；评估者
even function 偶函数
event 事件
event anticipation 事件预期
event appraisal 事件评估
event-appraisal model 事件评价模型
event-based 基于事件的
event-based agent simulation 基于事件的智能体仿真
event-based discrete simulation 基于事件的离散仿真
event-based language 基于事件的语言
event-based program 基于事件的程序
event-based programming 基于事件的程序设计
event-based programming paradigm 基于事件的程序设计范式
event-based simulation 基于事件的仿真

event-based task 基于事件的任务
event causality 事件因果关系
event-centered 以事件为中心
event-centered approach 以事件为中心的方法
event class 事件类
event code 事件代码
event communication 事件通信
event condition 事件条件
event-condition-action 事件-条件-动作
event-condition-action rule 事件-条件-动作规则
event coordination 事件协调
event delivery 事件交付
event density 事件密度
event description section 事件描述部分
event descriptor 事件描述符
event desirability 事件可取性
event-directed programming 事件导向的程序设计
event-driven 事件驱动
event-driven application 事件驱动的应用程序
event-driven architecture 事件驱动架构
event-driven logic 事件驱动逻辑
event-driven program 事件驱动程序
event-driven programming 事件驱动的程序设计
event-driven simulation 事件驱动的仿真
event exchange 事件交换
event-exchange format 事件交换格式
event extraction 事件抽取
event file 事件文件
event flag 事件标志
event following 事件跟踪
event-following simulation 事件跟踪仿真
event generator 事件发生器
event granularity 事件粒度
event graph 事件图
event handler 事件处理程序
event history 事件历史
event-history approach 事件历史方法
event horizon 事件视界
event instantiation 事件实例化
event interpreter 事件解释器
event language 事件语言
event learning 事件学习
event line 事件线
event list 事件列表
event node 事件节点
event notice 事件通知
event occurrence 事件发生
event-oriented 面向事件的
event-oriented model 面向事件的模型
event-oriented rule 面向事件的规则
event-oriented simulation 面向事件的仿真
event processing 事件处理
event processor 事件处理器
event queue 事件队列
event recorder 事件记录器
event retraction 事件撤回
event scheduling 事件调度
event-scheduling approach 事件调

度法
event-scheduling simulation 事件调度仿真
event segment 事件段
event selection 事件选择
event sequence 事件序列
event-sequence diagram 事件序列图
event sequencing 事件排序
event space 事件空间
event synchronization 事件同步
event time 事件时间
event trace 事件跟踪
event tracing 事件追踪
event-triggered 事件触发
event-triggered control 事件触发控制
event type 事件类型
event validity 事件有效性
event-validity test 事件有效性测验；事件合理性测验
evolution 演变；进化
evolutionary algorithm 进化算法
evolutionary computation 进化计算
evolutionary data 进化数据
evolutionary emergence 进化涌现
evolutionary game 进化博弈
evolutionary holon 子整体进化
evolutionary maintenance 进化维护
evolutionary method 进化方法
evolutionary model 进化模型
evolutionary multimodel 进化多模型
evolutionary reality 进化的现实
evolutionary simulation 进化仿真
evolutionary-simulation algorithm 进化仿真算法
evolutionary solution 进化解
evolutionary system 进化系统
evolutionary-system simulation 进化系统仿真
evolutionary validation 进化验证
evolvability 可进化性
evolvable 可进化的
evolvable metamodel 可进化的元模型
evolvable model 可进化模型
evolvable software architecture 可演进的软件构架
evolvable system 可进化系统
evolve (v.) 进化
evolving environment 演化环境
evolving model 演化模型
evolving phenomenon 演化现象
evolving reality 不断演化的现实
evolving system 演变系统
ex ante 事前
ex ante analysis 事前分析
ex ante evaluation 事前评估
ex ante prediction 事前预测
ex ante simulation 事前仿真
ex ante value 事前价值
ex post 事后的
ex post calculation 事后计算
ex post facto 事后的
ex post facto variable 事后变量
ex post prediction 事后预测
ex post simulation 事后仿真
ex situ experiment 异地实验
exact 精确的
exact model 精确的模型
exact-scale replica 同比例副本
exactness 正确；精确

examination 核查
exascale computer 超速计算机
exascale computing 超速计算
exascale simulation 超速仿真
exchange 交换
exchange (v.) 交换
exchange data 交换数据
exchanged data 被交换的数据
exchanged information 被交换的信息
exchanged model 被交换的模型
excitation 激励
exclusive execution 独占执行
executable 可执行的
executable architecture 可执行构架
executable cognitive-model 可执行的认知模型
executable error 可执行的错误
executable experiment 可执行的实验
executable hypothesis 可执行的假设
executable model 可执行的模型
executable scenario 可执行的场景
executable simulation model 可执行的仿真模型
execute (v.) 执行
execution 执行
execution control 执行控制
execution detail 执行细节
execution efficiency 执行效率
execution environment 执行环境
execution error 执行错误
execution model 执行模型
execution phase 执行阶段
execution profiling 执行分析
execution testing 执行测试
execution-time efficiency 运行时间效率
execution-time management 执行时间管理
execution tracing 执行跟踪
execution visualization 运行可视化
executive 执行的；执行程序
executive council for M&S 建模与仿真执行委员会
executive software 执行软件
exercise 练习；演练
exercise management 演练管理
exercise manager 演练管理器
exhaustive discipline 详尽的规则
existential assumption 存在的假设
existential dependence 存在的依赖
existential dependency 存在的依赖关系
existing condition 存在条件；现存条件
existing reality 存在的现实
existing system of interest 现有的收益系统
exogenous 外源；外生的
exogenous assumption 外生假设
exogenous covariate 外生协变量
exogenous event 外生事件
exogenous fault detection 外源故障检测
exogenous input 外源性输入
exogenous variable 外生变量
expanded time 膨胀的时间；扩展的时间
expanded-time simulation 扩展时间仿真
expansion 扩展

expectancy-value model　期望值模型
expectation-driven　期望驱动
expectation-driven reasoning　期望驱动型推理
expected behavior　预期的行为
expected input　期望输入
expected state　期望状态
expected value　希望值，期望值
expected-value model　期望值模型
experience　经验
experience (v.)　体验
experience-aimed　针对经验的
experience-aimed simulation　针对经验的仿真
experience- and nonexperience-based knowledge　基于经验的和非经验的知识
experience base　经验库
experience-based　基于经验的
experience-based knowledge　基于经验的知识
experience constraint　经验约束
experience-directed　经验导向的
experience management　经验管理
experience model　经验模型
experience parameter　经验参数
experience perspective　体验视角
experience-related parameter　经验相关的参数
experienced　经历过的
experiencing　正在经历的
experiential　经验的
experiential concept　经验的概念
experiential context　经验上下文
experiential knowledge　经验知识
experiential learning　经验学习
experiential model　经验模型
experiential planning　体验规划
experiential study　经验研究
experiential tool　经验工具
experiential variable　经验变量
experientialism　经验主义
experientialist　经验论者
experientially　经验上地
experiment　试验；实验
experiment (v.)　做实验；进行试验
experiment agent　实验代理
experiment-aimed　针对实验的
experiment-aimed simulation　针对实验的仿真
experiment-based　基于实验的
experiment constraint　实验约束
experiment design　实验设计
experiment management　实验管理
experiment model　实验模型
experiment planning　实验规划
experiment-related parameter　与实验（试验）相关的参数
experiment reproducibility　实验重现性
experiment space　实验空间
experimental　实验的
experimental bias　实验偏差
experimental code　实验的代码
experimental concept　实验的概念
experimental condition　实验条件
experimental-conditions interoperability　实验条件互操作性
experimental context　实验情境
experimental data　实验数据

experimental design 实验设计
experimental-design methodology 实验设计方法学
experimental error 实验误差
experimental frame 实验框架
experimental frame acceptability 实验框架的可接受性
experimental frame base 实验框架库
experimental frame composability 实验框架的可组合性
experimental frame composition 实验框架组合
experimental frame decomposition 实验框架分解
experimental frame derivability 实验框架可导性
experimental frame interoperability 实验框架互操作性
experimental frame realization 实验框架的实现
experimental knowledge 实验知识
experimental method 实验方法
experimental methodology 实验方法学
experimental model 实验模型
experimental observation 实验观察
experimental procedure 实验步骤；实验程序
experimental result 实验结果
experimental scenario interoperability 实验场景互操作性
experimental science 实验科学
experimental set-up 实验布置
experimental simulation 实验仿真
experimental study 实验研究

experimental system 实验系统
experimental technique 实验技术
experimental tool 实验工具
experimental validity 实验有效性
experimental variable 实验变量
experimentalism 实验主义
experimentalist 实验论者
experimentalize (*v.*) 做实验
experimentally 实验式地；实验上
experimentarian 依赖实验的
experimentation 实验
experimentation bias 实验偏差
experimentation challenge 实验挑战
experimentation credibility 实验可信度
experimentation ease 实验轻松
experimentation error 实验误差
experimentation interoperability 实验互操作性
experimentation investigation 实验调查
experimentation manager 实验管理器
experimentation parameter 实验参数
experimentation perspective 实验视角
experimentation policy 实验策略
experimentation specification 实验规范
experimentation support 实验支持
experimentation tool 实验工具
experimentation variable 实验变量
experimentative 实验性的
experimentator 实验员
experimented 被实验的
experimenter 实验者
experimenter bias 实验者偏见

experimenting 进行实验
experimentist 实验专家
expert 专家
expert model 专家模型
expert modeling 专家建模
expert system 专家系统
expert system embedded within simulation 嵌入式仿真中的专家系统
expert system shell 专家系统外壳
expertise-driven 专业知识驱动的
expertise-driven decision 专业知识驱动的决策
explain (v.) 解释
explained emergent behavior 解释紧急行为
explained variable 解释变量
explanation 解释
explanatory 解释的
explanatory model 解释性模型
explanatory simulation 解释性仿真
explanatory variable 解释变量
explicit 显式
explicit Adams-Bashforth method 显式的 Adams-Bashforth 方法
explicit algorithm 显式算法
explicit assumption 显性的假设
explicit constraint 显式约束
explicit coordination 显式协调
explicit declaration 显式声明
explicit dependency 显式依存关系
explicit event 显式事件
explicit first-order algorithm 显式一阶算法
explicit goal 明确目标
explicit integration 显式积分
explicit integration algorithm 显式积分算法
explicit knowledge 显性知识
explicit method 显式法
explicit midpoint rule 显式中点法则
explicit model checking 显式模型检查
explicit numerical integration 显式数值积分
explicit numerical integration formula 显式数值积分公式
explicit Nyström technique 显式奈斯特龙技术
explicit single-step method 显式单步法
explicit state 显式状态
explicit value 确切值
exploded logit 分对数分解
exploitative modeling 开发性建模
exploration model 探测模型
exploration simulation 探索性仿真
exploratory analysis 探索性分析
exploratory experiment 探索性实验
exploratory model 探索性模型
exploratory modeling 探索性建模
exploratory multimodel 探索式多模型
exploratory multisimulation 探索性多仿真
exploratory-multisimulation methodology 探索性的多仿真方法学
exploratory simulation 探索性仿真
exploratory testing 探索性测试

exponential distribution 指数分布
exponential random variable 指数随机变量
exposure 曝光；暴露；陈列
exposure variable 暴露变量
expression 表达；表达式
expression-oriented SL 面向表达式的仿真语言
expressive 表现的
expressive power 表现力
expressiveness 可表达性
expressiveness formalism 表现力形式主义
extendable finite-state-machine 可扩展有限状态机
extendable finite-state-model 可扩展有限状态模型
extendable model 可扩展模型
extended knowledge-attributed Petri net 扩展知识属性佩特里网
extended state-transition diagram 扩展状态转移图
extended stochastic Petri net 扩展随机佩特里网
extensibility 可扩展性
extensible 可延伸的
extensible federation 可扩展联邦
extensible finite-state-machine 可扩展有限状态机
extensible framework 可扩展框架
extensible M&S framework 可扩展建模与仿真框架
extensible model 可扩展模型
extensible multimodel 可扩展多模型
extensible simulation 可扩展仿真
extensible simulation-infrastructure 可扩展仿真的基本结构
extension 扩展
extension mechanism 扩展机制
extension notion 扩展概念
extensive variable 外延变量
external activity 外部活动
external agent 外部智能体；外部代理
external cloning of a federate 联邦成员的外部克隆
external coupling 外部耦合
external error 外部错误；外在误差
external event 外部事件
external evolution 外部演化
external excitation 外部激励
external goal 外部目标
external inactivity 外部闲置
external input 外部输入
external input coupling 外部输入耦合
external memory 外部记忆
external model 外部模型
external output coupling 外部输出耦合
external output coupling function 外部输出耦合函数
external performance evolution 外部性能演化
external reality 外部现实
external schema 外部模式
external scheme 外部方案
external state 外部状态
external system 外部系统
external time management 外部时间管理
external transition 外部变换

external transition-function 外部转换函数
external validity 外部有效性
external variable 外部变量
externalized knowledge 外部化知识
externally activated multimodel 外部激活多模型
externally coupled 外部耦合的
externally generated goal 外部生成的目标
externally generated input 外部生成的输入
extract (v.) 提取
extraction 提取
extraneous variable 外部变量
extrapolation error 外推误差
extrapolation method 外推法
extrapolation technique 外推法，外推技术
extreme event 极端事件
extreme input testing 极端输入测试
extreme programming 极限编程
extreme scale computer 极限规模计算机
extreme scale computing 极限规模计算
extreme scale simulation 极限规模仿真
extrinsic model 外部模型
extrinsic parameter 外部参数
extrinsic simulation 外部仿真
extrinsic simulism 外部仿真主义

Ff

fabrication error 工艺误差
face-to-face learning 面对面的学习
face validation 外观验证
face validity 表面效度
face value 面值
facial animation markup language 面部动画标记语言
fact 事实
fact-based 基于事实的
fact-based decision making 基于事实的决策
fact detection 事实探测
factor 因素
factor screening 因素筛选
factorial design 因子设计
factual error 事实错误
failure 失效；失败；故障
failure analysis 故障分析；失效分析
failure avoidance 故障避免
failure avoidance paradigm 避障范式
failure detection 故障检测
failure detection mechanism 故障检测机制
failure insertion testing 故障插入测试
fair simulation 公平仿真
faithfully stable 切实稳定的
faithfully stable integration algorithm 稳定整合算法
fake input 假输入
fake news 假消息
fake reality 假现实
falcification 伪造
falcify (v.) 伪造
fallacies in logic 逻辑错误
fallacious 谬误的
fallacy 谬论

false 否；假的；错误
false alarm error 假警报错误
false error 虚假错误
false knowledge 虚假知识
falsifiability 可证伪性
falsifiable hypothesis 可证伪假设
falsification 伪造；证伪
falsify (v.) 伪造
falsity 虚假
family 家族
family of systems 系统家族
fantasy 幻想
fast simulation 快速仿真
fast simulation modeling 快速仿真建模
fast time 快速时间
fast time constant 快速时间常数
faster than real-time system 超实时系统
fatal error 致命误差
father model 父模型
fault 故障
fault analysis 故障分析
fault behavior 故障行为
fault detection 故障检测
fault insertion testing 错误插入测试
fault management 故障管理
fault simulation 故障仿真
fault tolerance 故障容限；容错
fault tolerant 容错的
fault-tolerant simulation 容错仿真
fault tree 故障树
faulty 有故障的
faulty behavior 故障的行为
faulty component 错误的组件；故障组件
faulty model 故障模型
faulty modeling assumption 错误的建模假设
faulty simulation 故障仿真
FD (Finite Deterministic) DEVS 有限确定性离散事件系统规范
FD-DEVS 有限确定性离散事件系统规范
feasibility 可行性
feasibility study 可行性研究
feasible 可行的
feasible reality 可行的现实
feasible state 可行状态
feasible value 可行的数值
feature 特征
feature volatility 特征波动性
FEDEP 联邦开发与执行过程
federate 联邦成员
federate cluster 联邦成员集群
federate composition 联邦成员组成
federate computer 联邦计算机
federate instantiation 联邦实例化
federate interface specification 联邦接口规范
federate interoperability 联邦成员的互操作性
federate markup 联邦标记
federate model 联邦成员模型
federate ownership management 联邦所有权管理
federate qualification 联邦成员资格认证
federate simulation 联邦仿真
federate simulation output data 联邦仿

真输出数据
federate time 联邦时间
federated 联邦的
federated simulation 联邦仿真
federated simulation architecture 联邦仿真体系结构
federated simulation environment 联邦仿真环境
federated simulation service 联邦仿真服务
federated simulation system 联邦仿真系统
federation 联邦
federation certification 联邦认证
federation component 联邦组件
federation design 联邦设计
federation design methodology 联邦设计方法学
federation development 联邦开发
federation element 联邦元素
federation exchange data 联邦交换数据
federation execution 联邦执行
federation execution control 联邦执行控制
federation execution data 联邦执行数据
federation execution sponsor 联邦执行发起者
federation functionality 联邦功能
federation initialization 联邦初始化
federation manager 联邦管理器
federation model 联邦模型
federation object model 联邦对象模型
federation object model template 联邦对象模型模版
federation objective 联邦目标
federation of federations 联邦联合
federation required execution detail 联邦所需的执行细节
federation reusability 联邦可重用性
federation reusable 联邦可重用的
federation reuse 联邦重用
federation structure transformation 联邦结构变换
federation time 联邦时间
federation time axis 联邦时间轴
feedback 反馈
feedback control connection 反馈控制连接
feedback coupling 反馈耦合
feedback linearization 反馈线性化
feedback loop 反馈环
fictitious reality 虚拟现实
fidelity 保真度
fidelity and resolution 保真度和分辨率
fidelity management 保真度管理
fidelity modeling 保真度建模
fiducial error 基准误差
fiducial localization error 基准点定位误差
fiducial registration error 基准注册错误
field 场；域
field behavior 场行为
field experiment 现场实验
field instrumentation 现场仪器
field markup language 场标记语言
field method 场方法

field-of-view 视场
field simulation 场仿真
field testing 现场测试
figurative comparison 比喻比较
file 文件
file creation error 文件生成错误
file error 文件错误
file management 文档管理
filing error 归档错误
filter 滤波器
filter (v.) 滤波
filtered data 过滤的数据
filtered value 过滤的值
filtering 滤波
filtering model 滤波模型
final condition 最终条件
final model 终结模型
final state 终结状态，终态，末态
final value 终值
fine-grain approach 细粒度方法
fine-grain granularity 细粒度
fine-grain model 细粒度模型
fine-grained component 细粒度的组件
finer-grained component 细粒度的组件
finite 有限
finite difference 有限差分
finite-difference approximation 有限差分近似
finite-difference method 有限差分法
finite-difference model 有限差分模型
finite element 有限元素
finite-element analysis 有限元分析
finite-element approximation 有限元近似
finite-element method 有限元法
finite-element model 有限元模型
finite game 有限对策
finite-memory DEVS 有限存储离散事件系统规范
finite model 有限的模型
finite-state automaton 有限状态自动机
finite-state automaton model 有限状态自动机模型
finite-state-machine 有限状态机
finite-state machine simulation 有限状态机仿真
finite-state system 有限状态系统
fire 开火；激发
fired rule 触发规则
firing 开火；激发
firing delay 发射延时
firing of transition 触发变迁
firing phase 触发阶段
firing sequence 触发序列
firmware 固件
firmware error 固件错误
first 首先
first-degree simulation 一级仿真
first-order accurate 一阶精度
first-order control 一阶控制
first-order cybernetics 一阶控制论
first-order effect 一阶效应
first-order model 一阶模型
first-order quantizer 一阶数字转换器
first-order synergy 一阶协同作用
fit 拟合
fit (v.) 拟合

fit indicator 拟合指标
fitness 适应度
fitness function 适应度函数
fitness to purpose 目标适用性
fitting 拟合
fixed error 固有误差
fixed input 固定输入
fixed metamodel 固定的元模型
fixed-point iteration 定点迭代
fixed-structure automaton 固定结构自动机
fixed time-step 固定时间步长
fixed-topology simulation 固定拓扑仿真
fixed value 固定值
fixed variable 固定变量
flat sequential mapping 扁平顺序映射
flat sequential simulator 扁平顺序仿真器
flattened conservative parallel simulator 扁平化保守并行仿真器
flattened G-DEVS simulation 扁平化广义离散事件系统规范仿真
flattened simulation 扁平化仿真
flattened simulation structure 扁平化仿真结构
flaw 缺陷
flawed assertion 有瑕疵的断言
flawed model 有瑕疵的模型
flexibility 柔性；灵活性
flexible dynamic model updating 柔性动态模型更新
flexible point 柔点
flexible system 柔性系统
flight training 飞行训练

flow 流
flow variable 流量变化
fluid simulation 流体仿真
focused attention 集中注意
fog computing 雾计算
folded Petri net 折叠佩特里网
following 以下；后面的
following event 后续事件
forbidden situation 被禁止的情境
forbidden transition 禁止越迁
forcast (v.) 预测
forcasting 预测的
forced event 强制事件
forced input 强制输入
forecast 预测
forecasting 预测的
forecasting model 预测模型
forehead retina 前额视网膜
forehead retina system 前额视网膜系统
formal 正式的；形式上的
formal aspect of model transformation 模型变换的形式方面
formal behavior 正式行为
formal composability 形式上的组合性
formal conceptual model 形式概念模型
formal context model 形式上下文模型
formal contextual model 形式情境模型
formal control 形式控制
formal definition 形式定义
formal evaluation 形式评价

formal language 形式语言
formal method 形式方法
formal-method approach 形式化方式方法
formal model 形式模型
formal model evaluation 形式模型评价
formal modeling 形式建模
formal modeling approach 形式建模方法
formal ontology 形式化本体
formal property 形式化特性
formal property verification 形式化特性校核
formal review 形式评审
formal review process 形式评审过程
formal semantic system 形式语义系统
formal specification 形式规范
formal system 形式系统
formal technique 形式化技术
formal variable 形式变量
formal verification 形式化校核
formal verification technique 形式化校核技术
formal VV&T technique 形式化校核、验证与测试技术
formalism 形式主义
formalism comparison 形式主义比较
formalism expressiveness 形式主义表现力
formalism extensibility 形式的可扩展性
formalism extension 形式扩展
formalism primitive 形式本原
formalism's expressive power 形式化表现力
formalism's granularity 形式主义粒度
formalism's strength 形式主义力量
formalization 形式化
formalization technique 形式化技术
format 格式
format (v.) 格式
format description layer 格式描述层
formation 编队；构成
formatting 格式化
formatting error 格式化错误
formatting layer 格式化层
formula 公式
formulated problem quality 构思出的问题质量
formulation 公式化
forward difference operator 前向差分算子
forward engineering 正向工程
forward Euler algorithm 前向欧拉算法
forward Euler integration 前向欧拉积分
forward multisimulation 前向多仿真
forward multisimulation gaming 正向多仿真博弈
forward reasoning 正向推理
forward-reasoning model 正向推理模型
foundation 基础
foundational knowledge 基础知识
Fourier analysis 傅里叶分析
fourth-order accurate 四阶精度
fractional error 相对误差；部分误差
fractional factorial design 部分析因

设计
fragmentation 碎片
frame 框架；帧
frame applicability 框架的适用性
frame base 框架库
frame-based 基于框架的
frame-model applicability 框架模型的适用性
frame rate 帧速率
framework 框架；架构
free-form game 自由形式博弈
free game 无规博弈
free-lock 无锁的
free variable 自由变量
freedom 自由
frequency 频率
frequency error 频率误差
frequency of documentation 文档频率
frequency plot 频率图
frequency ratio 频率比
frequent user 常用用户
front-end 前端
front-end analysis 前端分析
front-end interface 前端接口
FRT (Finite and Real-Time) DEVS 有限实时离散事件系统规范
FRT-DEVS 有限实时离散事件系统规范
full cosimulation 全协同仿真
full factorial design 完全析因设计
full-immersive simulation 完全沉浸式仿真
full order 全阶
full-order model 全阶模型
full-scale error 满量程误差

full-system simulation 全系统仿真
full validity 完全有效性
fully controllable plant 完全可控装置
fully controllable system 完全可控系统
fully coupled simulation 全耦合仿真
fully distinguishable system 完全可区分系统
fully-implicit Runge-Kutta algorithm 全隐式龙格-库塔算法
fully observable plant 完全可观工厂
fully observable system 完全可观系统
fun game 趣味游戏
function 函数；功能
function approximator 函数逼近
function modeling 功能建模
function value 函数值
functional 功能的
functional analysis 功能分析
functional area 功能区域
functional attribute 功能属性
functional data administrator 功能数据管理员
functional decomposition 功能分解
functional decomposition diagram 功能分解图
functional design space 功能设计空间
functional domain area 功能域区
functional evolution 功能演化
functional fidelity 功能保真度
functional game 功能游戏
functional interface 功能接口
functional model 功能模型
functional model attribute 功能模型属性

functional modeling 功能建模
functional process 功能过程
functional process improvement 功能过程的改进
functional property 功能特性
functional requirement 功能需求
functional similarity 功能相似性
functional simulation 功能仿真
functional structure 功能结构
functional system 功能系统
functional system design 功能系统设计
functional testing 功能测试
functional verification 功能校核
functionality 功能性
functionality encapsulated 功能封装的
functionality encapsulation 功能封装
functionally 功能上地
functionally correct 功能上正确的
fundamental principle 基本原理
fusion 融合
future behavior 将来的行为
future events list 未来事件列表
future trend 未来趋势
future value 将来值
fuzzificate (v.) 模糊化
fuzzificated 模糊化的
fuzzificated value 模糊值
fuzzification 模糊化
fuzzification method 模糊化方法
fuzzifier 模糊器
fuzzy 模糊的
fuzzy calculus 模糊演算
fuzzy causal directed graph 模糊因果有向图
fuzzy clustering defuzzification 模糊聚类去模糊化
fuzzy cognitive map 模糊认知图
fuzzy data 模糊数据
fuzzy decision 模糊决策
fuzzy-DEVS 模糊离散事件系统规范
fuzzy differential equation 模糊微分方程
fuzzy digraph 模糊有向图
fuzzy equation 模糊方程
fuzzy expert system 模糊专家系统
fuzzy finite-state-machine 模糊有限状态机
fuzzy function 模糊函数
fuzzy inductive reasoning 模糊归纳推理
fuzzy inference 模糊推理
fuzzy inference algorithm 模糊推理算法
fuzzy inference engine 模糊推理引擎
fuzzy inference network 模糊推理网络
fuzzy initial condition 模糊初始条件
fuzzy input 模糊输入
fuzzy membership function 模糊隶属度函数
fuzzy model 模糊模型
fuzzy modeling 模糊建模
fuzzy number 模糊数
fuzzy parameter 模糊参数
fuzzy Petri net 模糊佩特里网
fuzzy qualitative value 模糊定性值
fuzzy quantum state 模糊量子状态
fuzzy regression 模糊回归

fuzzy relation 模糊关系
fuzzy restriction 模糊约束
fuzzy rule 模糊规则
fuzzy rule-base 模糊规则库
fuzzy rule-base class 模糊规则库类
fuzzy rule-based model 基于模糊规则的模型
fuzzy rule class 模糊规则类
fuzzy rule object 模糊规则对象
fuzzy set 模糊集合
fuzzy set class 模糊集类
fuzzy simulation 模糊仿真
fuzzy state transition graph 模糊状态迁移图
fuzzy system 模糊系统
fuzzy system modeling 模糊系统建模
fuzzy system simulation 模糊系统仿真
fuzzy truth value 模糊真值
fuzzy value 模糊值
fuzzy variable 模糊变量

Gg

gain error 增益误差
game 博弈；游戏；对策
game against nature 反自然博弈
game architecture 游戏架构
game asset generation 游戏资产生成
game asset management 游戏资产管理
game-based 基于游戏的
game-based education technique 基于游戏的教育技术
game-based learning 游戏式学习
game-based tool 基于游戏的工具
game-based training 基于游戏的训练
game control room 博弈控制空间
game cycle 博弈周期
game design 游戏设计
game developer 游戏开发者
game development 游戏开发
game development curriculum 游戏开发路线
game development tool 游戏开发工具
game engine 游戏引擎
game-engine design 游戏引擎设计
game experience 博弈经验
game graph 博弈图
game-like simulation 博弈式仿真
game management 博弈管理
game mechanics 博弈机制
game of chance 机会游戏
game of life 生存博弈
game parameter 博弈参数
game-playing strategy 博弈策略；游戏比赛策略
game program 游戏程序
game programming 游戏程序设计
game project management 游戏项目管理
game reliability 博弈可靠性
game scenario 博弈场景
game scenario management 博弈场景管理
game security 博弈安全
game shell 游戏外壳
game simulation 博弈仿真
game software 游戏软件
game technology 游戏技术

game-theoretic approach 博弈论方法
game-theoretic simulation 博弈论仿真
game-theoretical 博弈论的
game-theoretical model 博弈论模型
game theory 博弈论；对策论
game tree 博弈树；对策树
game update 游戏更新
game with agenda 议程上的游戏
game with intelligent entity 与智能实体进行的博弈
gameable 可玩的
gamer 玩家
gameware 游戏软件
gamification 游戏化
gamified 游戏化的
gamified module 游戏化模块
gamified training 游戏化训练
gamify (v.) 游戏化
gaming 博弈
gaming association 对策关联
gaming community 游戏社区
gaming device 游戏设备
gaming education 游戏教育
gaming engine 游戏引擎
gaming experience 游戏体验
gaming for training 用于训练的游戏
gaming group 博弈群
gaming hardware 游戏硬件
gaming in entertainment 娱乐游戏
gaming multisimulation 博弈多仿真
gaming organization 博弈组织
gaming paradigm 博弈范式
gaming process 博弈过程
gaming simulation 博弈仿真

gaming technique 博弈技术
gamist 博弈论学者
gamma distribution 伽马分布
gamma random variable 伽马随机变量
Gane-Sarson data flow modeling 加恩-萨尔松数据流建模
gap 间隙；缺口
gap analysis 缺口分析
gate-level simulation 门级仿真
gated discipline 选通规则
gateway 网关；通道
Gaussian copula simulation 高斯copula仿真
Gaussian distribution 高斯分布
Gaussian distribution simulation 高斯分布仿真
Gaussian elimination 高斯消元法
Gaussian error 高斯误差
Gaussian random function simulation 高斯随机函数模拟
Gaussian random variable 高斯随机变量
Gaussian sequential simulation 高斯序列模拟
Gaussian simulation 高斯仿真
Gaussian variable 高斯变量
GDEVS 广义离散事件系统规范
gedanken simulation 理想化仿真
gedanken experiment 理想实验
general accreditation 一般认证
general boundary condition 一般边界条件
general equilibrium model 一般平衡模型

general error 一般性错误
general interoperability 一般互操作性
general M&S knowledge 一般性建模与仿真知识
general model 一般性模型
general purpose distributed simulation 通用分布仿真
general purpose simulation system 通用仿真系统
general purpose SL 通用仿真语言
general semantic graph 通用语义图
general simulation knowledge 通用仿真知识
general use M&S 通用建模与仿真
general use M&S application 通用建模与仿真应用
generalization error 泛化误差
generalized coordinate 广义坐标
generalized discrete-event 广义离散事件
generalized level set defuzzification 广义水平集去模糊化
generalized linear mixed model 广义线性混合模型
generalized linear model 广义线性模型
generalized mixed-mode simulation 广义混合模式仿真
generalized model 广义模型
generalized model simulation 通用模型仿真
generalized multimodeling methodology 广义多重建模方法学
generalized queue 广义队列
generalized simulation 通用仿真
generalized stochastic Petri net 广义随机佩特里网
generalized synchronization 广义同步
generalized variable 广义变量
generated 生成
generated reality 生成现实
generation 产生
generative 生成的
generative behavior 生成行为
generative model 生成模型
generative modeling 生成建模
generative multisimulation 生成的多种仿真
generative parallax simulation 生成视差仿真
generative simulation 生成仿真
generator 发生器
generic 泛化；一般的；通用的
generic abstraction 泛化抽象
generic agent framework 通用智能体框架
generic algorithm 通用算法
generic coupling template 通用耦合模版
generic domain 通用域
generic element 通用元素
generic error 一般错误
generic failure 一般故障
generic library 通用库
generic methodology 通用方法论
generic model 通用模型
generic model component template 通用模型组件模板
generic model template 通用模型模版
generic modeling environment 通用建

模环境
generic ontology　通用本体
generic simulation　通用仿真
generic simulation model　通用仿真模型
generic-type variable　通用类型变量
genetic algorithm　遗传算法
genetic-algorithm simulation　遗传算法仿真
genotype　基因型
genotypic modeling　基因型建模
genotypic representation　基因型表达
geometric distribution　几何分布
geometric model　几何模型
geometric modeling　几何建模
geometric parameter　几何参数
geometric random variable　几何随机变量
geometric shape modeling　几何形状建模
geometry　几何学
geopositional input　地理位置输入
GEST coupling　GEST 耦合
gesture　姿态；手势
gesture input　手势输入
gesture markup language　姿态标记语言
ghost model　虚幻的模型
GK-DEVS　几何学与运动学离散事件系统规范
glass box　玻璃箱
glass box model　玻璃箱模型
glass box testing　玻璃箱测试
global　全局的
global approximation　全局近似

global behavior　全局行为
global constraint　全局约束
global error　全局误差
global error bound　全局误差界
global information　全局信息
global integration error　全局积分误差
global knowledge base　全局知识库
global minimum time　全域最小时间
global model　全局模型
global optimization　全局优化
global optimization problem　全局最优化问题
global position sensing input　全球位置感知输入
global property　整体特性
global property of a system　系统的全局（综合）特性
global relative accuracy　全局相对精度
global relative error　全局相关误差
global schedular　全局时间表
global state-transition function　全局调度器
global time　全局时间
global time manager　全局时间管理器
global transition function　全局转移函数
global variable　全局变量
goal　目标
goal component　目标组件
goal-determined　目标确定的
goal-determined simulation　目标确定的仿真
goal-directed　目标导向的
goal-directed activity　目标导向的

活动
goal-directed agent 目标导向的智能体
goal-directed behavior 目标导向行为
goal-directed experiment 目标导向实验
goal-directed experimentation 目标导向实验
goal-directed knowledge processing 目标导向的知识处理
goal-directed multimodel 目标导向的多模型
goal-directed system model 目标导向的系统模型
goal-directed system simulation 目标导向的系统仿真
goal-driven 目标驱动的
goal-driven activity 目标驱动的活动
goal-driven decision 目标驱动的决策
goal-free simulation 无目标仿真
goal-generating system simulation 目标生成系统仿真
goal of the study 研究目标
goal-oriented system simulation 面向目标的系统仿真
goal parameter 目标参数
goal-processing system simulation 目标处理系统仿真
goal programming 目标规划
goal-programming method 目标规划方法
goal-regression simulation 目标回归仿真
goal-seeking simulation 目标搜索仿真

goal-seeking system 目标搜索系统
goal-setting system 目标设置系统
goal-setting system simulation 目标设置系统仿真
goal softening strategy 目标软化策略
goal state 目标状态
goal structure 目标结构
goal variable 目标变量
goal with fixed structure 固定结构目标
goal with variable structure 变结构目标
god game 上帝模拟游戏
golden section method 黄金分割法
good enough solution 足够好的解决方案
good modeling practice 好的建模实践
Goore game Goore游戏
GOTS (Government Off-the-Shelf) software 政府成品软件
government off-the-shelf model library 政府成品模型库
government off-the-shelf simulation library 政府成品仿真库
gradient search method 梯度搜索法
gradual validity 渐近有效性
Graeco-Latin square design 希腊拉丁方设计
grammar-based 基于语法的
grammar-based program generator 基于语法的程序生成器
grammatical error 语法错误
granularity 粒度
graph 图
graph-based 基于图的

graph-based model 基于图的模型
graph-based model transformation 基于图的模型转换
graph-based multimodel specification formalism 基于图的多模型规范形式
graph-based representation 基于图的表征
graph-based specification formalism 基于图的规范形式
graph-based transformation 基于图的变换
graph database 图数据库
graph-driven 图驱动的
graph-driven approach 图驱动的方法
graph-driven methodology 图驱动的方法论
graph model 图模型
graph primitive 图形原语
graph structure 图结构
graphic coupling 图形耦合
graphic-oriented SL 面向图形的仿真语言
graphic processing unit 图形处理单元
graphic representation 图形表达
graphic user interface 图形用户界面
graphical 图形的
graphical comparison 图形化的比较
graphical interface 图形界面
graphical model 图形模型
graphical modeling tool 图形建模工具
graphical object 图形对象
graphical object-oriented model 面向图形对象的模型
graphical simulation 图形仿真
graphical specification 图形规范
graphical tool 图形工具
graphics 图形学
graphics engine 图形引擎
graphing 图表；制图
gray box 灰箱
gray box approach 灰箱方法
gray box approach in modeling 灰箱建模法
gray box testing 灰盒测试
Greenwich mean-time 格林尼治标准时间
grid 网格
grid and cluster computing 网格与集群计算
grid-based 基于网格的
grid-based collaborative simulation environment 基于网格协同仿真环境
grid-based computing 基于网格的计算
grid-based environment 基于网格环境
grid-based simulation 基于网格的仿真
grid computing 网格计算
grid federation 网格联邦
grid resource driver 网格资源驱动器
grid router 网格路由器
grid simulation 网格仿真
grid simulation environment 网格仿真环境
grid simulation technology 网格仿真技术

grid simulation tool   网格仿真工具
grid topology   网格拓扑
grid width   网格宽度
ground truth   地面实况；地面真值
group   群；组
group behavior   群体行为
growth   增长
growth error   增长误差
GUI (Graphic User-Interface)   图形用户界面
guide   指南
guideline   指南
guise   外观；伪装
Gumbel random variable   Gumbel 随机变量

Hh

Hamiltonian   哈密顿函数
Hamiltonian field   哈密顿场
hand-gesture input   手势输入
hand simulation   手仿真
handheld game   掌上游戏
handheld gaming   掌上游戏
handler   处理程序；处理器
handling   处理
hands-on simulation   操作仿真
hands-on training   操作训练
haptic   触觉的
haptic controller   触觉控制器
haptic data   触觉数据
haptic device   触觉设备
haptic feedback   触觉反馈
haptic feedback hardware   触觉反馈硬件

haptic input   触觉输入
haptic interfacing   触觉接口
haptic perception   触觉
haptic sensation   触觉感知
haptic simulator   触觉仿真器
haptic technology   触觉技术
haptics   触觉论
hard constraint   触觉约束
hardcore gamer   硬核玩家
hard coupled   硬耦合的
hard coupling   硬耦合
hard error   硬错误
hard real-time constraint   硬实时约束
hard real-time system   硬实时系统
hard simulation   硬仿真
hard system thinking   硬系统思维
hardware   硬件
hardware development   硬件开发
hardware environment   硬件环境
hardware error   硬件错误
hardware implementation   硬件实现
hardware-in-the-loop   半实物；硬件在回路
hardware-in-the-loop simulation   半实物仿真；硬件在回路仿真
hardware installation error   硬件安装错误
hardware integrity   硬件完整性
hardware model   硬件模型
hardware simulator   硬件仿真器
hardware socket interface technique   硬件套接字接口技术
hard-wired   硬接线的
hard-wired data   硬连线数据
harmful error   有害误差

harmless error 无害化误差
head-down display 下视显示
head-down instrument 下视仪表
head-mounted display 头盔式显示器；头戴显式
head-up display 平视显示
head-up instrument 平视仪表
help 帮助；说明
hermeneutic 解释的
hermeneutics 解释学
Hessian 黑塞形式
Hessian matrix 黑塞矩阵
heterogeneity 异质性
heterogeneous 异质的
heterogeneous agent 异构智能体；异构代理
heterogeneous condition 异质条件
heterogeneous data 异构数据
heterogeneous distributed system 异构分布系统
heterogeneous model 异构型模型
heterogeneous network 异构网络
heterogeneous simulation 异构仿真
heterogeneous simulation technique 异构仿真技术
heterogeneous simulator 异构仿真器
heterogeneous system 异构型系统
heterogenous modeling formalism 异构建模形式
heterogenous multiscale method 异构多尺度方法
heteroscedastic error 异方差误差
heteroscedasticity 异方差性
Heun's integration method 霍因积分法

heuristic 启发式的
heuristic error 启发式误差
heuristic evaluation 启发式评价
heuristic knowledge 启发式知识
heuristic learning 启发式学习
heuristic method 启发式方法；试探法
heuristic procedure 启发过程
heuristic simulation 启发式仿真
heuristic value 启发式的数值
heuristics 启发法；试探法
hidden attribute 隐属性
hidden layer 隐含层
hidden Markov chain 隐马尔可夫链
hidden value 隐含值
hidden variable 隐含变量
hierarchic 分层的；递阶
hierarchical 递阶的；分层的；层次
hierarchical clustering 分层聚类
hierarchical colored Petri net 层次着色佩特里网
hierarchical complex system 递阶复杂系统
hierarchical complexity measure 层次复杂性测量
hierarchical concept 递阶概念
hierarchical coupling 层次化耦合
hierarchical decision 分层决策
hierarchical decomposition 层次分解
hierarchical DEVS 递阶离散事件系统规范
hierarchical DEVS simulator 层次结构离散事件系统规范仿真器
hierarchical federation 层次化联邦
hierarchical form 分层形式
hierarchical model 层次模型

hierarchical model calibration 递阶模型校正
hierarchical model composition 层次模型组合
hierarchical model structure 层次化模型结构
hierarchical modeling 层次化建模
hierarchical modular DEVS 递阶模块化离散事件系统规范
hierarchical multimodel 层次结构多模型
hierarchical organization 层次组织
hierarchical sequential mapping 层次结构顺序映射
hierarchical sequential simulator 层次结构顺序仿真器
hierarchical simulation 层次式仿真
hierarchical simulation engine 层次仿真引擎
hierarchical structure 层次结构；递阶结构
hierarchical structuring formalism 层次结构形式
hierarchically coupled 层次耦合的
hierarchy 层次结构
hierarchy of models 模型的层次结构
hierarchy of simulations 多仿真系统层次结构
high cognitive complexity 高认知的复杂性
high-cognitive-complexity individual 高认知复杂性个体
high-fidelity simulation 高保真的仿真
high-fidelity simulator 高保真仿真器

high granularity 高粒度
high-granularity model 高粒度模型
high-integrity system 高完整性系统
high level 高层次
high-level abstraction 高层抽象
high-level architecture 高层体系结构
high-level component 高层次的组件
high-level conceptual model 高级概念模型
high-level language 高级语言
high-level model 高级的模型
high-level modeling environment 高层建模环境
high-level Petri net 高级佩特里网
high-level simulation 高级仿真
high-level simulation environment 高级仿真环境
high-level stimulus 高级（层）激励
high-level symbolic knowledge 高级符号知识
high-level system model 高级系统模型
high-level tool 高级工具
high order 高阶
high-order algorithm 高阶算法
high-order effect 高阶影响
high-order forward-difference operator 高阶前向差分算子
high-order method 高阶方法
high-order model 高阶模型
high-order nonlinear system 高阶非线性系统
high-order Runge-Kutta algorithm 高阶龙格-库塔算法
high-order synergy 高阶协同

high-order system 高阶系统
high-performance computer 高性能计算机
high-performance computing 高性能计算
high-performance simulation 高性能仿真
high-performance simulation system 高性能仿真系统
high-resolution combat modeling 高分辨率战斗建模
high-resolution model 高分辨率模型
high-resolution modeling 高分辨率建模
high-resolution simulation 高分辨率仿真
high situational complexity 高情境复杂性
high-speed computer 高速计算机
high-speed data transfer 高速数据传送
high-speed experimentation 高速实验
high-speed simulation 高速仿真
higher-index model 高指数模型
higher-index problem 高指标问题
higher level 更高层次
higher-level model 更高水平模型
higher-order model 更高阶次模型
higher-order transformation 更高阶变换
higher-resolution model 更高分辨率模型
highly distributed system 强分布系统
highly interactive system 高交互系统
highly nonlinear system 强非线性系统
historic data 历史数据
historical data 历史数据
historical-data validation 历史数据验证
historical-data validity 历史数据有效性
historical overview of M&S 建模与仿真的历史回顾
historical simulation 历史仿真
historical test 历史测试
historical validity 历史有效性
history 历史
HLA (High-Level Architecture) 高层体系结构
HLA awareness 高层体系结构意识
HLA-based simulation 基于高层体系结构的仿真
HLA compatible 高层体系结构兼容的
HLA compliance 高层体系结构合格
HLA-compliance certification facility 高层体系结构合格的认证设施
HLA-compliance test 高程体系结构合格测试
HLA-compliant 高层体系结构相容的
HLA-compliant interface 高层体系结构相容的接口
HLA-compliant simulation 高层体系结构相容的仿真
HLA-compliant simulation study 高层体系结构相容的仿真研究
HLA-compliant simulation system 高层体系结构相容的仿真系统
HLA federation 高层体系结构联邦

HLA interface specification 高层体系结构接口规范
HLA parameter 高层体系结构参数
HLA protocol data 高层体系结构协议数据
HLA rule 高层体系架构规则
holacratic system 全息透镜系统
holarchic model 合弄结构模型
holarchy 合弄结构
holarctic modeling 泛北极区建模
holdout validation 保持验证
holistic behavior 整体行为
holistic concept 整体概念
holistic effect 整体效应
holistic macro-level behavior 整体宏观行为
holistic model 整体性模型
holistic simulation 整体仿真
holistic training 整体训练
holistic variable 整体变量
holographic display 全息显示
holographic simulation 全息仿真
holon 子整体
holon-based model 基于子整体的模型
holonic 整体的
holonic agent simulation 完整智能体仿真；完整代理仿真
holonic modeling 完整建模
holonic simulation 完整仿真
holonic system 完整系统
holonic system simulation 完整系统仿真
holonization 全息化
holonize (v.) 全息化
holonized 全息化的
holonomic constraint 完整约束
holonomic model 完整模型
holonomic system 完整系统
homeostatic variable 稳态变量
homogeneous 同质的
homogeneous agent 同构代理
homogeneous boundary condition 齐次边界条件
homogeneous condition 齐次条件
homogeneous equation 齐次方程
homogeneous model 齐次模型
homogeneous network 同构网络
homogeneous system 同构型系统；均匀系统
homology 同源
homomorph 同态
homomorphic 同态的
homomorphic mapping 同态映射
homomorphic model 同态模型
homomorphic relation 同型关系
homomorphism 同态
homoscedastic error 同方差误差
homoscedasticity 同方差性
homothetic 同形的
homothety 同形
host 主机
host computer 主计算机
host model 宿主模型
host simulation computer 主仿真计算机
host simulation system 主仿真系统
host variable 宿主变量
hot emotional trigger 激情触发
HPC simulation 高性能计算仿真

human 人
human-agent communication 人-智能体通信
human-agent interaction 人-智能体交互
human-agent system 人-智能体系统
human behavior 人类行为
human behavior modeling 人类行为建模
human behavior representation 人类行为表示（法）
human-centered 以人为中心的
human-centered activity 以人为中心的活动
human-centered environment 以人为中心的环境
human-centered process 以人为中心的过程
human-centered simulation 以人为中心的仿真
human-centered system 以人为中心的系统
human-centric 以人为中心的
human-centric approach 以人为中心的方法
human-centric model 以人为中心的模型
human-computer interface 人机界面
human-defined activity 人为定义的活动
human error 人为误差
human factors 人为因素
human-in-the-loop 人在环，人在回路
human-in-the-loop simulation 人在环（回路）仿真
human-in-the-loop simulator 人在环（回路）仿真器
human-in-the system 人在系统
human-initiated simulation 人为仿真
human-interpretable descriptive model 人类可解释的描述模型
human-machine interface 人机界面
human-machine simulation 人机交互仿真
human-made system 人造系统
human modeler 人体建模器
human modeling 人体建模
human perception 人的感知
human performance measurement 人的功效测量
human-readable model 人类可读的模型
human-simulation interaction 人与仿真交互
human transcription error 人类转录错误
human view 人类视角
hybrid 混合
hybrid agent 混合型主体
hybrid automaton 混合自动机
hybrid computer 混合计算机
hybrid computer simulation 混合计算机仿真
hybrid continuous-system SL 混合连续系统仿真语言
hybrid factorial design 混合析因设计
hybrid gaming simulation 混合博弈仿真
hybrid model 混合模型
hybrid optimization 混合优化

hybrid reality 混合现实
hybrid simulation 混合仿真
hybrid simulation language 混合仿真语言
hybrid simulation model 混合仿真模型
hybrid simulation software 混合仿真软件
hybrid simulation study 混合仿真研究
hybrid simulation system 混合仿真系统
hybrid simulator 混合仿真器
hybrid SL 混合仿真语言
hybrid source SL 混合源仿真语言
hybrid synchronization 混合同步
hybrid system 混合系统
hybrid systemic solution 混杂系统解
hyper-federation 超级联邦
hyper-heuristics 超启发式
hyperactive step-size adjustment strategy 动步长调整策略
hyperbolic partial differential equation 双曲线偏微分方程
hyperbolic PDE 双曲型偏微分方程
hypergame 超对策
hypergame theory 超对策理论
hypergeometric distribution 超几何分布
hypergeometric random variable 超几何随机变量
hyperreality 超现实
hypertext-supported documentation 超文本支持文档
hypervariable 超变量
hypervariable data 超变量数据

hypervariate 超变量
hypotheses space 假设空间
hypothesis 假设
hypothesis error 假设误差
hypothesis test 假设检验
hypothesis testing 假设检验
hypothesis validity 假说有效性
hypothesize (v.) 假设
hypothetical event 假设事件
hypothetical proposition 假设命题
hypothetical system 假设系统
hysteresis 滞后
hysteresis error 滞后误差
hysteresis width 滞环宽度
hysteretic 滞后的
hysteretic quantization 滞后量化
hysteretic quantization function 迟滞量化函数

Ii

I/O 输入/输出
I/O-based 基于输入/输出的
I/O-based model 基于输入/输出的模型
I/O-based system model 基于输入/输出的系统模型
icon 图标
iconic model 图像模型
ideal point model 理想点模型
ideal-seeking simulation 理想搜索仿真
idealized model 理想化模型
identical marking 完全相同的标记
identifiable 可辨识的

identifiable system 可辨识系统
identification 辨识
identification error 辨识误差
identified 被辨识的
identified model 被辨识的模型
identified parameter 被辨识的参数
identified state 辨识的状态
identified state-variable 辨识的状态变量
identified structure 被识别的结构
identified system 已识别的系统
identify (v.) 确定；识别
identity 身份
identity simulation 特性仿真
identity simulation game 特性仿真游戏
illegal 非法的
illegal access error 非法进入错误
illegal error 非法错误
illegal seek error 非法查找错误
illusion 错觉
image 图像
image-based 基于图像的
image-based model 基于图像的模型
image-based modeling 基于图像的建模
image generation 图像生成
image interpolation 图像插值
imagery 意象
imaginary 虚构的
imaginary axis 虚轴
imaginary experience 虚构的经验
imaginary part 虚部
imaginary state 虚态
imagined reality 想象的现实

imagistic simulation 意象仿真
imitated reality 模仿现实
imitation 模拟
imitation-based 基于模仿的
imitation-based gaming 基于模仿的博弈
imitation perspective 模仿视角
imitative representation 模仿表现
immature model 不成熟的模型
immediate transition 瞬时变迁
immediate transition priority rule 即时过渡优先规则
immersion 沉浸
immersive 沉浸式的
immersive environment 浸入式环境
immersive experience 沉浸式体验
immersive learning experience 沉浸式学习体验
immersive modeling 沉浸式建模
immersive simulation 沉浸式仿真
immersive training 沉浸式训练
immersive visualization 沉浸式可视化
imminent event 临近事件
impact 影响；碰撞
impact modeling 碰撞建模；影响建模
impact of simulation 仿真影响
impact time 碰撞时间
imperative 命令的；命令式
imperative-driven 命令驱动的
imperative-driven simulation 命令驱动的仿真
imperfect model 有缺陷的模型
implement (v.) 实施；实现
implementability 可实现性

implementability of simulation recommendation 仿真推荐的可实现性
implementable design 可实施设计
implementable design space 可实施的设计空间
implementation 实现
implementation dependent 依赖于实现
implementation independent 自主实现
implementation-independent model 独立执行的模型
implementation mechanism 实现机制
implementation option 可执行选项
implementation phase 实现阶段
implementation platform 实施平台
implementation verification 实施校验
implemented 实施的
implemented object 实现的对象
implications of achieving the goal 实现目标的含意
implicit 隐式的
implicit assumption 隐性的假设
implicit boundary condition 隐边界条件
implicit constraint 隐式约束
implicit coordination 隐式协调
implicit dependency 暗含依存关系
implicit event 隐式事件
implicit extrapolation method 隐式外推法
implicit goal 隐性目标
implicit integration 隐式积分
implicit integration algorithm 隐式积分算法
implicit interaction 隐式交互
implicit knowledge 隐性知识
implicit method 隐式方法
implicit midpoint rule 隐式中点规则
implicit Milne method 隐式密尔恩法
implicit single-step method 隐式单步法
implicit solution 隐式解
implicit state 隐式状态
implicit value 隐含值
importance sampling 重要性采样
importance-sampling-based 基于重要性采样的
importance-sampling-based simulation 基于重要性采样的仿真
importance sampling-based simulation method 基于重要性采样的仿真方法
importance-sampling-based simulation technique 基于重要性采样的仿真技术
importance-sampling estimator 重要性采样估计器
imposed 施加的
imposed goal 施加的目标
imposed input 强加性输入
imprecision 不精确
improved model 改进的模型
improved simulation algorithm 改进的仿真算法
improvement 改进
impulse 脉冲
impulse variable 脉冲变量
in-basket simulation 公文处理法仿真
in context simulation 上下文仿真

in silico 硅片上的，计算机中的
in silico analog 计算机模拟
in silico data 电子数据
in silico experience 电子经验
in silico experiment 计算机实验
in silico experimental set-up 计算机实验设置
in silico experimentation 计算机实验
in silico model 计算机模型
in silico observation 计算机观测
in silico set-up 计算机设置
in silico simulation 计算机仿真
in situ 在原位的
in situ analog 原位模拟
in situ data 原位数据
in situ experience 现场经验
in situ experiment 现场试验；原位实验
in situ experimental set-up 现场实验装置
in situ experimentation 现场试验；原位实验
in situ model 原位模型
in situ observation 就地观测
in situ simulation 原位仿真
in the large simulation 在大型仿真中
in the small simulation 在小型仿真中
in vitro 体外的
in vitro analog 在体外模拟
in vitro data 体外数据
in vitro experience 体外体验
in vitro experiment 体外试验
in vitro experimental set-up 体外实验装置
in vitro experimentation 体外试验
in vitro model 体外模型
in vitro observation 体外观察
in vitro set-up 体外装置
in vitro simulation 体外仿真
in vivo 体内
in vivo analog 体内模拟
in vivo data 体内数据
in vivo experience 体内体验
in vivo experiment 体内试验
in vivo experimental set-up 体内实验装置
in vivo experimentation 体内试验
in vivo model 体内模型
in vivo observation 体内观察
in vivo set-up 体内装置
in vivo simulation 体内仿真
inaccessible prediction 无法获取的预测
inaccuracy 不准确
inaccurate 不准确的
inaccurate model 不准确的模型
inaccurately 不准确地
inactive behavior 不活跃的行为
inactive model 闲置的模型
inactive state 不活动状态
inactivity 闲置
inadequate data 不当数据
inadequate model 不完全模型
inadequate stimulus 不充分激励
inadvertent error 疏忽造成的误差
inappropriate simulation 非适宜仿真
incompatibility principle 不相容性原理
incompatible formalism 不兼容形式
incompatible interface 不兼容接口

incompatible property　不兼容属性
incomplete data　不完整数据
incomplete information　不完整信息
incomplete model　未完成的模型
inconsistent emergence　不一致涌现
inconsistent emergent behavior　不一致的涌现行为
inconsistent formula error　不一致公式误差
inconsistent model　不一致的模型
incorrect　错误的
incorrect assumption　错误的假设
incorrect output　错误输出
incorrect prediction　错误的预测
incorrect result　不正确结果
increased value　增量
incredible solution　不可信解
incremental　增加的
incremental energy　增加的能量
incremental simulation　增量仿真
incremental transformation　增量变换
indecomposable system　不可分的系统
independence　独立性
independent　独立的
independent assumption　独立的假设
independent state variable　独立状态变量
independent time advancement　独立的时间推进
independent V&V　独立的校核与验证
independent variable　自变量，独立变量
independent verification and validation　独立的校核与验证

indeterminate state　不确定的状态
index　指数；索引
index error　索引错误
indexed　索引的
indicator　指示器
indicator variable　指示变量
indirect communication　间接的通信
indirect coordination　间接协调
indirect coupling　间接耦合
indirect input　间接输入
indirect input device　间接输入设备
indirect interaction　间接交互
indirect measurement technique　间接测量技术
indirect simulation　间接仿真
indirect understanding　间接理解
indirectly coupled　间接耦合的
individual　个体
individual-based model　基于个体的模型
individual-based modeling　基于个体的建模
individual-based simulation　基于个体的仿真
individual behavior　个体的行为
individual control　个体控制
individual holon　单个子整体
individual model　个体模型
individual simulation　个体仿真
individual variable　个体变量
induced　诱导；感应的
induced emergence　诱导出现
induction　诱发；归纳；感应
induction variable　归纳变量
inductive　归纳的；感应的；诱导的

inductive assertion 归纳断言
inductive method 归纳法
inductive modeling 归纳建模
inductive reasoning 归纳推理
inductive simulation 归纳法仿真
inductive simulation proof 归纳仿真证明
industrial classification code 工业化分类代码
industrial scale 工业规模
industrial-scale simulation 工业规模仿真
industry 产业
industry fragmentation 产业断裂
industry identity 行业认同
industry scope 工业领域
ineffective decision 无效决策
inessential game 非本质博弈
inference 推论
inference engine 推理引擎
inference rule 推理规则
infinite game 无限对策，无限博弈
infinite-state system 无限状态系统
infinite stiffness 无限刚度，无穷刚度
inflow 流入
influence diagram 影响图
info trace 信息追踪
infohabitant 信息居民
informal conceptual model 非正式化概念模型
informal definition 非正规定义
informal model 非正式模型
informal model description 非正式模型描述
informal specification 非正式化描述

informal technique 非正式化方法
informal VV&T technique 非正式化校核、验证与测试方法
informatics 信息论
information 信息
information animation 信息动画
information asymmetry 信息不对称
information exchange 信息交换
information hiding 信息隐藏
information management 信息管理
information model 信息模型
information processing 信息处理
information processing model 信息处理模型
information resource dictionary 信息资源词典
information resource dictionary system 信息资源词典系统
information source 信源，信息源
information system 信息系统
information systems modeling 信息系统建模
information technology 信息技术
information visualization 信息可视化
information warfare 信息战
informative data 信息数据
infrastructure 基本结构；基础设施
infrastructure perspective 基础设施的视角
infrastructure system 基础设施系统
inherit ($v.$) 继承
inheritance 继承
inherited error 继承的误差
inhibitor Petri net 抑制佩特里网
inhibitory arc 抑制弧

inhomogeneous boundary condition 非齐次边界条件
inhomogeneous system 非齐次系统
initial 初始的
initial condition 初始条件
initial marking 初始标记
initial section 起始段
initial state 初始状态，初态
initial transient 初始瞬值
initial value 初始值
initial-value problem 初值问题
initialization 初始化
initialization condition 初始化条件
initialization error 初始化误差
initialization mechanism 初始化机制
initialization of random number seed 随机数种子初始化
initialization of state variable 状态变量初始化
initialization phase 初始化阶段
initialization section 初始化部分
initialization value 初始化值
initialize (v.) 初始化
initialized variable 初始化的变量
inline integration 内联集成
inlining implicit Runge-Kutta algorithm 内联隐式龙格-库塔算法
inlining partial differential equation 内联偏微分方程
innovative simulation approach 新型仿真方法
innovative simulation technique 新型仿真技术
innovative simulation tool 新型仿真工具

input 输入
input acceptability 输入的可接受性
input angle 输入角
input area 输入区
input assertion 输入断言
input assessment 输入评估
input axis 输入轴
input block 输入块
input buffer 输入缓冲区
input buffer register 输入缓冲区寄存器
input channel 输入通道
input characteristic 输入字符
input condition 输入条件
input control 输入控制
input data 输入数据
input-data analysis 输入数据分析
input device 输入设备
input distribution 输入分布
input-driven validation 输入驱动的验证
input equation 输入方程
input equipment 输入设备
input event 输入事件
input file 输入文件
input format 输入格式
input formatting 输入格式化
input instruction 输入指令
input inverter 输入逆变器
input limit 输入限制
input limited 输入限制的
input limiter 输入限制器
input linearity 输入线性度
input loading 输入加载
input log 输入记录

input matrix　输入矩阵
input media　输入媒体
input message　输入消息
input node　输入节点
input observation port　输入观察端口
input/output　输入/输出
input/output analysis　输入/输出分析
input/output buffer　输入/输出缓冲区
input/output bus　输入/输出总线
input/output channel　输入/输出通道
input/output control　输入/输出控制
input/output control system　输入/输出控制系统
input/output controller　输入/输出控制器
input/output coprocessor　输入/输出协处理器
input/output coupling　输入/输出耦合
input/output device　输入/输出设备
input/output diagram　输入/输出示意图
input/output error　输入/输出误差
input/output instruction　输入/输出指令
input/output interrupt　输入/输出中断
input/output matrix　输入/输出矩阵
input/output model　输入/输出模型
input/output pair　输入/输出对
input/output port　输入/输出端口
input/output port specification　输入/输出端口规范
input/output process　输入/输出处理
input/output processor　输入/输出处理器
input/output register　输入/输出寄存器

input/output requirements　输入/输出要求
input/output routine　输入/输出程序
input/output specification　输入/输出规范
input/output statement　输入/输出语句
input/output switching　输入/输出转换
input/output system　输入/输出系统
input/output table　输入/输出表
input port　输入端口
input power　输入功率
input preparation　输入准备
input process　输入过程
input program　输入程序
input quantization error　输入量化误差
input queue　输入队列
input range　输入范围
input rate　输入率
input reader　输入阅读器
input restriction　输入限制
input section　输入段
input segment　输入片段
input sensitivity　输入灵敏度
input signal　输入信号
input space　输入空间
input span　输入跨度
input specification　输入规格
input specification ease　输入规格简易
input station　输入站
input storage　输入存储
input stream　输入流
input synchronization　输入同步

input terminal 输入终端
input threshold 输入阈值
input trajectory 输入轨迹
input unit 输入装置；输入单元
input validation 输入确认
input validity 输入有效性
input value 输入值
input value set 输入值集合
input variable 输入变量
input vector 输入向量
input verification 输入校核
inscription error 铭文标定误差
insightful model 富有洞察力的模型
insignificant event 不重要事件
inspect (v.) 检验
inspection 检查
inspirational reality 鼓舞人心的现实
instability 不稳定
installability 可安装性
installation 安装
installation error 安装误差
installer error 安装程序错误
instance 实例
instance data 实例数据
instance variable 实例变量
instantiated 实例化
instantiated model 实例化的模型
instantiated operation 实例化操作
instantiated variable 实例化变量
instantiation 实例化
instantiation operation 实例化操作
instruction 指令
instructional simulation 教学模拟
instructional system 教学系统
instructional system design 教学系统设计
instrument 仪器
instrument error 仪表误差
instrumentable variable 可测量变量
instrumental error 仪器误差
instrumental variable 辅助变量
instrumentation 仪器，仪表
instrumentation error 仪器误差
instrumentation system 测量系统
instrumented 仪表化的
instrumented entity 仪表化的实体
instrumented model 仪表化的模型
instrumented variable 辅助化的变量
integer value 整数值
integer-valued variable 整数值变量
integer variable 整型变量
integral equation 积分方程
integral problem 积分问题
integrate (v.) 集成
integrated architecture 集成架构
integrated environment 集成环境
integrated model 集成模型
integrated modeling 集成建模
integrated modeling and simulation framework 集成建模和仿真框架
integrated simulation 一体化仿真
integration 综合，集成；积分
integration accuracy 积分精度；集成精度
integration algorithm 积分算法；集成算法
integration cost 整合费用；集成费用
integration data 集成数据
integration error 积分误差；集成误差
integration framework 集成框架

integration method 积分法；集成方法
integration step size 积分步长
integration variable 积分变量
integrative modeling 综合建模；一体化建模
integrative modeling methodology 综合建模方法学；一体化建模方法学
integrative multimodeling 一体化多重建模
integrator 积分器
integrator equation 积分方程
integrity 完整性
integrity checking 完整性检查
integrity grid 完整性网格
intelligence 智能
intelligence community coordinating group 智能社会协调群
intelligent agent 智能代理
intelligent behavior 智能行为
intelligent complex adaptive system 智能的复杂适应系统
intelligent control 智能控制
intelligent entity 智能实体
intelligent game 智能博弈
intelligent hybrid systemic solution 智能混合系统解
intelligent interactive system 智能交互系统
intelligent interface 智能接口
intelligent model 智能化模型
intelligent simulation 智能仿真
intelligent simulation environment 智能仿真环境
intelligent system 智能系统
intelligent-system simulation 智能系统仿真
intelligent systemic solution 智能系统解
intelligent user interface 智能用户界面
intelligibility 可理解性
intended behavior 预期行为
intended consequence 预期后果
intended emerging behavior 预期出现的行为
intended outcome 预期结果
intended reality 预期现实
intensive 密集的
intensive variable 强度变量
intent model 意图模型
intent modeling 意图建模
intentional coupling 有意的耦合
intentional knowledge processing 有意图的知识处理
intentional state 意图状态
inter-action 交互作用
inter-agent 智能体间；代理间
inter-agent communication 智能体间的通信；代理间的通信
inter-cloud federation 云间联邦
inter-holonic cooperation 跨整体合作
interacting agent 交互代理
interacting entity 交互实体
interaction 交互，互动
interaction-based simulation 交互式仿真
interaction-based system 交互式系统
interaction capability 交互能力
interaction coherence 交互一致性
interaction graph 交互图

interaction model 交互模型
interaction modeling 交互建模
interaction parameter 交互参数
interaction point 交互点
interaction structure 交互性结构
interaction style 交互方式
interaction variable 交互变量
interactional complexity 交互复杂性
interactional simplicity 交互简化
interactive 交互式的
interactive behavior 交互性的行为
interactive capability 交互能力
interactive complexity measure 交互复杂性测量
interactive computation 交互式计算
interactive computer-based training 基于计算机的交互式训练
interactive demonstration 交互示范
interactive environment 交互环境
interactive experience 交互式体验
interactive game 交互式游戏
interactive graphical simulation 交互式图形化仿真
interactive help 交互式帮助
interactive intent modeling 交互式意图建模
interactive model 交互的模型
interactive modeling 交互式建模
interactive-movie-based learning 基于交互式电影的学习
interactive plot 交互绘图
interactive reality 交互现实
interactive simulation 交互式仿真
interactive simulation environment 交互式仿真环境
interactive simulation game 交互式仿真博弈
interactive simulation model 交互式仿真模型
interactive SL 交互式仿真语言
interactive system 交互式系统
interactive training 交互式训练
interactive transition system 交互式转换系统
interactive virtual reality 交互式虚拟现实
interactive visualization 交互式可视化
interchange 交换
interchange model 交换模型
interchange standard 交换标准
intercloud 云际；互联云
interconnected model 互联的模型
interconnectedness 连通度
interconnection 互连
interconnection technology 互连技术
interdisciplinary 跨学科
interdisciplinary modeling 学科交叉建模；跨领域建模
interdisciplinary topic 跨学科的主题
interest 收益
interest management 兴趣管理
interesting variable 兴趣变量
interface 接口；界面
interface agent 接口代理
interface analysis 界面分析
interface attribute 接口属性
interface component 接口组件
interface definition 接口定义
interface functionality 接口功能

interface issue 接口问题
interface layer 接口层
interface module 界面模块；接口模块
interface specification 接口规范
interface specification language 接口规范语言
interface technique 接口技术
interface technology 接口技术
interface testing 接口测试；界面测验
interfacing 接口，接口技术
interference 干扰
interlanguage 中介语
interlanguage mapping 人工辅助语言映射
interlanguage model mapping 人工辅助语言模型映射
intermediate error 中间误差
intermediate input 中间输入
intermediate model 中间模型
intermediate state 中间状态
intermediate value 中间值
intermediate variable 中间变量
intermittent 断断续续的，间歇的
intermittent generation 间断生成
intermittent simulation 间歇式仿真
intermittent SL 间歇式仿真语言
intermodular coupling 组件之间的耦合
internal activity 内在活动，内部活动
internal agent 内部智能体；内部代理
internal characteristic 内部的特性
internal completeness 内部完整性
internal consistency 内部一致性
internal coupling 内部耦合
internal error 内部误差

internal event 内部事件
internal evolution 内部演化
internal excitation 内部激励
internal goal 内部目标
internal inactivity 内部闲置
internal input 内部输入
internal model 内部的模型
internal parameter 内部参量
internal performance evolution 内部性能演变
internal scheme 内部方案
internal simulator 内部仿真器
internal state 内部状态
internal structure 内部构件
internal tacit model 内隐模型
internal time management 内部时间管理
internal transition 内部变换
internal transition function 内部转移函数
internal understanding 内部理解
internal validity 内部有效性
internal variable 内变量
internally activated multimodel 内部激活多模型
internally coupled 内部耦合的
internally generated goal 内部产生的目标
internally generated hypothesis 内部产生的假设
internally generated input 内部生成的输入
internally generated question 内部产生的问题
internally valid 内部有效的

international confederation 国际的联盟
Internet-based 基于互联网的
Internet-based data collection 基于因特网的数据收集
interoperability 互操作性
interoperability layer 互操作层
interoperability level 互操作性水平
interoperability model 互操作性模型
interoperability service 互操作服务
interoperability of simulation model 仿真模型的互操作性
interoperability standard 互操作标准
interoperable 可互操作的
interoperable federation 可互操作的联邦
interoperable game 可互操作的博弈
interoperable gaming 可互操作的博弈
interoperable model 可互操作的模型
interoperable simulation 可互操作的仿真
interoperable simulation environment 可互操作的仿真环境
interoperable simulation model 可互操作的仿真模型
interoperable war game 可互操作的战争游戏
interoperable war gaming 可互操作的战争游戏
interoperate (v.) 互操作
interoperation 互操作
interoperation adaptor 互操作适配器
interpersonal skill 人际技巧
interpolated variable 插值变量
interpolation 插值
interpolation error 插值误差
interpret (v.) 解释；说明
interpretation 释义
interpretation error 解释误差
interpretation model 解释模型
interpretational simulation 解释仿真
interpreted 解释
interpreted model 解释模型
interpretive simulation 解释型仿真
interpretive SL 解释型仿真语言
interpretivism 解释主义
interpretivist 解释主义者
interpretivist methodology 解释主义方法论
interpretivist paradigm 解释主义范式
interpretivist theory 解释主义理论
interrelated model 关联的模型
interrupt 中断
interrupt (v.) 中断；干扰
intersection entity 交叉实体
intersimulation data 交互仿真数据
interval 区间；间隔
interval-oriented 面向区间的
interval-oriented simulation 面向区间的仿真
interval scale 区间尺度
interval-scale variable 区间尺度变量
interval tolerance 区间公差
interval value 区间值
interval variable 区间变量
intervening variable 中介变量
intra-action 内在互动
intra-holonic cooperation 全方位合作，全息合作

intractable equation 难解方程
intractable simulation 难解仿真
intralanguage 内部语言
intralanguage mapping 内部语言映射
intralanguage model mapping 内部语言模型映射
intrasimulation data 内部仿真数据
intrinsic error 基本误差；固有误差
intrinsic model 本质模型
intrinsic simulation 内在仿真；本征模拟
intrinsic simulism 内在仿真
intrinsic value 内在价值
introductory phase 引入阶段
introspection 内省
introspection-based coupling 基于内省的耦合
introspective model 自省模型
introspective simulation 内省仿真
introspective simulation model 内省仿真模型
introspective system 内省系统
intrusive method 侵入法
invalid 无效的
invalid input testing 无效输入测试
invalid simplification 无效简化
invalidity 无效
invariable 不变的；常数
invariance 不变性
invariance property 不变性
invariant 不变的；不变量；不变条件
invariant embedding 不变嵌入
inverse 逆
inverse behavior 逆行为
inverse cubic polynomial 逆三次多项式
inverse Hermite interpolation 反赫米特插值
inverse Hessian 逆黑塞矩阵
inverse logit 逆分对数
inverse modeling 逆向建模
inverse multibody system dynamics 逆多体系统动力学
inverse ontomimetic simulation 逆模拟仿真
inverse problem 逆问题
inverse transformation 逆变换
inversion 逆
invert (v.) 反转
inverted input 反相输入
inverter 反相器
inverting input 反相输入
investigate (v.) 调查
investigation 调查
investment game 投资博弈
involuntary emotional behavior 不由自主的情绪行为
irrecoverable error 不可更正的错误
irreducible uncertainty 不可减少的不确定性
irreflexive constraint 非自反约束
irreflexive relationship 非自反关系
irregular domain 不规则论域
irregular error 不规则误差
irregular tessellation 不规则曲面细分
irregular variable 不规则变量
irrelevant 不相关的
irrelevant data 无关数据
irrelevant information 无关信息
irrelevant input 不相干输入

irrelevant model 不相干模型
irrelevant processing 非相关处理
irrelevant stimulus 不相干激励
irrelevant variable 不相干的变量
isomorph 同构
isomorphic 同构的
isomorphic mapping 同构映射
isomorphic model 同构的模型
isomorphic phenomenon 同构现象
isomorphic relation 同构关系
isomorphism 同构
issue 问题
iteration 迭代
iteration connection 迭代连接
iteration loop 迭代循环
iteration scheme 迭代方案
iteration variable 迭代变量
iterative method 迭代法
iterative modeling 迭代建模
iterative modeling methodology 迭代建模方法学
itinerant agent 流动智能体；流动代理

Jj

Jacobi linearization 雅可比线性化
Jacobian matrix 雅克比矩阵
JMASS (Joint M&S System) 联合建模与仿真系统
joint experiment 联合试验
joint experimentation 联合试验
joint M&S 联合建模与仿真
joint M&S executive panel 联合建模与仿真执行程序面板
joint M&S investment plan 联合建模与仿真投资计划
joint M&S proponent 联合建模与仿真提议者
joint M&S system 联合建模与仿真系统
joint multiresolution modeling 联合多分辨率建模
joint simulation 联合仿真
joint simulation system 联合仿真系统
joint synthetic battlespace 联合合成作战空间
joint training confederation 联合训练联盟
joint warfare system 联合作战系统
joint warfare training 联合作战训练
JSIMS (Joint Simulation System) 联合仿真系统
judgment error 判断误差
judgmental data 判断数据
jump discontinuity 跳跃不连续性
justified assumption 经证明的假设
justified belief 正当信念
justified reliance 正当信赖
JWARS (Joint Warfare System) 联合作战系统

Kk

Keller-Segel model Keller-Segel 模型
kernel function 核函数
kernel game 内核游戏
key 关键
key-process variable 关键过程变量
key value 关键值
key variable 关键变量

keyboard emulator 键盘仿真器
keyboard input 键盘输入
keyboard simulator 键盘仿真器
kinematic model 动态模型
kinetic model 运动模型
kinetic variable 动力学变量
Kiviat chart 基维亚特图表
knowledge 知识
knowledge acquisition 知识获取
knowledge area 知识领域
knowledge attributed 知识属性
knowledge-attributed Petri net 具有知识属性的佩特里网
knowledge base 知识库
knowledge-based 基于知识的
knowledge-based approach 基于知识的方法
knowledge-based assessment 基于知识的评估
knowledge-based decision 基于知识的决策
knowledge-based dynamic simulation composition 基于知识的动态仿真组合
knowledge-based experimentation support 基于知识的实验支持
knowledge-based MAS 基于知识的多代理系统
knowledge-based model-processing support 基于知识的模型处理支持
knowledge-based modeling support 基于知识的建模支持
knowledge-based program generation support 基于知识的程序生成支持
knowledge-based simulation 基于知识的仿真
knowledge-based simulator 基于知识的仿真器
knowledge-based support 基于知识的支持
knowledge-based system 基于知识的系统
knowledge-based visualization 基于知识的可视化
knowledge capture stage 知识获取阶段
knowledge compendium tool 知识概要工具
knowledge discovery 知识发现
knowledge elaboration 知识提炼
knowledge-generation activity 知识产生活动
knowledge horizon 知识视野；知识面
knowledge infrastructure 知识的基本结构
knowledge-intensive MAS 知识密集型多主体系统
knowledge-intensive system 知识密集型系统
knowledge management perspective 知识管理视角
knowledge-processing activity 知识处理活动
knowledge reconstruction 知识重构
knowledge transducer 知识转换器
knowledgeable behavior 知识渊博的行为
known object 已知的对象
known state 已知的状态
known value 已知的值

known variable 已知变量
Kriging 克里格
Kriging assumption 克里格假设
Kriging equation 克里格方程
Kriging error 克里格误差
Kriging interpolation 克里格插值
Kriging interpolation in simulation 仿真中的克里格插值
Kriging metamodel 克里格元模型
Kriging metamodeling 克里格元建模
Kriging metamodeling in discrete-event simulation 离散事件仿真中的克里格元建模
Kriging model 克里格模型
Kriging model algorithm 克里格模型算法
Kriging model optimization 克里格模型优化
Kriging model selection 克里格模型选择
Kriging model validation 克里格模型验证
Kriging modeling 克里格建模
Kriging modeling for global approximation 用于全局逼近的克里格建模
Kriging simulation 克里格仿真

Ll

L-system 林氏系统
L-system simulation 林氏系统仿真
labeled state-transition system 标记状态转移系统
labeled transition 标记迁移；标记转换
labeled-transition system 标记迁移系统
labor classification code 人工分类代码
laboratory 实验室
laboratory experiment 实验室试验
laboratory simulation 实验室仿真
lag 滞后
lag (v.) 滞后
lag variable 延迟变量
lagged 滞后的，延迟的
lagged endogenous variable 滞后内生变量
lagged variable 滞后的变量
Lagrangian model 拉格朗日模型
Lagrangian modeling 拉格朗日建模
lambda calculus λ演算
Lanchester equation 兰彻斯特方程
landscape model 景观模型
language 语言
language error 语言错误
language specification 语言规范
Laplace equation 拉普拉斯方程
Laplace transformation 拉普拉斯变换
large eddy simulation 大涡模拟
large-grained component 大粒度的组件
large scale 大尺度
large-scale analysis 大规模分析
large-scale data analysis 大规模数据分析
large-scale dataset 大规模数据集
large scale experiment 大规模试验
large scale model 大尺度模型
large-scale simulation 大规模仿真

large-scale simulation environment 大规模仿真环境
large-scale simulation experiment 大规模仿真试验
large-scale simulation model 大尺度仿真模型
large-scale simulation software development 大规模仿真软件开发
large-scale stiff system 大型刚性系统
large-scale system 大规模系统
large simulation 大型仿真
large simulation data set 大型仿真数据集
largest step size 最大步长
last state 上一个状态
latency 时延
latent 潜在的
latent aspect 潜在的方面
latent aspect of model 模型的潜在方面
latent error 潜在误差
latent model 潜在模型
latent phase 隐藏期
latent value 潜在值
latent variable 潜在变量
Latin square design 拉丁方设计
lattice-based model 基于格子的模型
lattice model 格子化模型
launch 发射
launch of the system 系统的开始
law 法则；法律
layer 层
layer-based 基于层的
layer-based segmentation 基于层的分割

layered approach 分层的方法
layered model 分层的模型
layout 布局
lazy DEVS event 懒惰离散事件系统规范事件
lazy DEVS evolution 懒惰离散事件系统规范演变
lazy evaluation 惰性计算
lazy simulation 懒惰仿真
lead variable 前置变量
lean simulation 精益沙盘模拟
learned behavior 经过训练学到的行为
learning 学习
learning algorithm 学习算法
learning automaton 学习自动机
learning data 学习数据
learning data set 学习数据集合
learning experience 学习经验
learning game 学习博弈
learning heuristic 学习启发式
learning system 学习系统
learning theory 学习理论
least training 最小规模的训练
least upper bound 最小上界
legacy 遗留，遗赠，遗产
legacy application 已有应用
legacy data 已有数据
legacy M&S 已有建模与仿真
legacy model 已有模型
legacy simulation 已有仿真
legacy simulation system 已有仿真系统
legacy system 已有系统
legacy tool 已有工具

legacy transformation 已有变换
legal value 合法值
leisure game 休闲游戏
length 长度
length of the simulation run 仿真运行长度
level 级；水平
level of abstraction 抽象层次
level of accuracy 精度级别
level of aggregation 聚集层次
level of detail 细节层次
level of fidelity 逼真度等级
level of game 博弈水平
level of M&S validatability 建模与仿真级别的可验证性
level of model specification 模型层规范
level of perception 感知水平
level of resolution 分辨率级别
level of service 服务水平
level of system specification 系统规范层
level of validity 有效性的等级
level variable 水平变量
lexical 词汇的
lexical error 词汇错误
lexical evaluation 词汇评价
lexically-bound variable 词汇约束变量
lexically-scoped variable 词汇作用域变量
lexicographic model 词典模型
lexicographic ordering 按词典排序
Liapunov stability 李雅普诺夫稳定性
library-driven simulation 库驱动的仿真
licenced model 许可的模型
life cycle 生命周期
life cycle cost 生命周期成本
life cycle cost model 生命周期成本模型
life cycle cost modeling 生命周期成本建模
life cycle management 生命周期管理
life cycle model 生命周期模型
life cycle modeling 生命周期建模
life cycle of a simulation study 一个仿真研究的生命周期
life cycle of simulation 仿真生命周期
life cycle support 全生命周期支持
Likert-scale data 李克特量表数据
Likert-scale data analysis 李克特量表数据分析
Likert-scale data interpretation 李克特量表数据解释
limit 限制
limit (v.) 限制
limit cycle 极限环
limit value 极限
limitation 局限性
limited 有限的
limited dependent variable 受限因变量
limited discipline 有限规则
limited environmental interaction 有限的环境交互
limited input 有限的输入
limited interaction 有限互动
limited solution space 有限解空间
limiter 限制器

limiting coupling　限制耦合
Lindenmayer system　林登梅耶系统，林氏系统
Lindenmayer system simulation　林氏系统仿真
line　线路
line-of-sight simulation　视线仿真
linear　线性
linear bisimilar system　线性互相似系统
linear connection　线性的连接
linear conservation law　线性守恒定律
linear constant coefficient system　线性常系数系统
linear coupling　线性耦合
linear interpolation　线性插值
linear mapping　线性映射
linear mixed model　线性混合模型
linear model　线性模型
linear optimization　线性优化
linear programming　线性规划
linear programming embedded simulation　线性规划嵌入式仿真
linear programming embedded within simulation　嵌入仿真中的线性规划
linear second derivative model　线性二阶导数模型
linear single-input single-output model　线性单输入/单输出模型
linear stability theory　线性稳定性理论
linear state-space model　线性状态空间模型
linear system　线性系统
linear system simulation　线性系统仿真
linear system solver　线性系统求解器
linear time-invariant continuous-time system　线性时不变时间连续系统
linear topology　线性拓扑
linearity　线性度
linearization　线性化
linearization error　线性化误差
linearize (v.)　线性化
linearly coupled　线性耦合的
linearly implicit method　线性隐式方法
line-of-sight　视线
linguistic value　语言值
linguistic variable　语言变量；语义变量
link　链接；链路；连接
link trainer　连接训练器
linkable code　可连接的代码
linkage　接合
linkage of live, virtual, and constructive simulations　实况仿真、虚拟仿真与构造仿真间的联接
linkage to live simulation　连接到实时仿真
linked data　关联的数据
linked list　关联列表
linked model　关联的模型
linking　连接
little endian　小端字节
little endian data format　低位优先数据格式
live data　实时数据
live entity　真实实体
live event learning　真实事件学习

live exercise 实装演习
live instrumented entity 实装仪表化的实体
live instrumented simulation 实装仪表化的仿真
live simulation 实况仿真
live system 实况系统
live system-enriching simulation 实况系统增强仿真
live system-supporting simulation 实况系统支撑仿真
live training 实兵实抗训练
live training environment 实兵实抗训练环境
live training simulation 实况训练仿真
live training system 实况训练系统
livelock 活锁
load (*v.*) 加载
load bearing assumption 承载假设
load model 负载模型
loading error 加载误差
local 局部的
local accuracy 局部精度
local area network 局域网
local complexity 局部的复杂性
local containment 局部包容性
local error 局部误差
local error tolerance 局部误差容限
local hidden variable 局部隐含变量
local information 局部信息
local integration error 局部积分误差
local knowledge base 局部知识库
local model 局部模型
local property 局部特性
local property of a system 系统的局部特征
local quantum variable 局部量子变量
local realism 局部现实主义
local relative accuracy 局部相对精度
local schedular 局部时间表
local stability 局部稳定性
local time 局部时；地方时
local transition function 局部转移函数
local value 局部值
local variable 局部变量
localizability 本地化
localization 局部化；定域化
localize (*v.*) 定位
location 定位；区位
location-based 基于定位的
location-based experience 基于定位的体验
location-based game 基于定位的游戏
location tracing 位置跟踪
locative game 定位游戏
log 运行日志
logarithmically quantized integrator 对数量化积分
logic 逻辑
logic error 逻辑错误
logic game 逻辑游戏
logic simulation 逻辑仿真
logic simulator 逻辑仿真器
logic variable 逻辑变量
logical boundary 逻辑上的边界
logical consequence 逻辑结果
logical data model 逻辑数据模型
logical deduction 逻辑推论
logical empiricism 逻辑经验主义

logical environment 逻辑环境
logical equivalence 逻辑等价
logical error 逻辑误差
logical fallacy 逻辑上的谬论
logical inference 逻辑推理
logical interface 逻辑接口
logical interoperability 逻辑互操作性
logical knowledge 逻辑知识
logical model 逻辑模型
logical relation 逻辑关系
logical schema 逻辑架构
logical simulation 逻辑仿真
logical simulation topology 逻辑仿真拓扑
logical technique 逻辑技术
logical time 逻辑时间
logical time axis 逻辑时间轴
logical time management 逻辑时间管理
logical validity 逻辑正确性
logical value 逻辑值
logical variable 逻辑变量
logical verification 逻辑校核
logically deduced behavior 逻辑推理行为
logically named variable 逻辑命名变量
login ID 登录ID
logistic distribution 逻辑分布
logistic growth model 逻辑斯谛增长模型
logistics 物流
logit 分对数
logit function 对数函数
logit model 对数单位模型

lognormal random variable 对数正太随机变量
logon right 登录权限
long-haul network 长距离传输网络
long-lived state 耐久状态
long-term predictability 长期可预测性
long term trajectory forecast 长期轨迹预测
look ahead 预测
look-ahead time 前瞻时间
loop 环路，回路，闭路；循环
loop-breaking DEVS model 拆环离散事件系统规范模型
loop testing 循环测试
loose association 松散联系
loose coupling 松耦合
loose temporal coupling 松散时间耦合
loosely coupled 松耦合的
loosely coupled federated simulation 松耦合的联合仿真
loosely coupled model 松耦合模型
loosely coupled system 松耦合系统
loss-of-activation error 激活损失误差
low cognitive complexity 低认知的复杂性
low cognitive-complexity individual 低认知复杂性个体
low coupling 低耦合
low-fidelity simulation 低保真度仿真
low-fidelity simulator 低保真度仿真器
low granularity 低粒度
low-granularity model 低粒度模型

low interactive system 低交互系统
low level 低水平，低层级
low-level component 低层级组件
low level data 低级数据
low-level language 低级语言；初级语言
low-level modeling environment 低层级建模环境
low-level simulation 低层级仿真
low-level stimulus 低层级激励
low-order effect 低位效应
low-order explicit method 低阶显式方法
low-resolution model 低分辨率模型
low situational complexity 低情境的复杂性
lower bound 下界，下限
lower bound on the time stamp 时间戳下界
lower level 下层；下级
lower limit 下限
lower-resolution model 低分辨率模型
lower-triangular form 下三角形式
ludic simulation 路德模拟
lumped 集中的；集总
lumped continuous-time Markov chain 集总连续时间马尔可夫链
lumped DEVS model 集总离散事件系统规范模型
lumped model 集总模型
lumped-parameter model 集总参数模型
lurking variable 隐含变量

Mm

M&S (Modeling and Simulation) 建模与仿真
M&S accreditation 建模与仿真认证
M&S adjunct tool 建模与仿真辅助工具
M&S application 建模与仿真应用
M&S application quality 建模与仿真应用质量
M&S application sponsor 建模与仿真应用出资者
M&S architecture 建模与仿真结构
M&S as a cloud service 作为云服务的建模与仿真
M&S basic training course 建模与仿真基础训练课程
M&S body of knowledge 建模与仿真的知识体系
M&S BOK (M&S Body of Knowledge) 建模与仿真知识体系
M&S capability 建模与仿真能力
M&S category 建模与仿真范畴
M&S course 建模与仿真课程
M&S curriculum 建模与仿真课程
M&S developer 建模与仿真开发者
M&S development tool 建模与仿真开发工具
M&S engineer 建模与仿真工程师
M&S executive agent 建模与仿真执行代理
M&S facility 建模与仿真设备
M&S federation 建模与仿真联邦
M&S fidelity 建模与仿真的逼真度
M&S framework 建模与仿真框架

M&S group 建模与仿真群
M&S history 建模与仿真历史记录
M&S information source 建模与仿真信息源
M&S infrastructure 建模与仿真基础设施
M&S interoperability 建模与仿真的互操作性
M&S life cycle 建模与仿真生命周期
M&S life cycle management 建模与仿真生命周期管理
M&S market 建模与仿真市场
M&S master plan 建模与仿真主计划
M&S ontology 建模与仿真本体
M&S organization 建模与仿真组织
M&S paradigm 建模与仿真范式
M&S planning 建模与仿真规划
M&S practitioner 建模与仿真从业者
M&S practitioner career 建模与仿真从业者职业生涯
M&S principle 建模与仿真原理
M&S process 建模与仿真过程
M&S professional 建模与仿真专家
M&S program manager 建模与仿真程序管理器
M&S proponent 建模与仿真提议者
M&S repository 建模与仿真库
M&S reproducibility 建模与仿真的可重复性
M&S requirements 建模与仿真需求
M&S resolution 建模与仿真分辨率
M&S scalability 建模与仿真的延展性
M&S service 建模与仿真服务
M&S shareware 建模与仿真共享软件
M&S software design 建模与仿真软件设计
M&S sponsor 建模与仿真赞助商
M&S theory 建模与仿真理论
M&S tool 建模与仿真工具
M&S user 建模与仿真用户
M&S working group 建模与仿真工作组
machine-based formal method 基于机器的形式化方法
machine-centered simulation 以机器为中心的仿真
machine-centric model 机器中心模型
machine error 机器误差
machine intelligence 机器智能
machine intelligible model 机器可理解的模型
machine learning 机器学习
machine simulation 机器仿真
macro behavior 宏观行为
macro error 宏观误差
macro-level behavior 宏观层行为
macro-level pattern 宏观层模式
macro-level phenomenon 宏观层现象
macro model 宏模型
macro-step size 宏观步长
macroscopic behavior 宏观行为
macroscopic emergent behavior 宏观紧急行为
macroscopic model 宏观模型
main effect 主要效果
maintainability 可维护性；可维修性
maintainable 可维修的
maintenance 维护

maintenance model 维护模型
maintenance simulation 维护仿真
maintenance training 维护训练
malfunction 功能失常；误动作
man-centered simulation 以人为中心的仿真
man-in-the-loop 人在环，人在回路
man-in-the-loop simulation 人在环（回路）仿真
man-in-the-loop simulator 人在环（回路）仿真器
man-machine simulation 人机仿真
man-machine system simulation 人机系统仿真
man-made system 人造系统
management 经营；管理；操纵
management game 管理对策
management knowledge 管理知识
management of game 对策管理
management of simulation game 仿真游戏管理
management risk 管理风险
management service 管理服务
manager 管理者
managing 管理
managing gaming community 管理游戏社区
manipulated variable 操作的变量
manipulation 操作
manual input 手动输入
manual simulation 人工模拟
many-to-many relation 多对多关系
many-to-one relation 多对一关系
map 映射
mapping 映射

margin of error 误差范围
marginal input 边缘化输入
marginal stability 边迹稳定性
marginally stable 边迹稳定的
marginally stable result 边迹稳定结果
marginally stable system 临界稳定系统
marginally stiff 临界刚性的
marginally stiff system 临界刚性系统
marked graph 标示图
market 市场
market-based 基于市场的
market place 市场
marking 标识，标记
Markov assumption 马尔可夫假设
Markov chain 马尔可夫链
Markov chain model 马尔可夫链模型
Markov chain simulation 马尔可夫链仿真
Markov game 马尔可夫博弈
Markov model 马尔可夫模型
Markov process 马尔可夫过程
Markov simulation 马尔可夫仿真
Markovian property 马尔可夫特性
markup 标记
markup language 标记语言
MAS (MultiAgent System) 多智能体系统
mash-up 混聚
mass matrix 质量矩阵
mass storage 海量存储器
massive-data analysis 大规模数据分析
massive parallel processing 大规模并行处理

massive scale　大规模
massive-scale model　大规模模型
massive-scale simulation　大规模仿真
massively multiagent system　大规模多智能体系统
massively multiplayer　大型多人
massively multiplayer game　大规模多用户游戏
massively multiplayer online game　大规模多用户在线游戏
massively multiplayer simulation　大规模多用户仿真
massively multiuser　大规模多用户
massively multiuser game　大规模多用户游戏
master plan　主计划
match (v.)　匹配
matching　匹配
matchmaking algorithm　匹配算法
matchmaking engine　匹配引擎
material reality　物质现实
mathematical game　数学游戏
mathematical knowledge　数学知识
mathematical language　数学语言
mathematical markup language　数学标记语言
mathematical model　数学模型
mathematical modeling　数学建模
mathematical programming　数学程序设计
mathematical simulation　数学仿真
mathematical variable　数学变量
matrix　矩阵
matrix inversion　矩阵的逆
matrix-valued variable　矩阵值变量

mature model　成熟模型
mature understanding system　成熟的理解系统
maximization　最大化
maximum likelihood model　最大化可能性模型
maximum payoff　最大收益
maximum value　最大值
MDE-based process model　基于模型驱动工程的过程模型
Mealy automata　米利自动机
Mealy automaton　米利自动机
Mealy machine　米利机
Mealy model　米利模型
mean error　平均误差
mean solar time　平太阳时
mean time to failure　出故障的平均时间
mean time to restore　平均恢复时间，到还原点的平均时间
meaning of a variable　变量的含义
measurable　可测量的
measurable variable　可测变量
measure　测度；度量
measure (v.)　量出；估量
measure of effectiveness　有效性度量
measure of merit　价值度量
measure of outcome　输出测量
measure of performance　性能测量
measured　测量的
measured behavior　可测量的行为
measured outcome variable　测量的输出变量
measured quantity　被测量
measured value　测量值

measured variable 被测变量
measurement 测量
measurement accuracy 测量精度
measurement error 测量误差
measurement technique 测量技术
measurement unit 测量单位
measuring instrument error 测量仪器造成的误差
mechanical design modeling 机械设计建模
mechanical error 机械误差
mechanical system 力学系统；机械系统
mechanism 机理；机制；机械装置
mechanistic model 机械学模型
mechanized reasoning 机械化推理
mechatronic system 机电系统
media 媒体；介质
mediated 仲裁；中介
mediated digraph model 间接有向图模型
mediated modeling 间接建模
mediated understanding 间接的理解
mediating agent 调解代理
mediation 中介
mediator variable 中间变量
medical training 医疗培训
medium-scale system 中规模系统
medium-term predictability 中期可预测性
member variable 成员变量
membership 成员身份，隶属关系
membership function 成员函数
membership value 隶属度值
memory 内存；存储器

memory error 内存错误
memory leak 内存泄漏
memory model 记忆模型
memory state model 记忆状态模型
memory variable 记忆变量
memoryless model 无记忆的模型
memoryless multimodel 无记忆的多模型
memoryless quantization 无记忆量化
memoryless random variable 无记忆随机变量
memoryless variable 无记忆变量
mental bias 心理偏差
mental complexity 心理复杂性
mental model 智力模型
mental modeling 智力建模
mental simulation 精神仿真
menu 选单
menu-based 基于选单的
menu-driven 选单驱动
menu-driven tool 菜单驱动的工具
merge (*v.*) 归并
merit 优点，价值
mesh 网格，网孔
mesh-based 基于网格的
mesh-based simulation 基于网格的仿真
mesh generation step 网格生成步骤
mesh refinement 网格细化
meshfree simulation 无网格仿真
mesoscale modeling 中尺度建模
mesoscale simulation 中尺度仿真
mesoscale system 中尺度系统
mesoscopic model 介观模型
message 消息

message delivery 消息传递
message delivery service 消息传递服务
message error 消息误差
message passing 消息传递
meta-analysis 元分析
meta-metamodel 元元模型
metaclass 元类
metacomputing 元计算
metadata 元数据
metadata standard 元数据标准
metadata template 元数据模板
metadiscursive value 不确实的数值
metagame 元对策
metaheuristics 共通启发式演算法
metaknowledge 元知识
metalevel component 元级的组件
metalevel concept 元级概念
metalinguistic 元语言的
metamechanism 元构件
metamodel 元模型
metamodel-based agent model 基于元模型的智能体模型
metamodel-based approach 基于元模型的方法
metamodel-based model 基于元模型的模型
metamodel-based simulation 基于元模型的仿真
metamodel evolution 元模型演变
metamodel validation 元模型验证
metamodel verification 元模型校核
metamodeling 元建模
metamodeling foundation 元建模基础
metamodeling technique 元模型构建技术
metamodeling theory 元建模理论
metamorph 变形，变质
metamorph model 变质模型
metamorphic 变质的
metamorphic model 变质模型
metamorphic multimodel 变质的多模型
metamorphic reality 变质现实
metamorphic simulation 变质仿真
metamorphism 变质作用
metamorphosis 变形
metaphor 隐喻
metaphorical expression 隐喻表示
metaphorical model 隐喻模型
metaprogramming 元编程
metascheduler 元调度器
metasimulation 元仿真
metasimulation study 元仿真研究
metastable state 亚稳态
metasyntactic variable 元语法变量
metasyntactical variable 元语法变量
metavariable 元变量
metavariate 元变量
method 方法
method error 方法误差
method-of-lines 直线法
method-of-lines solution 直线法解
method-of-lines technique 直线法技术
methodological constraint 方法论约束
methodological framework 方法论体系
methodological knowledge 方法论知识

methodological ontology　方法论本体
methodology　方法论
methodology-based　基于方法学的
methodology-based classification　基于方法论的分类
methodology-based software tool　基于方法学的软件工具
methodology tailoring　方法裁剪
metric　度量
metric-aware　度量意识
metric-awareness　度量意识
metric-driven　指标驱动
micro-level behavior　微观行为
micro step size　微观步长
microanalytic simulation　微观分析仿真
microbehavior　微观行为
microcomputer simulation　微型计算机仿真
microelectromechanical system　微机电系统
microgrid simulation　微网格仿真
micromodel　微小模型
microphone input　麦克风输入
microscopic model　微观模型
microsimulation　微观仿真
microsimulation model　微观仿真模型
microstep size　微步长
middleware　中间件
military game　军事游戏
military gaming　军事博弈
military simulation gaming　军事仿真博弈
mimesis　模仿
mimetic　模仿的

mimetic play　模拟游戏
mimetic relation　相似关系
mimetics　模仿
mind-reading helmet　意识读取头盔
minimal　最小的
minimal effect　最小影响
minimal effect model　最小影响模型
minimal model　最小模型
minimal value　最小值
minimization　最小化
mining　挖掘
mining data　挖掘数据
mirror simulation　镜像仿真
mirror world simulation　镜像世界仿真
misbehavior　错误行为
miscalculation　误算
misconception　误解
miscouple (v.)　错误耦合
miscoupled　错误耦合
miscoupling　错误耦合
misimplementation　错误实现
misinterpretation　误读
misleading　误导
misperceived reality　错误感知的现实
missed assumption　错过的假设
missing data　缺失的数据
missing value　缺失值；漏测值
mission level　任务级
mission-level modeling　任务等级建模
mission-level simulation　任务级仿真
mission rehearsal　任务演练
mission rehearsal simulation　任务排演仿真
mission space　任务空间

mission space entity 任务空间实体
mission system 任务系统
misspecification error 错误设定误差
mistake 错误
mistake detection 错误探测
mistake forgiving environment 容错环境
mistake prevention 错误预防
mistake proofing 错误证明
misunderstanding 误解
misunderstood reality 错误理解的现实
mixed coupling 混合式耦合
mixed coupling recipe 混合式耦合方法
mixed discrete and continuous system 离散连续混合系统
mixed formalism 混合形式
mixed-formalism model 混合形式模型
mixed-formalism SL 多范式仿真语言
mixed granularity 混合粒度
mixed-granularity model 混合粒度模型
mixed logit 混合分对数
mixed method 混合方法
mixed methodology 混合方法学
mixed methodology model 混合方法学模型
mixed-mode integration 混合模式集成
mixed-mode simulation 混合模式仿真
mixed model 混合模型
mixed partial and ordinary differential equations 常微分、偏微分混合方程
mixed reality 混合现实
mixed-reality system 混合现实系统
mixed-reality training simulator 混合现实训练仿真器
mixed-resolution model 混合分辨率模型
mixed-signal simulation 混合信号仿真
mixed simulation 混合仿真
mixed state 混合状态
mixed-state model 混合状态模型
mixed symbolic and numerical approach 混合符号和数值方法
mixed system 混合系统
mixed-time model 混合时间模型
ml-DEVS 建模语言离散事件系统规范
mobile collaboration 移动协同
mobile computing 移动计算
mobile device activated simulation 移动设备激活的仿真
mobile device initiated simulation 移动设备发起的仿真
mobile device-triggered simulation 移动设备触发的仿真
mobile game 手机游戏
mobile simulation 移动仿真
mobility model 移动模型
mock simulation 模拟仿真
mock-up 样机
mock-up model 样机模型
modal value 模式值
mode 模式
mode error 模式误差

mode switching 模式切换
model 模型
model (v.) 建模
model abstracting technique 模型抽象化技术
model abstraction 模型抽象
model acceptability 模型的可接受性；模型合格
model acceptability standard 模型可接受性标准
model accreditation 模型确认
model accuracy 模型精度
model acquisition 模型获取
model activation 模型激活
model adaptation 模型适应
model adaptivity 模型自适应性
model adequacy 模型充分性
model affinity 模型亲和性
model aggregation 模型聚合
model alignment 模型配准；模型对齐
model analysis 模型分析
model applicability 模型适用性
model appraisal 模型评估
model appropriateness 模型适当性
model archive 模型存档
model archiving 模型存档
model assessment 模型评估
model assumption 模型假设
model attribute 模型属性
model authoring 模型创作
model availability 模型可用性
model bank 模型（存储）体
model base 模型库
model base management 模型库管理
model base manager 模型库管理器

model-based 基于模型的
model-based activity 基于模型的活动
model-based analysis 基于模型的分析
model-based codesign 基于模型的协同设计
model-based complex system 基于模型的复杂系统
model-based development 基于模型的开发
model-based DEVS methodology 基于模型的离散事件系统规范方法学
model-based engineering 基于模型的工程
model-based experiment 基于模型的试验
model-based initiative 基于模型的主动性
model-based interoperability 基于模型的互操作性
model-based method 基于模型的方法
model-based methodology 基于模型的方法学
model-based prediction 基于模型的预估
model-based reality 基于模型的现实
model-based reasoning 基于模型的推理
model-based review 基于模型的评估
model-based review diagnosis 基于模型的评估诊断
model-based review process 基于模型的评估过程
model-based simulation 基于模型的仿真

model-based simulation monitor 基于模型的仿真监视器
model-based software 基于模型的软件
model-based solution 基于模型的解决方案
model-based system 基于模型的系统
model-based systems engineering 基于模型的系统工程
model-based technique 基于模型的技术
model-based testing 基于模型的测试
model-based testing technique 基于模型的测试技术
model-based tool 基于模型的工具
model-based validation 基于模型的验证
model-based verification 基于模型的校核
model behavior 模型行为
model behavior discontinuity 模型行为不连续性
model behavior fitting 模型行为拟合
model behavior ontology 模型行为本体
model benchmarking 模型基准
model boundary 模型边界
model brokering 模型代理
model building 模型建立
model building formalism 模型建立形式
model calibration 模型校准
model centric 模型中心
model-centric approach 模型中心方法

model certification 模型认证
model characterization 模型表征
model checking 模型检查
model checking algorithm 模型检测算法
model checking technique 模型检测技术
model checking tool 模型校验工具
model class 模型类
model classification 模型分类
model coarseness 模型粗糙度
model coevolution 模型协同进化
model comparison 模型比较
model compiler 模型编译器
model complexity 模型复杂性
model complexity assessment 模型复杂性评估
model component 模型组件
model component pragmatism 模型组件实用主义
model component semantics 模型组件语义
model component template 模型组件模板
model composability 模型可组合性
model composer 模型组合器
model composition 模型组合
model composition instance 模型组合实例
model composition rule 模型组合规则
model composition technology 模型组和技术
model comprehensibility 模型可理解性
model compression 模型压缩

model conceptualization 模型概念化
model confidence 模型置信度
model configuration 模型配置
model consistency 模型一致性
model constant 模型常数
model constraint 模型约束
model construction 模型构建
model construction management 模型构建管理
model content 模型内容
model continuity 模型的连续性
model correctness 模型正确性
model correctness analysis 模型正确性分析
model cost 模型成本
model coupling 耦合模型
model credibility 模型可信性
model data 模型数据
model database 模型数据库
model deactivation 模型失活
model decomposition 模型分解
model deployment 模型部署
model description 模型描述
model description language 模型描述语言
model design 模型设计
model developer 模型开发者
model development 模型开发
model development environment 模型开发环境
model development framework 模型开发框架
model-development infrastructure 模型开发设施
model differencing 模型差分化

model-directed 模型导向的
model-directed process 模型引导的过程
model-directed system 模型导向的系统
model discovery 模型发现
model docking 模型对接
model documentation 模型文档
model documenting 模型建档
model-driven 模型驱动
model-driven approach 模型驱动的方法
model-driven architecture 模型驱动架构
model-driven architecture tool 模型驱动构架工具
model-driven code generation 模型驱动的代码生成
model-driven development 模型驱动开发
model-driven development approach 模型驱动开发方法
model-driven development language 模型驱动开发语言
model-driven development methodology 模型驱动开发方法学
model-driven development of user interface 用户界面（接口）的模型驱动开发
model-driven development technique 模型驱动开发技术
model-driven engineering 模型驱动工程
model-driven interoperability 模型驱动互操作性

model-driven language 模型驱动语言
model-driven methodology 模型驱动方法学
model-driven process 模型驱动过程
model-driven reasoning 模型驱动推理
model-driven reproducibility 模型驱动的可重复性
model-driven scenario 模型驱动的想定
model-driven simulation engineering 模型驱动的仿真工程
model-driven system 模型驱动的系统
model-driven systems engineering 模型驱动的系统工程
model-driven technique 模型驱动技术
model-driven user interface 模型驱动的用户界面（接口）
model driver 模型驱动器
model dynamics 模型动力学
model editing environment 模型编辑环境
model editor 模型编辑器
model elaboration 模型精化
model element 模型元素
model engineering 模型工程
model engineering body of knowledge 模型工程知识体系
model engineering ontology 模型工程本体
model engineering process 模型工程过程
model engineering standard 模型工程标准
model engineering support 模型工程支持
model engineering tool 模型工程工具
model environment 模型环境
model epistemology 模型认识论
model equivalencing 模型等价
model error 模型误差
model error statistic 模型误差统计
model evaluation 模型评价
model evolution 模型演变
model exchange 模型交换
model execution 模型执行
model execution property 模型执行属性
model expert 模型专家
model extraction 模型提取
model family 模型体系；模型族
model federation 模型联邦
model fidelity 模型逼真度
model file 模型文件
model fitting 模型拟合
model flexibility 模型灵活性
model formalism 模型形式
model generation 模型生成
model generation problem 模型生成问题
model generator 模型生成器
model granularity 模型粒度
model graph 模型图
model heterogeneity 模型的异质性
model homogeneity 模型的同质性
model homomorphism 模型同态
model identification 模型辨识
model implementation 模型实现
model-in-the-loop 模型在环

model input/output 模型的输入/输出
model instance 模型实例
model instantiation 模型实例化
model integration 模型集成
model integrity 模型完整性
model integrity requirements 模型完整性需求
model intelligibility 模型的可理解性
model interaction 模型交互
model-interface analysis 模型接口分析
model interface error 模型接口错误
model interface testing 模型接口测试
model interoperability 模型互操作性
model interoperability standard 模型互操作标准
model interoperation 模型互操作
model isomorphism 模型同构
model latency 模型延迟
model library 模型库
model life cycle 模型生命周期
model life cycle cost 模型生命周期成本
model lifecycle engineering 模型生命周期工程
model life cycle process 模型生命周期过程
model location 模型定位；模板定位
model maintainability 模型可维护性
model maintenance 模型维护
model management 模型管理
model-management infrastructure 模型管理基础设施
model management tool 模型管理工具

model manager 模型管理器
model manipulation 模型操作
model mapping 映射模型
model market place 模型市场
model matching 模型匹配
model maturity 模型成熟度
model memory 模型记忆
model merging 模型归并，模型融合
model modifiability 模型可修改性
model modularity 模型模块化
model morphism 模型同型
model navigation 模型导航
model of self 自身模型
model of user 用户的模型
model ontology 模型本体
model operator 模型算子
model-oriented 面向模型的
model packaging 模型封装
model packaging tool 模型封装工具
model parameter 模型参数
model partitioning 模型分割
model pedigree 模型系谱
model plausibility 模型的似真性；模型合理性
model portability 模型可移植性
model postulate 模型假定
model pragmatics 模型语用学
model precision 模型精度
model process management 模型过程管理
model processing 模型处理
model processing environment 模型处理环境
model processing support 模型处理支持

model provider 模型提供方
model pruning 模型裁剪
model qualification 模型鉴定
model quality 模型质量
model quality assurance 模型质量保证
model query 模型查询
model query language 模型查询语言
model querying 模型查询
model realism 模型实在论
model recommender 模型推荐人
model reconstruction 模型重构
model reconstruction management 模型重构管理
model reduction 模型降阶
model reduction technique 模型降阶技术
model refactoring 模型重构
model referability 模型可交付性
model refinement 模型精炼
model registry 模型注册
model-related 模型相关的
model relation 模型关系
model relationship 模型关联
model relevance 模型关联
model relevant 模型相关的
model reliability 模型可靠性
model reliance 模型依赖
model replacement 模型替换
model replicability 模型可复制性；模型可重现性
model replication 模型复制；模型重现
model repository 模型库
model repository management 模型仓库管理
model representation 模型表示（法）
model requirements 模型需求
model requirements management 模型需求管理
model resolution 模型分辨率
model resource 模型资源
model retrieval 模型检索
model reusability 模型可重用性
model reusable 模型可重用
model reuse 模型重用
model reuse technology 模型重用技术
model robustness 模型鲁棒性
model scalability 模型可扩展性
model scaling 模型缩放
model security 模型安全
model selection 模型选择
model semantics 模型语义
model sensitivity 模型灵敏度
model shareability 模型可共享性
model sharing 模型共享
model simplification 模型简化
model size 模型规模
model-size limit 模型大小限制
model-size metric 模型尺寸度量
model space 模型空间
model specification 模型规范
model specification environment 模型规范环境
model specification error 模型设定误差
model specification language 模型规范语言
model stability 模型稳定性

model stage 模型级
model staging 模型分级
model structure 模型结构
model support 模型支架；模型支持
model supportability 模型保障性；模型可支持行
model switch 模型切换
model switching 模型切换
model synchronization 模型同步
model synthesis 模型综合
model synthesis error 模型综合误差
model systems engineering 模型系统工程
model taxonomy 模型分类
model template 模型模板
model test 模型试验；模型测试
model testing 模型测试
model theory 模型论
model-to-code transformation 模型代码变换
model-to-entity mapping 模型实体映射
model-to-model mapping 模型间映射
model-to-model transformation 模型间变换
model-to-text transformation 模型文本变换
model tool 模型工具
model topography 模型地貌
model traceability 模型可追溯性
model trade-off algorithm 模型权衡算法
model transferability 模型可传递性
model transformation 模型变换
model transformation approach 模型变换方法
model transformation engine 模型转换引擎
model transformation language 模型转换语言
model transformation operator 模型转换算子
model transformation paradigm 模型转换范式
model-transformation tool 模型变换工具
model transition 模型转换
model tree 模型树
model trustworthiness 模型可信度
model understandability 模型的可理解性
model update 模型更新
model updating 模型更新
model usability 模型的可用性
model use 模型使用
model usefulness 模型实用性
model user 模型使用者
model utility 模型效用
model validation 模型验证
model validity 模型有效性
model veracity 模型的真实性
model verification 模型校核
model verification, validation and certification 模型校核、验证与确认
model verisimilitude 模型逼真性
model version 模型版本
model version control 模型版本控制
model versioning 模型版本控制
model wrap 模型封装
model wrapping 模型封装

modeled dynamics　建模动力学
modeler　建模器；制造模型者
modeler's risk　建模器风险
modeling　建模
modeling activity　建模活动
modeling agent　建模代理
modeling algorithm　建模算法
modeling and simulation　建模与仿真
modeling and simulation framework　建模与仿真框架
modeling application　建模应用
modeling approach　建模方法
modeling assumption　建模假设
modeling capability　建模能力
modeling challenge　建模挑战
modeling concept　建模概念
modeling constraint　建模约束
modeling consultant　建模顾问
modeling convention　建模惯例
modeling domain　建模域
modeling environment　建模环境
modeling environment support　建模环境支持
modeling error　建模误差
modeling expert　建模专家
modeling for diagnosis　诊断建模
modeling formalism　建模形式化
modeling framework　建模框架
modeling function　建模函数
modeling infrastructure　建模基础设施
modeling interface　建模界面
modeling issue　建模问题
modeling language　建模语言
modeling layer　建模层
modeling literacy　建模素养
modeling maturity　建模完备性
modeling method　建模方法
modeling methodology　建模方法学
modeling natural environment　建模自然环境
modeling ontology　建模本体
modeling paradigm　建模范式
modeling perspective　建模视角
modeling practice　建模实践
modeling primitive　建模元语
modeling principle　建模原理
modeling process　建模过程
modeling relation　建模关系
modeling relationship　建模关系
modeling scenario　建模场景
modeling semantics　建模语义
modeling, simulation and visualization　建模、仿真和可视化
modeling software　建模软件
modeling stage　建模阶段
modeling standard　建模标准
modeling structure　建模结构
modeling support　建模支持
modeling system　建模系统
modeling system structure　建模系统结构
modeling technique　建模技术
modeling term　建模术语
modeling theory　建模理论
modeling tool　建模工具
modeling tool design　建模工具设计
modeling tool evaluation　建模工具评价
modeling tool implementation　建模工具实现

modeling tool validity 建模工具有效性
modeling tool usability 建模工具可用性
model's behavior 模型行为
moderate modeling noise 中度建模噪声
moderate noise 中等强度噪声
moderated emergence 适度的涌现
moderated stable emergence 适度稳定的涌现
moderated stable emergent behavior 缓和稳定的突现行为
moderated unstable emergence 适度不稳定的涌现
moderated unstable emergent behavior 缓和不稳定的突现行为
moderating variable 调节变量
moderator 调节器；仲裁员
moderator variable 调节变量；干涉变量
modifiability 可修改性
modifiable 可更改的
modifiable goal 可变更的目标
modifiable property 可修改属性
modification model 修正模型
modified Euler integration method 改进的欧拉积分法
modified game 修正对策
modified Koch L-system 改进的科赫-林氏系统
modified model 修正模型
modified Newton iteration 改进牛顿迭代法
modified screen 修正屏幕

modifier 调节器
modify (v.) 修改，更改
modular 模块化的
modular component 模块化组件
modular DEVS form 模块化的离散事件系统规范形式
modular form 模块化形式
modular model 模块化模型
modular modeling 模块化建模
modular simulation 模块化仿真
modular simulation program 模块化仿真程序
modular system 模块化系统
modular system modeling 模块化系统建模
modularity 模块化；模块性
modularization 模块化；组合化
modularize (v.) （使）模块化
modulation 调制
module 模块；组件
module boundary 模块边界
module interface 模块接口
module testing 模块测试
modus pollens rule 正断法规则
modus tollens rule 否定后件律规则
molecular communication 分子通信
moment closure 矩封闭
moment-closure approximation 矩封闭近似
moment-closure technique 矩封闭技术
momentum 动量；动力
monadic model 一元模型
monitor 监视器
monitor software 监控软件

monitored 监控的
monitored variable 监控变量
monitoring 监测，监控
monitoring data 监测数据
monitoring program 监控程序
monogenic system 单演系统
monolithic model 单体模型
monotonic evolutionary simulation algorithm 单调进化仿真算法
monotonic Kriging 单调克里格
monotonous input 单一输入
Monte Carlo 蒙特卡罗
Monte Carlo algorithm 蒙特卡罗算法
Monte Carlo experiment 蒙特卡罗实验
Monte Carlo method 蒙特卡罗方法
Monte Carlo simulation 蒙特卡罗仿真
Monte Carlo test 蒙特卡罗测试
Moore automata 摩尔自动机
Moore automaton 摩尔自动机
Moore machine 摩尔机
Moore model 摩尔模型
Moore neighborhood 摩尔型邻域
moral error 道德错误
moral principle 道德原则
moral reasoning 道德推理
morphable model 形变模型
morphic 形态学的；同型的
morphism 同型
morphological assessment 形态学评估
morphological understanding 形态理解
morphology 形态学

motivational state 激发状态
motive 动机
motor control skill 运动控制技术
motor skill 运动控制
motor skill enhancing 运动技术增强
mouse emulator 鼠标仿真器
mouse simulator 鼠标仿真器
moving-boundary simulation 移动边界仿真
MS&V (Modeling Simulation and Visualization) 建模仿真和可视化
MSMP (Modeling and Simulation Master Plan) 建模与仿真总体规划
mulfunction 故障
mulfunctioning 出故障的
multi 多；多面；多重
multiagent 多智能体
multiagent-based simulation 基于多智能体的仿真
multiagent model 多智能体模型
multiagent orthogonality 多智能体正交性
multiagent participatory-simulation 多智能体参与式仿真
multiagent simulation 多智能体仿真
multiagent-supported simulation 多智能体支持的仿真
multiagent system 多智能体系统
multiaspect 多方面
multiaspect model 多方面模型
multiaspect modeling 多方面建模
multiaspect multimodel 多方面多模型
multiaspect multisimulation 多层面多仿真

multiaspect simulation 多方面仿真
multiaspect simulation model 多方面仿真模型
multiaspect system 多方面系统
multiaspect understanding 多方面理解
multiattribute 多属性
multibody simulation 多体仿真
multibody system 多体系统
multibody system dynamics 多体系统动力学
multicast 多点传输；组播
multicomponent 多元的；多组件的
multicomponent DEVS 多分量离散事件系统规范
multicomportment model 多元模型
multicore cache simulator 多核缓存仿真器
multicoupling 多耦合
multicriteria decision 多准则决策
multicriteria decision making 多准则决策
multidimensional 多维的；多面的
multidimensional data model 多维数据模型
multidimensional data set 多维数据集合
multidimensional data visualization 多维数据可视化
multidimensional modeling 多维度建模
multidimensional space 多维空间
multidisciplinary system 多学科系统
multidomain 多领域
multidomain modeling 多域建模
multidomain modeling language 多域
multidomain simulation system 多域仿真系统
multiechelon 多级
multiechelon model 多级模型
multiengineering 多个工程
multievent 多个事件
multifacet 多层面性；多面的
multifaceted 多层面的，多方面的
multifaceted model 多层面模型
multifaceted modeling 多层面建模
multifaceted modeling formalism 多层面建模形式
multiformalism 多形式的
multiformalism model 多形式模型
multiformalism modeling 多形式建模
multi-input 多输入
multi-input multi-output 多输入多输出
multi-input multi-output model 多输入多输出模型
multi-input multi-output system 多输入多输出系统
multi-input single-output 多输入单输出
multi-input single-output model 多输入单输出模型
multi-input single-output system 多输入单输出系统
multi-instance event 多实例事件
multilanguage 多语言
multilayer 多层次
multilayer network 多层网络
multilayer system 多层系统
multilevel 多层的；多级的

multilevel approach　多级方法
multilevel clustering　多级聚类
multilevel control　多级控制
multilevel dynamic coupling　多级动态耦合
multilevel interaction　多级交互
multilevel model　多级模型
multilevel modeling　多级建模
multilevel multimodel framework　多级多模型框架
multilevel security　多层安全
multilevel simulation　多级仿真
multilevel structure　多级结构
multilingual programming　多语言编程
multilingual simulation　多语言仿真
multimedia documentation　多媒体文件
multimedia-enriched simulation　多媒体强化仿真
multimedia model　多媒体模型
multimedia modeling　多媒体建模
multimedia modeling language　多媒体建模语言
multimedia simulation　多媒体仿真
multimethod simulation　多方法仿真
multimodal　多模态的
multimodal distribution　多模态分布
multimodal input　多模态输入
multimodal interfacing　多模态接口技术
multimodal model　多模态模型
multimodal user interface　多通道用户界面
multimode method　多模态方法

multimodel　多模型
multimodel agent coupling　多模型智能体耦合；多模型代理耦合
multimodel configuration　多模型配置
multimodel coupling　多模型耦合
multimodel framework　多模型框架
multimodel simulator　多模型仿真器
multimodeling　多重建莫
multimodeling formalism　多重建模形式
multimodeling methodology　多重建模方法学
multimodeling paradigm　多重建模范式
multinomial　多项式；多项的
multinomial distribution　多项式分布
multinomial logit　多项式分对数
multiobjective model　多目标模型
multiobjective optimization problem　多目标优化问题
multi-output　多输出
multiparadigm　多范式
multiparadigm model　多范式模型
multiparadigm modeling　多范式建模
multiparadigm simulation　多范式仿真
multiparadigm simulation system　多范式仿真系统
multiperspective　多视角
multiperspective model　多视角模型
multiperspective modeling　多角度建模
multiperspective simulation　多视角仿真
multiphase process　多相过程
multiphysics model　多物理场模型

multiphysics modeling 多物理场建模
multiphysics simulation 多物理场仿真
multiplatform 多平台
multiplatform model 多平台模型
multiplatform modeling 多平台建模
multiplayer 多人模式
multiplayer game 多参与者游戏
multiplayer gaming technique 多参与者游戏（博弈）技术
multiplayer model 多参与者模型
multiplayer online game 多人在线游戏
multiplayer simulation 多用户仿真
multiplayer support 多用户支持
multiple 多重的；多种的
multiple access 多路存取
multiple anticipation 多重预期
multiple control 多重控制
multiple data model 多数据模型
multiple decomposition 多重分解
multiple emergence 多重涌现
multiple emergent behavior 多重突现行为
multiple experimental frame 多重试验框架
multiple fidelity simulation 多保真度仿真
multiple input 多输入
multiple input multiple-output 多输入多输出
multiple input multiple output model 多输入多输出模型
multiple input single output 多输入单输出
multiple input single output model 多输入单输出模型
multiple instantiation 多重实例化
multiple measure 多重测量
multiple modeling paradigm 多重建模范式
multiple output 多输出
multiple perception 多重感知
multiple queues 多队列
multiple response variable 多响应变量
multiple run 多次运行
multiple run simulation study 多运行仿真研究
multiple scenario instantiation 多重想定实例化
multiple simulator 多路仿真器
multiple spatial scales 多重尺度空间
multiple states 多状态
multiple synchronization point 多同步点
multiple temporal scales 多时间尺度
multiple time-scale variable 多时间尺度变量
multiple transition 多重转换
multiple transition subnet 多重转换子网
multiple zero-crossing 多零交点
multiplicity 多样性
multiplism 多重性
multipoint modeling 多点建模
multiprocess 多进程
multiprocessor simulation 多处理器仿真
multirate 多速率

multirate integration 多速率集成
multirate sampling 多速率采样
multirate simulation 多速率仿真
multiresolution 多分辨率
multiresolution approach 多分辨率方法
multiresolution federate 多分辨率联邦成员
multiresolution M&S 多分辨率建模与仿真
multiresolution model 多分辨率模型
multiresolution modeling 多分辨率建模
multiresolution multimodel 多分辨率多模型
multiresolution multiperspective modeling 多分辨率多视角建模
multiresolution multisimulation 多分辨率多重仿真
multiresolution simulation 多分辨率仿真
multiresolution simulation model 多分辨率仿真模型
multirun experiment 多次运行实验
multirun simulation study 多运行仿真研究
multisampling system 多采样系统
multiscale 多尺度
multiscale analysis 多尺度分析
multiscale hierarchical model 多尺度层次模型
multiscale M&S 多尺度建模与仿真
multiscale method 多尺度法
multiscale model 多尺度模型
multiscale modeling 多尺度建模
multiscale modeling technique 多尺度建模技术
multiscale modeling tool 多尺度建模工具
multiscale phenomenon 多尺度现象
multiscale problem 多尺度问题
multiscale simulation 多尺度仿真
multisensory I/O 多传感器输入/输出
multisensory input 多传感器输入
multisensory interaction 多传感器交互
multisim 一种电路仿真软件
multisimulation 多仿真
multisimulation engine 多仿真引擎
multisimulation formalism 多仿真形式
multisimulation framework 多仿真框架
multisimulation gaming 多仿真博弈
multisimulation gaming strategy 多仿真博弈策略
multisimulation management game 多仿真管理对策
multisimulation methodology 多仿真方法学
multisimulator 多仿真器
multisourse data 多源数据
multistability 多重稳定性
multistable state 多稳态
multistage integration 多级集成
multistage model 多级模型
multistage modeling 多阶段建模
multistage modeling formalism 多阶段建模形式
multistage multimodel 多级多模型

multistage multisimulation 多阶段多仿真
multistage simulation 多阶段仿真
multistage simulation model 多阶段仿真模型
multistage validation 多级验证
multistage validity 多级有效性
multistate 多状态
multistate object 多态对象
multistep integration algorithm 多步积分算法
multistep integration method 多步积分法
multistep method 多步法
multitasked 多任务的
multitasking 多任务处理
multitasking behavior 多任务处理行为
multitasking model 多任务处理模型
multitouch emulation 多点触控仿真
multiunderstanding 多重理解
multiuser 多用户
multivalued function 多值函数
multivariable 多变量
multivariate 多元
multivariate analysis 多元分析
multivariate Gaussian random variable 多元高斯随机变量
multivariate random variable 多元随机变量
multivision understanding 多视觉理解
mutable variable 可变变量
mutation 突变，变异
mutation analysis 突变分析
mutation-based 基于变异的
mutation for model-driven approach 模型驱动方法的变异
mutation operator 变异算子
mutation testing 变异测试
mutation tool for 变异测试工具
mutational model 突变模型
mutational multimodel 突变多模型
mutual simulation 交互仿真

Nn

n-body simulation n 体仿真
n-person game n 人博弈
n-way model merging n 路模型融合
nanoscale simulation 纳米级仿真
nanoscale system 纳米级系统
nanosimulation 纳米仿真
narrative model 叙述模型
narrative simulation 叙述仿真
national federation 国家联邦
national model 国家模型
natural complexity 自然复杂度
natural computing 自然计算
natural experiment 自然实验
natural group variable 自然组变量
natural model 自然模型
natural system 自然系统
naturalistic experiment 自然主义实验
nature-inspired 自然启发
nature-inspired algorithm 自然启发算法
nature-inspired paradigm 自然启发范式
near-optimal 近似最优

near real-time 近实时，准实时
need-driven goal 需求驱动的目标
negative binomial random variable 负二项随机变量
negative constraint 负约束
negative emergence 负涌现
negative emergent behavior 负涌现行为
negative-feedback loop 负反馈回路
negative hypergeometric random variable 负超几何随机变量
negative thought experiment 消极思想实验
negative training 消极训练
negative value 负值
negligible error 可忽略误差
negotiation 谈判，协商
nested coupled multimodel 嵌套耦合多模型
nested coupling 嵌套耦合
nested game 嵌套博弈
nested logit 嵌套分对数
nested model 嵌套的模型
nested multisimulation 嵌套多仿真
nested simulation 嵌套仿真
net 网；网络
netcentric 网络中心
netcentric gaming 网络中心博弈
netcentric simulation 网络中心仿真
netcentric system 网络中心系统
netcentric wargaming 网络中心战争博弈
network 网络
network-based trust game 基于网络的信任博弈

network communication service 网络通信服务
network device error 网络设备故障
network error 网络错误
network filter 网络滤波器
network game 网络游戏
network game experiment 网络游戏实验
network game theory 网络博弈论
network management 网络管理
network manager 网络管理器
network model 网络模型
network node 网络节点
network-oriented 面向网络的
network-oriented simulation 面向网络的仿真
network permission error 网络权限错误
network security 网络安全
network theory 网络理论
network-type hypergame 网络型超对策
networked game 网络化游戏
networked simulation 网络化仿真
networked system 网络化系统
neural-level modeling 神经系统级建模
neural network 神经网络
neural network-based model 基于神经网络的模型
neural network metamodel 神经网络元模型
neuro-fuzzy inference algorithm 神经模糊推理算法
neurotrust game 神经信任游戏

neutral emergence 中心涌现
neutral input 中性输入
news game 新闻游戏
Newton-Gregory backward polynomial 牛顿-格雷戈里后向多项式
Newton-Gregory forward polynomial 牛顿-格雷戈里前向多项式
Newton-Gregory polynomial 牛顿-格雷戈里多项式
Newton iteration 牛顿迭代（法）
next event 下一事件
next event formalism 下一事件形式化
next external event 下一外部事件
next internal event 下一内部事件
nociceptive input 伤害性输入
nociceptive stimulus 伤害性刺激
node 节点
noise 噪声
noisy data 噪声数据
noisy input 噪声输入
noisy model 有噪声的模型
noisy value 噪声值
nominal emergence 标称涌现
nominal intentional emergence 标称的有意涌现
nominal scale 标称尺度
nominal-scale variable 标称尺度变量
nominal step size 标称步长
nominal value 标称值
nominal variable 标称变量
non-equation-oriented simulation 面向非方程式的仿真
non-experience-based knowledge 基于非经验的知识
non-HLA-compliant simulation 不符合高层体系结构的仿真
non-line-of-sight simulation 非视线仿真
non-load-bearing assumption 非承重假设
non-strictly-competitive game 非严格竞争博弈
non-zero-sum game 非零和博弈
non-zero-sum simulation 非零和博弈模拟
non-zero-sum simulation game 非零和仿真博弈
nonabsorbing state 非吸收状态
nonadverse event 非逆事件
nonanticipatory model 非预期模型
nonanticipatory system 非预期系统
nonautonomous model 非自治模型
nonbasic variable 非基本变量
noncentral Chi-Squared random variable 非中心卡方随机变量
noncognitive variable 非认知变量
noncompensatory model 非补偿模型
noncontrollable assumption 不可控假设
noncontrolled state 非控制状态
nonconvergent simulation 非收敛仿真
noncooperative 非合作的
noncooperative game 非合作博弈
nondeterministic DEVS 非确定性离散事件系统规范
nondeterministic experiment 非确定性试验
nondeterministic function 非确定性函数
nondeterministic model 非确定性

模型

nondeterministic problem　不确定性问题

nondeterministic simulation　非确定性仿真

nondeterministic Turing machine simulation　非确定性图灵机仿真

nondeterministic variable　非确定性变量

nondominated solution　非主导解决方案

nondynamic behavior　非动态行为

nonemergent　非突发的

nonemergent property　隐性，没有表现的特性

nonempirical knowledge　非经验知识

nonentertainment game　非娱乐游戏

nonergodic behavior　非遍历行为

nonessential event condition　非必要事件条件

nonfoundational knowledge　非基础知识

nonfunctional property　非功能性

nonfuzzy value　非模糊值

nongame application　非应用程序

nonhierarchical model　非层次的模型

nonholonomic constraint　非完整约束

nonholonomic model　非完整模型

nonholonomic system　非完整系统

nonhomeostatic variable　非稳态变量

nonintrusive　非侵入性的

nonintrusive method　非侵入性方法

noninverting input　同相输入

noniterative co-simulation　非迭代协同仿真

nonkinetic model　非动力学模型

nonlinear　非线性的

nonlinear affine input　非线性仿射输入

nonlinear affine input system　非线性仿射输入系统

nonlinear causality　非线性因果关系

nonlinear connection　非线性连接

nonlinear control　非线性控制

nonlinear coupling　非线性耦合

nonlinear dynamical system　非线性动力系统

nonlinear function　非线性函数

nonlinear induction variable　非线性归纳变量

nonlinear mapping　非线性映射

nonlinear model　非线性模型

nonlinear optimization　非线性优化

nonlinear-second derivative model　非线性二阶微分模型

nonlinear stability　非线性稳定性

nonlinear state-space model　非线性状态空间模型

nonlinear statistical coupling　非线性统计耦合

nonlinear system　非线性系统

nonlinear system simulation　非线性系统仿真

nonlinearity　非线性特性

nonlinearly coupling　非线性耦合

nonlocality　非定域性

nonmanipulated variable　非操作变量

nonmembership　非隶属度

nonmutational multimodel　非突变的多模型

nonnative code   非本地代码
nonnominal emergence   非标称涌现
nonnominal intentional emergence   非标称有意涌现
nonnumeric   非数值的
nonnumeric processing   非数字处理
nonnumerical computation   非数值计算
nonnumerical method   非数值方法
nonnumerical simulation   非数值仿真
nonnumerical variable   非数值变量
nonobservable event   不可观事件
nonobservable variable   不可观变量
nonparametric method   非参数方法
nonparametric random variate   非参数随机变量
nonparametric random variate generation   无参随机变量生成
nonparametrized experimentation   非参数化实验
nonperiodic   非周期性的
nonperiodic output   非周期性输出
nonpropositional knowledge   非命题知识
nonrecoverable error   不可恢复的错误
nonreflexive emergence   非反射性涌现
nonregular error   不规则误差
nonrelational data   非关系数据
nonsampling error   非采样误差
nonsimulatable model   不可仿真模型
nonsimulation   非仿真
nonsimulation model   非仿真模型
nonspeech audio input   非语音音频输入
nonstandard cell   非标准单元
nonstandard data element   非标准数据元素
nonstationary   非平稳的
nonstationary data   非平稳数据
nonstiff simulation   非刚性仿真
nonterminating simulation   非终止仿真
nontraditional data   非传统数据
nontrial simulation   非实验仿真
nonunique random variable   非唯一随机变量
normal distribution   正态分布
normal Petri net   正规佩特里网
normal priority Petri net   正常优先级佩特里网
normalised value   归一化数值
normality of simulation data   仿真数据正态性
normalized stability region   规一化稳定性区域
normative   标准的；规范的；正常的
normative assessment   标准评估
normative behavior   规范的行为
normative decision   规范性决策
normative model   标准模型
normative simulation   规范仿真
normative system   规范系统
norms of experimentation   试验规范
norms of modeling methodology   建模方法学规范
norms of simulation methodology   模拟方法学规范
not perceptual simulation   非感知仿真

not-so-minimal 小尺度
not-so-minimal model 小尺度模型
notation 记号；标记法
notice 声明
notion 概念；想法，意图
notional data 概念数据
NP-complete NP 完全
NP-complete problem NP 完全问题
NP-hard NP 困难
NP-hard problem NP 难题
nth order central difference scheme N 阶中心差分格式
nth order scheme N 阶方案
nuisance variable 滋扰变量
null hypothesis 零假设
null Turing machine simulation 零图灵机仿真
null value 空值
number 数
number of replication 复制次数
number of runs 运行次数
number of simulation runs 仿真运行次数
numeric function 数值函数
numeric integration 数值积分
numeric range validation 数值范围验证
numerical 数值的
numerical analysis 数值分析
numerical approach 数值方法
numerical computation 数值计算
numerical computing 数值计算
numerical differential equation solver 数值微分方程求解器
numerical experiment 数值试验

numerical experimentation 数值实验
numerical input 数字输入
numerical instability 数值不稳定性
numerical integration 数值积分
numerical integration algorithm 数值积分算法
numerical integration method 数值积分法
numerical integration requirements 数值积分需求
numerical method 数值方法
numerical model 数值模型
numerical ODE solver 数值常微分方程求解器
numerical Petri net 数值佩特里网
numerical simulation 数值仿真
numerical solution 数值解
numerical stability 数值稳定性
numerical state-variable 数值状态变量
numerical value 数值
numerical variable 数值变量
numerically stable region 数值稳定区域
numerically unstable region 数值不稳定区域
Nyström algorithm 奈斯特龙算法

Oo

object 对象
object attribute 对象属性
object-based 基于对象的
object-based model 基于对象的模型
object class 对象类

object code 目标代码
object computer 目标计算机
object database 对象数据库
object definition 对象定义
object diagram 对象图
object file 目标文件
object-flow testing 对象流测试
object interaction 对象交互
object interaction control 对象交互控制
object-management architecture 对象管理架构
object model 对象模型
object model framework 对象模型框架
object model template 对象模型模版
object model transformation 对象模型转换
object modeling 对象建模
object-oriented 面向对象的
object-oriented design 面向对象的设计
object-oriented implementation 面向对象的实现
object-oriented language 面向对象的语言
object-oriented model 面向对象的模型
object-oriented modeling 面向对象的建模
object-oriented Petri net 面向对象的佩特里网
object-oriented programming 面向对象的程序设计
object-oriented representation 面向对象的表示（法）
object-oriented simulation 面向对象的仿真
object-oriented technology 面向对象技术
object ownership 对象所有权
object simulation program 对象仿真程序
object SL 对象仿真语言
object type 对象类型
objective 目标；客观的
objective complexity 客观复杂性
objective data 客观数据
objective experience 客观经验
objective function 目标函数
objective reality 客观现实
objective validation 客观验证
objective value 目标值
objective variable 目标变量
objectives-driven methodology 目标驱动的方法学
objectivity 客观性
observability 可观测性
observable 可观测的
observable behavior 可观测行为
observable event 可观测事件
observable plant 可观测装置
observable system 可观测系统
observable variable 可观测变量
observance 遵守；惯例
observation 观测
observation error 观测误差
observation error statistic 观测误差统计
observation event 观测事件

observation function 观测函数
observation mechanism 观测机制
observation model 观测模型
observation state 观测状态
observational data 观测数据
observational error 观测误差
observational knowledge 观测知识
observational variable 观测变量
observe (v.) 观测
observed 观测到的
observed behavior 观测到的行为
observed frequency 观测频率
observed information 观测信息
observed value 观测值
observed variable 观测变量
observer-dependent complexity 观察者依赖的复杂性
observer effect 观察者效应
observer-independent complexity 独立于观察者的复杂性
observer model 观察者模型
obsolete data 过时数据
obsolete model 过时的模型
Occam's razor 奥卡姆剃刀理论
occlusion 阻塞；遮挡
occurrence 发生
odd function 奇函数
ODE (Ordinary Differential Equation) 常微分方程
ODE solver 常微分方程求解器
off-line 离线的
off-line simulation 离线仿真
off-line storage device 离线存储设备
off-site backup 异地备份
off-the-shelf model 现成的模型
off-the-shelf model library 现成的模型库
off-the-shelf simulation package 现成的仿真程序包
offset error 偏移误差
omission 省略；遗漏
omission error 遗漏误差
omitted variable 遗漏变量
on-demand simulation 按需仿真
on-the-job training 在职培训
one-dimensional wave equation 一维波动方程
one-sided game 单边博弈
one-stage understanding 一阶段理解
one-to-many relation 一对多关系
one-to-one relation 一对一关系
one-way coupling 单向耦合
online 在线的
online decision support 在线决策支持
online diagnosis 在线诊断
online experience 在线体验
online game 在线游戏
online help 在线帮助
online learning 在线学习
online mode 联机模式
online model documentation 在线模型文档
online model recommender 在线模型推荐器
online role-play simulation 在线角色扮演仿真
online simulation 在线仿真
online simulation update 在线仿真更新
online SL 在线仿真语言

online storage device 在线存储设备
onsite backup 本地备份
ontological dictionary 本体词典
ontological emergence 本体涌现
ontological markup language 本体标记语言
ontological perspective 本体论视角
ontological representation 本体表达
ontological shift 本体论偏移
ontology 本体；本体论
ontology-based 基于本体的
ontology-based agent simulation 基于本体的智能体仿真
ontology-based description 基于本体的描述
ontology-based dictionary 基于本体的词典
ontology-based framework 基于本体的框架
ontology-based methodology 基于本体的方法学
ontology-based model specification 基于本体的模型规范
ontology-based multiagent simulation 基于本体的多智能体仿真
ontology-based simulation 基于本体的仿真
ontology-based simulation methodology 基于本体的仿真方法学
ontology-based tool 基于本体的工具
ontology fusion 本体融合
ontology language 本体语言
ontology model 本体模型
ontology of emergence 涌现本体
ontomimetic model 本体模型

ontomimetic simulation 本体仿真
opaque thought experiment 不透明的思维实验
open federation 开放联邦
open-form model 开放式模型
open-form simulation 开放式仿真
open game 开型博弈
open hierarchical system 开放的分级系统
open knowledge 公开的知识
open loop 开环
open loop connection 开环连接
open loop control system 开环控制系统
open loop game 开环博弈
open loop simulation 开环仿真
open source 开源
open source game technology 开源游戏技术
open source model 开源模型
open source multicore functional cache memory simulator 开源多核功能高速缓存仿真器
open source programming 开源编程
open source simulation 开源仿真
open source simulator 开源仿真器
open source software 开源软件
open system 开放系统
openness to experience 经验开放性
operability 可操作性
operate 操作
operating value 操作值
operation 操作；运算；运行；军事行动
operation phase 运行阶段

operational analysis　运行分析
operational environment　操作环境
operational game　操作对策
operational gaming　运筹博弈
operational model　操作模型
operational risk　操作风险
operational scenario　操作想定
operational validation　操作验证
operational validity　操作有效性
operational view　操作视图
operations peace　和平行动
operations research　运筹学
operations time alignment　操作时间对齐
operative error　操作误差
operative management knowledge　操作性管理知识
operator　算子；操作员
opportunity　时机，机会
optical computing　光学计算
optimal　最优的
optimal condition　最优条件
optimal reduced order　最佳降阶
optimal reduced-order model　最优降阶模型
optimal solution　最优解
optimality　最优性
optimistic event　乐观事件
optimistic execution　乐观执行
optimistic optimization　乐观优化
optimistic parallel simulation　乐观并行仿真
optimistic simulation　乐观仿真
optimistic simulation algorithm　乐观仿真算法
optimistic simulation architecture　乐观仿真架构
optimistic synchronization　乐观同步
optimistic synchronization protocol　乐观同步协议
optimistic time management　乐观时间管理
optimization　最优化
optimization approach　最优化方法
optimization embedded within simulation　优化嵌入仿真
optimization problem　最优化问题
optimization technique　最优化技术
optimization theory　最优化理论
optimization within simulation　仿真优化
optimizing simulation　优化仿真
optimum　最优
option　可选项
OR input　"或"输入
order　命令；次序
order control　指令控制
order of differential equation　微分方程的阶次
order reduction　降阶
ordered　有序的
ordered logit　有序分对数
ordered set　有序集合
ordering　排序
ordinal optimization　序优化
ordinal optimization algorithm　序优化算法
ordinal optimization theory　序优化理论
ordinal scale　顺序量表；分类量表

ordinal scale-variable 序数尺度变量
ordinal value 序数值
ordinal variable 顺序变量
ordinary differential equation 常微分方程
ordinary differential equation M&S 常微分方程建模与仿真
ordinary differential equation simulation 常微分方程仿真
ordinary Kriging 普通克里格法
ordinary Kriging simulation 普通克里格仿真
organic system-of-systems 有机体系
organization 组织；机构
organizational behavior 组织行为
organizational input 组织输入
organized 有组织的
oriented 面向……的
oriented graph 有向图
original 原始的
original data 原始数据
original notion 原始概念
orthogonal 正交的
orthogonal matrix 正交矩阵
orthogonal multiagent model 正交多智能体模型
orthogonality 正交性
orthogonality of goal components 目标分量的正交性
oscillating state 振荡状态
oscillation 振荡
outbound 出站
outcome 结果
outcome-based error detection 基于结果的误差检测
outcome-driven 结果驱动的
outcome-driven simulation 结果驱动的仿真
outcome metric 结果度量
outcome-oriented simulation 面向结果的仿真
outcome variable 结果变量
output 输出
output analysis 输出分析
output channel 输出通道
output condition 输出条件
output data 输出数据
output device 输出设备
output-driven validation 输出驱动的验证
output duplication 输出复制
output equation 输出方程
output event 输出事件
output file 输出文件
output interpolation 输出插值
output layer 输出层
output limited 输出限制
output limiter 输出限制器
output log 输出日志
output matrix 输出矩阵
output module manager 输出模块管理器
output observation port 输出观测端口
output periodicity 输出周期性
output port 输出端口
output process 输出过程
output segment 输出部分，输出段
output space 输出空间
output terminal 输出端
output time 输出时间

output trajectory 输出轨迹
output understanding 输出理解
output validation 输出验证
output value 输出值
output value set 输出值集合
output variable 输出变量
over-implicit numerical differentiation formula 超隐式数值微分公式
overdetermination of state 状态超定
overdetermined differential algebraic equation 超定微分代数方程
overdetermined linear system 超定线性系统
overdetermined linear-system solver 超定线性系统求解器
overestimation error 高估误差
overflow error 溢出错误
overimplicit numerical integration scheme 超隐式的数值积分法
overloading 过载；重载
overloading attribute 重载属性
override error 覆盖错误
oversimplification 过度简化
oversimplified 过于简单化的
overwrite error 重写错误
owned attribute 自有属性
ownership 所有权

Pp

P-DEVS 并行离散事件系统规范
package 包
package diagram 包图
packaging 组装；封装，包装
Padé approximation 帕德逼近
Padé approximation method 帕德近似法
pair 对；一对；配对
paltry input 微小输入
Pantelides algorithm 庞泰利代斯算法
parabolic PDE 抛物线型偏微分方程
paradigm 范式，模式；范例；方式
parallax simulation 视差仿真
parallax view 视差图
parallel 并行
parallel behavior 并行行为
parallel computing 并行计算
parallel configuration 并行配置
parallel connection 并行连接
parallel DEVS 并行离散事件系统规范
parallel discrete-event simulation 并行离散事件仿真
parallel distributed simulation 并行分布式仿真
parallel execution 并行执行
parallel expansion 平行扩展
parallel process 平行过程
parallel processing 平行处理
parallel processor 并行处理器
parallel simulation 并行仿真
parallel simulation engine 并行仿真引擎
parallel simulator 并行仿真器
parallel visualization 并行可视化
parallel world 平行世界
parallelization 并行化
parallelization of sequential program 顺序程序的并行化
parallelization of sequential simulation

顺序仿真的并行化

parallelization of simulation 仿真的并行化

parallelization pattern 并行化模式

parallelization speedup 并行加速

parallelized simulation 并行仿真

paralogism 谬论；逻辑倒错

parameter 参数，参量

parameter acceptability 参数可接受性

parameter base 参数库

parameter base manager 参数库管理器

parameter calibration 参数校准

parameter consistency 参数一致性

parameter error 参数错误

parameter estimation 参数估计

parameter identification 参数辨识

parameter identification methodology 参数辨识方法学

parameter manager 参数管理器

parameter morphism 参数同态

parameter sensitivity 参数灵敏度

parameter sensitivity analysis 参数敏感度分析

parameter sensitivity analysis methodology 参数灵敏度分析方法学

parameter set 参数集

parameter smoothing 参数平滑

parameter structure 参数结构

parameter tuning 参数调整

parameter validity 参数有效性

parameter value 参数值

parameter variability 参数可变性

parameterization 参数化

parameterize (v.) 用参数表示；确定……的参数

parameterized 参数化的

parameterized experimental frame 参数化试验框架

parameterized robustness 参数化鲁棒性

parametric 参数的，参量的；参数化

parametric control 参数控制

parametric experimental frame 参数化试验框架

parametric model 参数化模型

parametric random variate 参数随机变量

parametric random variate generation 有参随机变量生成

parametric uncertainty 参数不确定性

parametric variable 参数变量

parametrization error 参数化错误

parametrized experimentation 参数化实验

parasitic behavior 寄生行为

Pareto distribution 帕累托分布

Pareto frontier approach 帕累托前沿方法

Pareto optimal approach 帕累托最优法

Pareto optimal solution 帕累托最优解

Pareto random variable 多输入多输出系统

parity error 奇偶检验误差

parse 解析

parser 解析器

part 部分

part-whole relationship 部分整体关系

partial decomposability 部分可分解性
partial differential equation 偏微分方程
partial differential equation M&S 偏微分方程建模与仿真
partial differential equation simulation 偏微分方程仿真
partial equilibrium model 局部均衡模型
partial equilibrium simulation 局部均衡仿真
partial evaluation 部分求值
partial evaluator 部分求值程序
partial model 局部模型
partial ordering 偏序
partial simulation 局部仿真
partial specification 部分规范
partial state 部分状态
partial truth 部分真实
partial validity 部分有效性
partial value 偏值
partially causalized 部分因果关系的
partially-causalized algebraic equation system 部分因果代数方程系统
partially-causalized equation system 部分因果方程系统
partially-colored structure digraph 部分着色结构有向图
partially false 部分错误的
partially ordered set 部分有序集合
partially true 部分正确的
participant 参与者
participant variable 参与者变量
participation-based 基于参与的
participative 参与式；参与的
participative modeling 参与式建模
participative simulation 参与式仿真
participatory agent simulation 参与式代理仿真
participatory design 参与式设计
participatory experiment 参与式试验
participatory modeling 参与式建模
participatory simulation 参与式仿真
participatory simulation modeling 参与式仿真建模
partition 分区，划分；隔离物
partition (v.) 区分；隔开；分割
partition analysis 分区分析
partition testing 划分测试
partitioned 分段的
partitioning 划分
partitioning metric 分区度量
pass (v.) 通过；传递；忽略
passing 经过，通过；经过的，短暂的
passive component 无源元件
passive entity of a model 模型的被动实体
passive multimodel 被动多模型
passive state 被动状态
passive stigmergic system 被动式群体激发工作现象的系统
passive stigmergy 被动式群体激发工作
passive system 无源系统；被动系统
passively accepted input 被动接受的输入
password error 密码错误
past behavior 过去的行为
past experience 过去的经验
past trend 过去的趋势

path planner 路径规划器
path testing 路径测试
pathway simulation 路径仿真
pattern 模式
pattern-based 基于模式的
pattern-based transformation 基于模式的转换
pattern-directed multimodel 模式导向的多模型
pattern effect 模式效应
pattern language 模式语言
pattern-oriented 面向模式的
pattern-oriented modeling 面向模式的建模
payoff 报偿；支付
payoff function 支付函数
payoff matrix 支付矩阵
PC game PC 博弈
PDE (Partial Differential Equation) 偏微分方程
PDE modeling 偏微分方程建模
PDEVS model 平行离散事件系统规范模型
peace game 和平博弈
peace simulation 和平仿真
peace support 和平支持
peace support game 维和博弈
pedagogical simulation 教学仿真
pedigree 系谱
peer-to-peer model 对等的模型，点对点模型
peer-to-peer simulation 点对点仿真
pen and paper simulation 笔纸仿真
pen input 笔控输入
perceivable 可感知的

perceive (v.) 感知
perceived 感知到的
perceived correctness 感知到的正确性
perceived data 感知数据
perceived emergence 感知到的涌现
perceived endogenous input 感知到的内源输入
perceived event 感知事件
perceived exogenous input 感知到的外源输入
perceived external fact 感知的外部事实
perceived external input 感知的外部输入
perceived fact 感知的事实
perceived goal 感知的目标
perceived input 感知的输入
perceived internal fact 感知的内部事实
perceived internal input 感知的内部输入
perceived model 感知的模型
perceived payoff function 感知的支付函数
perceived reality 感知现实
perceived situation 感知到的情况
perceived suitability 适用性感知
perceived truth 感知到的真理
percentage error 百分比误差
percept 感知的对象
perception 感知，感觉
perception-based coupling 基于感知的耦合
perception-based modeling 基于感知

的建模
perception bias 知觉偏差
perception error 感知误差
perceptual 感知的
perceptual input 感官输入
perceptual knowledge 感性认识
perceptual simulation 感知仿真
perceptual variable 感知变量
perfect information game 完备信息博弈
perfect model 完美的模型
perfect rationality 完全理性
perfective maintenance 完善性维护
perfective model maintenance 完善的模型维护
performance 性能
performance analysis 性能分析
performance assessment 性能评估
performance attribute 性能属性
performance evolution 性能演化
performance indicator 性能指标
performance measure 性能测量
performance measurement 性能测量
performance metric 性能度量
performance model 性能模型
performance parameter 性能参数
performance prediction 性能预测
performance requirements 性能需求
performance testing 性能测试
performance tracking 性能跟踪
period 周期；时期
periodic 周期性的
periodic behavior 周期性行为
periodic boundary condition 周期边界条件
periodic function 周期函数
periodic output 周期性输出
periodic system 周期系统
periodicity 周期性
permanent error 固定误差
permissible output 允许的输出
permutation matrix 置换矩阵
persistence 持久性
persistent 持续的
persistent data 持续数据
persistent environment 持久性环境
persistent error 持续误差
persistent node 持久节点
persistent object 持久对象
persistent state 持久状态
persistent Turing machine 持久图灵机
personal data 个人资料
personal error 人为误差
personal game 个人游戏
personal knowledge 个人知识
personality filter 个性过滤器
personalization technology 个性化技术
personalized game 个性化游戏
personalized system 个性化系统
personalized user interface 个性化用户界面（接口）
perspective 透视；视角
persuasion model 劝导模型
persuasive media simulation 劝导媒体仿真
persuasive simulation 劝导仿真
persuasive technology 劝导技术
perturbation index 摄动指数
pervasive 普遍的

pervasive computing 普适计算
pervasive game 随境游戏
pervasive grid simulation 普适化网格仿真
pervasive simulation 普适仿真
pervasive system 普适系统
pervasive technology 普适技术
petascale 千兆级
petascale computer 千兆级计算机
petascale computing 千兆级计算
petascale simulation 千兆级仿真
Petri net 佩特里网
Petri net model 佩特里网模型
Petri net simulation 佩特里网仿真
phase 阶段；相位
phase-based 基于相位的
phase-based model 基于相位的模型
phase difference 相位差
phase model 相位模型
phase plane plot 相平面图
phase synchronization 相位同步
phase transition 相变
phasor 相量
phenomenal experience 现象性经验
phenomenological 现象的，现象学的
phenomenological approach 现象学方法
phenomenological error 现象学误差
phenomenological experience 现象学经验
phenomenological model 唯相模型
phenomenon 现象
phenotype 表型；显型
phenotypic 表型的
phenotypic modeling 表型建模
phenotypical 表型的
phenotypical change 表型变化
phenotypically 表型地
philosophical foundation 哲学基础
physical aspect 物理方面
physical attribute 物理属性
physical data model 物理数据模型
physical environment 物理环境
physical experiment 物理试验
physical fidelity 物理逼真度
physical interface 物理接口
physical model 物理模型
physical model attribute 物理模型属性
physical modeling 物理建模
physical network 物理网络，实体网络
physical perception 物理感知
physical reality 物理现实
physical realization 物理实现
physical schema 物理模式
physical simulation 物理仿真
physical system 物理系统
physical system simulation 物理系统仿真
physical time 物理时间
physical variable 物理变量
physically-based 基于物理的
physically-based modeling 基于物理的建模
physics-based 基于物理的
physics-based design 基于物理的设计
physics-based model 基于物理的模型
physics-based modeling 基于物理的建模

physics-based simulation 基于物理的仿真
physics engine 物理引擎
piecewise continuous segment 分段连续段
piecewise linear 分段线性
piecewise polynomial trajectory 分段多项式轨迹
pipeline topology 管线拓扑
pixel 像素
pixel-based model 基于像素的模型
plan 计划，规划；方案
planned behavior 计划行为
planner 规划者；规划器
planning 规划，计划
platform 平台
platform dependence 平台依赖性
platform dependent 平台相关的
platform implementation 平台实现
platform independence 平台独立性
platform independent 平台无关的
platform-independent model 与平台无关的模型
platform-independent modeling 与平台无关的建模
platform integration 平台集成
platform level 平台级
platform requirements 平台需求
platform-specific 平台特定的
platform-specific model 平台特定的模型
platform-specific modeling 平台特定的建模
Platonic reality 柏拉图式的现实
plausibility 似真性

plausible experiment 似真试验
plausible model 似真模型
plausible scenario 似真场景
plausible state 似真状态
plausible trajectory 似真轨迹
plausible value 似真值
play 播放
play room 游戏室
player 比赛者；演奏者；演员；游戏者
player profile 选手简介
player room 游戏者空间
playing 游戏；比赛
plot 绘图
plot interval 绘图间隔
plug-and-play composability 即插即用的可组合性
plug-and-play functionality 即插即用功能
plug-in 插件
plug-in-based 基于插件的
plug-in model 插件模型
plural modeling 多元建模
plus value 附加值
poiesis 创制
point 点
point behavior 点特征
point distribution model 点分布模型
point of view 论点；见解；视点
pointing mechanism 对准机理
Poisson distribution 泊松分布
Poisson random variable 泊松随机变量
Poisson stability 泊松稳定性
polar coordinate 极坐标

polarization error 极化误差
polarized network 极化网络
pole 极点
policy comparison 策略比较
policy experimentation 策略实验
political game 政治博弈
political variable 政治变数
polyformalism composition 多形式构成
polyformalism model 多形式模型
polyformalism model composition 多形式模型构成
polyformalism modeling 多形式建模
polyformic model 多形态模型
polygon 多边形
polygon-oriented 面向多边形的
polygon-oriented modeling 面向多边形的建模
polymorphism 多态性
polynomial 多项式
polynomial approximation 多项式逼近
polynomial model 多项式模型
polynomial trajectory 多项式轨迹
poor design 欠设计
Popperian falsification test 波普尔伪造试验
port 端口
port-based 基于端口的
port-based modeling 基于端口的建模
port-based modeling paradigm 基于端口的建模范式
port interconnection 端口互连
portability 可移植性
portable 便携式；可移植的
portable model 轻量级模型
portable simulation 轻量级仿真
portable simulation system 轻量级仿真系统
portal 门户
position error 位置误差
positional input 位置输入
positive constraint 正约束
positive emergence 积极的涌现
positive emergent behavior 积极的涌现行为
positive-feedback loop 正反馈环
positive spatial error 正空间误差
positivism 实证论
positivist methodology 实证方法论
positivist paradigm 实证主义范式
positivist theory 实证理论
postevent 后事件
postexecution state 后执行状态
postgame analysis 赛后分析
postmodernism 后现代主义
postprocessing 后处理
postprocessing step 后处理步骤
postprognostic analysis 预后分析
postrun 运行后的
postrun activity 运行后活动
postrun analysis 运行后分析
postrun computation 运行后计算
postrun output 运行后输出
postsilicon 流片后的；晶片
postsilicon validation 流片后验证
postsilicon verification 流片后校核
postsimulation analysis 仿真后分析
postsimulation phase 仿真后阶段
postsimulation report 仿真后报告

poststudy 研究之后的
poststudy activity 研究后活动
poststudy analysis 研究后分析
poststudy computation 研究后计算
poststudy output 研究后输出
postulate 前提；假设
potential defect 潜在缺陷
potential field 势场
power flow 功率流
power law 幂次法则
power law function 幂律函数
power split (*v.*) 功率分流
practical knowledge 实践知识
practical model 实用模型
practicality of use 实际用途
practice 实践
practitioner 实践者；从业者
Practopoiesis 普拉克珀斯
Practopoietic 普拉克珀斯的
Practopoietic cycle 普拉克珀斯周期
Practopoietic system 普拉克珀斯系统
pragmatic 实用主义的
pragmatic approach 语用方法
pragmatic assessment 语用评估
pragmatic assessment of the goal 目标的语用评估
pragmatic assessment of the goal of the study 研究目标的语用评估
pragmatic evaluation 语用评价
pragmatic interoperability 语用互操作性
pragmatic interoperability level 语用互操作性水平
pragmatic level 语用层次
pragmatic model 语用模型
pragmatic perspective 语用视角
pragmatic understanding 语用理解
pragmatics 语用论
pragmatism 实用主义
precise 精确的
precise coupling 精确的耦合
precise value 精确值
precision 精度
precocial system 前期系统
predator-prey model 猎食模型
predefined activity 预定义活动
predefined hypothesis 预定义假设
predicate 断言；声明
predicate calculus 谓词演算
predicate transformation 谓词转换
predicate variable 谓词变量
predict (*v.*) 预测，预估
predictability 可预测性
predictability of system quality characteristic 系统质量特性的可预测性
predictable 可预测的
predictable behavior 可预测行为
predictable data 可预测数据
predictable emergent behavior 可预测突现行为
predictable system 可预测系统
predicted behavior 预测的行为
predicted data 预测的数据
predicted trajectory 预测的轨迹
predicted value 预测值
predicted variable 预测变量
prediction 预测，预估
prediction error 预测误差
prediction model 预测模型

prediction-model maintainability 预测模型的可维护性
prediction-model traceability 预测模型的可追溯性
prediction simulation 预测仿真
prediction system 预测系统
predictive 预测性
predictive biosimulation 预测性仿生仿真
predictive decision 预测性决策
predictive display 预测性显示
predictive model 预测模型
predictive model validity 预测模型有效性
predictive modeling 预测建模
predictive simulation 预测仿真
predictive system 预测系统
predictive technique 预测技术
predictive validation 预测性验证
predictive validity 预测效度
predictively valid 预测有效的
predictor 预测；预测器
predictor-corrector 预测校正
predictor-corrector method 预估测校正法
predictor model 预测模型
predictor variable 预测变量
preexecution state 预执行状态
preference variable 首选项变量
preferred behavior 首选行为
preferred state variable 首选状态变量
preliminary phase 初步阶段
preparation 准备
preprocessing 预处理
prerequisite condition 先决条件

prerun activity 预运行活动
prescribe (v.) 规定
prescription 说明；法则
prescriptive knowledge 说明性知识
prescriptive model 说明性模型
prescriptive simulation 规定式仿真
presence 出席，到场；呈现
presentation layer 表示层
presentation manager 演示管理器
presentation of result 结果演示
preservation of robustness 鲁棒性维持
presimulation phase 预仿真阶段
prestudy activity 预研究活动
presumed behavior 假定的行为
pretend (v.) 假装
pretended reality 虚假的现实
pretended self-model 虚假的自模型
pretending 假装
prevent (v.) 预防
preventability 可预防性
preventable 可预防的
prevention 预防
preventive maintenance 预防性维护
preventive model-maintenance 预防性模型维护
primary assumption 原始假设
primary input 初始输入
primary task 主要任务
primitive 原始的；基元；原语
primitive activity 元活动；初始活动
primitive model 原始的模型
primitive modeling 原始建模
primitive transformation 原语转换
principle 原则，原理
principle of simplicity 简化原则

print interval 打印间隔
printer error 打印机错误
priority Petri net 优先佩特里网
priority queue data structure 优先队列数据结构
Prisk identification 普利斯科识别
prisoner's dilemma 囚徒困境
privacy grid 私密网络
private event 私有事件
private state event 私有状态事件
proactive 前摄的；主动式
proactive approach 主动式方法
proactive behavior 前摄行为
proactive decision 前摄决策
proactive interaction 前摄交互
proactive maintenance 前摄维护
proactive mechanism 主动机制
proactive model maintenance 前摄模型维护
proactive system 前摄系统
proactively 前摄地；主动地
probabilistic 概率的，概率性的
probabilistic analysis 概率分析
probabilistic assumption 概率假设
probabilistic bisimilarity 概率双相似
probabilistic bisimulation 概率互仿真
probabilistic database 概率数据库
probabilistic event 概率事件
probabilistic function 概率函数
probabilistic Ising model 概率伊辛模型
probabilistic model 概率模型
probabilistic modeling 概率建模
probabilistic process 概率过程
probabilistic risk assessment 概率风险评估
probabilistic sensibility analysis 概率灵敏度分析
probabilistic simulation process 概率仿真过程
probabilistic variable 概率变量
probability 概率；可能性
probability density function 概率密度函数
probability distribution 概率分布
probability distribution function 概率分布函数
probability error 概率误差
probability of acceptance 接受概率
probability of occurrence 事件概率
probability value 概率值
probable 概率的；可能的
probable error 概率误差
problem 问题
problem analysis 问题分析
problem complexity 问题复杂性
problem context 问题情境
problem documentation 问题文件
problem domain 问题域
problem domain requirements 问题域需求
problem formulation 问题公式化
problem solving 问题求解
problem solving environment 问题求解环境
problem solving paradigm 问题解决范式
problem solving process 问题解决过程
problem space 问题空间

problem specific knowledge 特定问题知识
problem specification 问题说明
problem specification phase 问题规格说明阶段
problem-specification-time efficiency 问题-规范-时间的效率
problematique 相互关联的问题群
procedural knowledge 过程性知识
procedural knowledge processing 过程性知识处理
procedural model 程序模型
procedure 程序；过程
procedure error 程序误差
procedure-related knowledge 程序相关的知识
process 过程；进程；处理
process abstraction 过程抽象化
process-based 基于过程的
process-based discrete-event simulation 基于过程的离散事件仿真
process control 过程控制
process error 过程误差
process improvement 过程改进
process improvement modeling 过程改进建模
process integration 过程集成
process interaction 过程交互
process interaction model 进程交互模型
process interaction simulation 过程交互仿真
process maturity model 过程成熟度模型
process model 过程模型
process modeling 过程建模
process-oriented 面向过程的
process-oriented DEVS 面向过程的离散事件系统规范
process-oriented model 面向过程的模型
process-oriented modeling 面向过程建模
process-oriented simulation 面向过程仿真
process-oriented world view 面向过程的世界观
process quality 过程质量
process reliabilism 过程可靠性
process simulation 过程仿真
process template 流程模板
process variable 过程变量
processable 适合处理的，可处理的
processed 加工过的
processing 处理；加工
processing error 处理误差
processing node 处理节点
processor 处理器
produce (v.) 生成
producer 制造者，制造机；发生器
product 产品；成果
product definition 产品定义
product development 产品开发
product evaluation 成果评估
product model 产品模型
product quality 产品质量
product requirements 产品需求
product specification 产品规格说明
product specification requirements 产品规格需求

product specification standard 产品规格标准
product testing 产品测试
production code 生产代码
productivity metric 生产力度量
professional 专业的
professional certification 专业认证
professional certification commission 专业认证委员会
professional simulation 专业仿真
professional simulation engineer 专业仿真工程师
professional simulation systems engineer 专业仿真系统工程师
professional simulationist 专业仿真者
profile 剖面图；侧面；轮廓；形象
profiling 压型；形貌测量；剖面测量；仿形
prognosed value 预测值
prognostic analysis 预后分析；后相关因素分析
prognostic model 预后模型
prognostic variable 预后变量
program 程序；规划
program (v.) 编程
program acceptability 程序可接受性
program analysis 程序分析
program-based 基于程序的
program-based simulation 基于程序的仿真
program certification 程序认证；程序检验
program check 程序检查
program composition 程序组合
program comprehensibility 程序可理解性
program correctness 程序正确性
program documentation 程序文件
program efficiency 程序效率
program error 程序错误
program generation support 程序生成支持
program generator 程序生成器
program group 程序组
program management 程序管理
program manager 程序管理器
program modularity 程序模块化
program-oriented simulation 面向程序的仿真
program reliability 程序可靠性
program robustness 程序鲁棒性
program-sensitive error 程序敏感性误差
program transformation 程序转换
program variable 程序变量
program verification 程序校核
programmable analysis 可编程分析
programmed input/output 程序输入/输出
programmed model 程序化模型
programming 编程，程序设计；规划
programming by questionnaire 问卷式编程
programming error 编程错误
programming language 编程语言
programming paradigm 程序设计范式
programming technique 编程技术
progressive model fitting 渐进模型拟合

project 工程，项目；计划，规划；课题
project management 项目管理
project planning 项目规划
project-specific 特定项目的
project-specific knowledge 特定项目知识
projection error 投影误差
proof 证明，论证，验证
proof-of-concept 概念验证，概念证明
proof-of-concept game 概念验证博弈
proof-of-concept simulation 概念验证仿真
proof of correctness 正确性证明
proofing 证明；试验；验算
propagated error 传播误差
property 性能，特性
property classification 性质分类
property selection 属性选择
property verification 特性校核
proponent 提议者；支持者
proportional error 比例误差；相对误差
proposition 命题；建议
proposition of an assumption 假设的命题
proposition variable 命题变量，命题变元
propositional knowledge 命题性知识
propositional model 命题模型
propositional variable 命题变量
protection mechanism 保护机制
protocol 协议
protocol converter 协议转换器
protocol data unit 协议数据单元

protocol data unit standard 协议数据单元标准
protocol entity 协议实体
protocol suite 协议套组
prototype 样机；原型
prototype-based model 基于原型的模型
prototype SL 原型仿真语言
prototypical model 原型模型
prototyping 原型设计；样机研究
provider 提供者
proximate cause 近因
proximate system 近似系统
proxy 代理；代理人；代理服务器；委托书
proxy simulation 代理仿真
pruning 修剪；剪枝；删除
pseudo-observation 伪观测
pseudoanalytical simulation 伪分析仿真
pseudocode 伪码
pseudoderivative 伪导数
pseudoindependent variable 伪自变量
pseudorandom number 伪随机数
pseudorandom number generator 伪随机数发生器
pseudosimulation 伪仿真
pseudosimulator 伪仿真器
psychological fidelity 心理逼真度
public behavior 公众行为
public data 公用数据
public domain 公共领域
public domain game 公共领域博弈
public domain simulation 公共领域仿真

publication 发布
publicly available 公开可用的
publicly available game technology 公开提供的游戏技术
publicly available model 公开可用的模型
publicly available simulation 公开可用的仿真
publicly available simulator 公开可用的仿真器
publicly available software 公开可用的软件
publish (*v.*) 发布
pull factor 拉力因素
pure coordination game 纯协调博弈
pure feedback coupling 纯反馈耦合
pure feedback coupling recipe 纯反馈耦合配方
pure software simulation 纯软件仿真
purpose 目的；宗旨
purpose of simulation 仿真目的
purpose perspective 目的视角
purposeful experiment 目的性的试验
purposeful simulation 有目的仿真
push factor 推动因素
puzzle game 益智游戏

Qq

quadratic error 平方误差
quadruple pole 四极
qualification 资格；条件；限制
qualification certificate 资格证书
qualification criterion 资格标准
qualified variable 限定变量
qualifier condition 限定符条件
qualitative 定性的,定质的；性质上的；质量的
qualitative algorithm assessment 定性算法评估
qualitative analysis 定性分析
qualitative assessment 质量评估；定性评估
qualitative assumption 定性假设
qualitative causal model 定性因果模型
qualitative concept 定性概念
qualitative constraint 定性约束
qualitative data 定性数据
qualitative decision-variable 定性决策变量
qualitative digraph model 定性有向图模型
qualitative evaluation 定性评价
qualitative measure 定性测量
qualitative mental model 定性心理模型
qualitative metric 定性度量
qualitative mixed simulation 定性混合模拟
qualitative model 定性模型
qualitative model analysis 定性模型分析
qualitative model validation 定性模型验证
qualitative modeling 定性建模
qualitative reasoning 定性推理
qualitative rule 定性规则
qualitative simulation 定性仿真
qualitative simulation model 定性仿真

模型
qualitative simulation system 定性仿真系统
qualitative term 定性术语
qualitative value 定性值
qualitative variable 定性变量
quality 质量；特性
quality assurance 质量保证
quality assurance paradigm 质量保证范式
quality characteristic 质量特征
quality control 质量控制
quality estimation 质量估计
quality indicator 质量指标
quality issue 品质问题
quality measure 质量测量
quality metric 品质度量
quality model 质量模型
quality prediction 质量预测
quantifiable measure 可量化测度
quantifiable relationship 可量化关系
quantification 量化，定量
quantified variable 量化变量
quantify (v.) 量化，定量
quantitative 定量的
quantitative algorithm assessment 定量算法评估
quantitative analysis 定量分析
quantitative assessment 定量的评估
quantitative assumption 定量的假设
quantitative causal model 定量因果模型
quantitative concept 定量的概念
quantitative constraint 定量约束
quantitative data 定量数据

quantitative decision variable 定量决策变量
quantitative digraph model 定量有向图模型
quantitative evaluation 定量评价
quantitative measure 定量测量
quantitative measurement 定量测量
quantitative metric 定量度量
quantitative mixed simulation 定量混合仿真
quantitative model 定量模型
quantitative model analysis 定量模型分析
quantitative model validation 定量模型验证
quantitative modeling 定量建模
quantitative rule 定量规则
quantitative simulation 定量仿真
quantitative simulation model 定量仿真模型
quantitative term 定量术语
quantitative value 定量值
quantitative variable 定量变量
quantity 数量
quantity metric 数量度量
quantization 量化
quantization-based 基于量化的
quantization-based approximation 基于量化的逼近
quantization-based integration 基于量化的积分
quantization-based method 基于量化的方法
quantization function 量化函数
quantization of state variable 状态变量

的量化
quantization with hysteresis 量化滞后
quantized DEVS simulator 量化离散事件系统规范仿真器
quantized input 量化输入
quantized input variable 量化输入变量
quantized integrator 量化积分器
quantized output variable 量化输出变量
quantized simulator 量化仿真器
quantized state 量化状态
quantized state system 量化状态系统
quantized state system method 量化状态系统方法
quantized state system solver 量化状态系统求解器
quantized state variable 量化状态变量
quantized system 量化系统，量子化系统
quantized value 量化数值
quantized variable 量化变量
quantizer 量化器；数字转换器
quantizing 量化
quantum 量子
quantum algorithm 量子算法
quantum computer 量子计算机
quantum computing 量子计算
quantum continuous variable 量子连续变量
quantum information processing 量子信息处理
quantum-inspired intelligent system 量子启发的智能系统
quantum logic 量子逻辑

quantum object 量子对象
quantum photonic 量子光子
quantum processor 量子处理器
quantum property 量子特性
quantum simulation 量子仿真
quantum simulator 量子仿真器
quantum state 量子态
quantum superposition 量子叠加
quantum system 量子系统
quantum system architecture 量子系统结构
quantum variable 量子变量
quantum walk 量子游走
quasi-analytic simulation 准分析（解析）仿真
quasi-analytic simulation technique 准分析（解析）仿真技术
quasi-continuous simulation 准连续仿真
quasi experiment 准实验
quasi experimentation 准实验
quasi-identity simulation 类同式仿真
quasi-linear PDE 准线性偏微分方程
quasi-Monte Carlo 准蒙特卡罗
quasi-Monte Carlo simulation 准蒙特卡罗仿真
quasi-natural experiment 准自然实验
quasi optimization 准优化
questionnaire 问卷
queue 队列
queue error 队列误差
queue simulation 队列仿真
queueing model 排队模型
queueing network model 排队网络模型

queueing system 排队系统
queueing theory 排队论
queuing discipline 排队规则
quiescent state 静态
quiz game 问答游戏

Rr

radar input 雷达输入
random 随机
random error 随机误差
random experiment 随机实验
random failure 随机失效
random function 随机函数
random number 随机数
random number generator 随机数发生器
random number seed 随机数种子
random process 随机过程
random sample 随机样本
random sampling 随机采样
random search 随机搜索
random simulation 随机仿真
random state 随机状态
random variable 随机变量
random variate 随机变量
random variate generation 随机变量生成
random walk 随机游走
random walk simulation 随机游走仿真
randomization 随机化
randomized balanced incomplete block design 随机平衡不完全区组设计
randomized complete block design 随机完全区组设计
randomized design 随机设计
range 范围
range of a variable 变量范围
range of applicability 适用性范围
range of model accuracy 模型精度范围
range set 区域集合
range set of a variable 变量范围集
ranging error 测距误差
ranked model 排列模型
ranked relation 排列关系
ranked transformation 排列转换
rapid prototyping 快速原型法
rare event 稀少事件
rare-event analysis 稀有事件分析
rare-event probability 稀有事件概率
rare-event sampling 稀有事件采样
rare-event simulation 稀有事件仿真
rare-event simulation methodology 稀有事件仿真方法
rare-event simulation technique 稀有事件仿真技术
rate 速率
rate-based 基于速率的
rate-based simulation 基于速率的仿真
rate data 速率数据
rate of change 变化率
rate variable 速率变量
ratio 比率
ratio scale 比率尺度
ratio scale variable 比率尺度变量
ratio simulation 比率仿真
ratio value 比率值

ratio variable 比率变量
rational behavior 合理的行为
rational knowledge 理性知识
rational model 理性模型
rational response 理性响应
rationality 合理性
raw information 原始信息
ray tracing 光线追踪
ray-tracing model 光线追踪模型
re-emergent state 重新出现的状态
re-evaluate 再评估
re-instantiation 再实例化
reachability analysis 可达性分析
reachability graph 可达图
reachability tree 可达树
reactive agent 反应式智能体，反应式代理
reactive approach 反应方法
reactive behavior 反应的行为
reactive decision 反应式决策
reactive interaction 反应性交互
reactive knowledge processing 反应式知识处理
reactive maintenance 反应式维护
reactive mechanism 反应机制
reactive model maintenance 反应式模型维护
reactive modeling 反应建模
reactive planning 反应规划
reactive relationship 反应型关系
reactive system 反应系统
reactive technique 反应技术
read error 读错误
readability 可读性
readable screen 可读屏幕
reader 读者
real 真实的
real axis 实轴
real battlefield 真实的战场
real environment 真实环境
real experience 真实经验
real experiment 真实实验
real knowledge 真实知识
real-life experience opportunity 真实体验机会
real-life-like experience 真实生活相似的经验
real-life simulation 真实仿真
real object 实际对象
real-part 实部
real system 真实系统；实际系统
real-system-based training 基于实际系统的训练
real-system data 实际系统数据
real-system data acceptability 实际系统数据可接受性
real-system-enriching simulation 实际系统增强仿真
real system experimentation 实际系统实验
real-system-support simulation 实际系统支持仿真
real-time 实时
real-time clock 实时时钟
real-time clock synchronization impulse 实时时钟同步脉冲
real-time constraint 实时约束
real-time continuous simulation 实时连续仿真
real-time data 实时数据

real-time data-driven simulation 实时数据驱动仿真
real-time decision making simulation 实时决策仿真
real-time domain 实时域
real-time environment 实时环境
real-time execution 实时执行
real-time input 实时输入
real-time input testing 实时输入测试
real-time modeling 实时建模
real-time object-oriented modeling 实时面向对象的建模
real-time output 实时输出
real-time platform reference 实时平台参考
real-time prediction 实时预测
real-time programming 实时程序设计
real-time service 实时服务
real-time simulation 实时仿真
real-time solution 实时解决方案
real-time system 实时系统
real-time training 实时训练
real-time validation 实时验证
real-time visualization 实时可视化
real value 真值
real-valued variable 真值变量
real variable 实变量
real world 真实世界
real-world data 真实世界数据
real-world time 真实世界的时间
realism 现实主义；实在论
realistic game 现实博弈
realistic model 现实的模型
realistic output 实际输出
realistic virtual reality 真实感虚拟现实
realistic virtual reality entertainment 逼真的虚拟现实娱乐
reality 实在；现实
reality-based 基于现实的
reality-based constraint 基于现实的约束
reality horizon 现实视野
reality-in-the-loop 现实在环（回路）
realization 实现
realization context 实现情境
realization of experimental frame 实验框架的实现
realized value 实现的价值
reasonable 适当的，合理的
reasonable model 合理模型
reasonable simulation 合理仿真
reasoning 推理
reasoning about models 关于模型的推理
reasoning-based intelligent system 基于推理的智能系统
reasoning error 推理误差
reasoning model 推理模型
reasoning simulation 推理仿真
reasoning under uncertainty 不确定性推理
receive operation 接收操作
received operation 接收到的操作
receptor 接收器
reckoning 计算；清算
recognition 识别
recognized assumption 公认假设
recommendation 推荐
recommended 被推荐的

recommender 推荐人
reconfigurable 可重构的
reconfigurable simulation 可重构仿真
reconfigurable simulator 可重构仿真器
reconfigurable software architecture 可重构的软件架构
reconfigurable synthetic environment 可重构合成环境
reconfigurable system 可重构系统
reconfigurable training 可重构训练
reconfiguration 重配置
reconfiguration facility 重配置设备
reconstruction 重建
recoverability 可恢复性
recoverable 可恢复的
recoverable error 可恢复的误差
recovery 恢复
rectification error 校正误差
recurrent environment 周期性环境
recursion 递归
recursive 递归的
recursive definition 递归定义
recursive model 递归模型
recursive simulation 递归仿真
redemption value 递归值
reduced-complexity model 简化复杂性的模型
reduced model 简化的模型
reduced nodeset 简化的节点集
reduced order 降阶
reduced-order model 降阶的模型
reduced system 简化系统
reducibility 可简化性；可还原性
reducibility of goal components 目标分量的可还原性
reducible 可简化的
reducible multimodel 可简化的多模型
reducible uncertainty 可简约的不确定性
reduction 还原；简化；规约
reductionism 还原论；简化论
reductionism paradigm 还原论范式；简化论范式
reductionist 简化（还原）论者
reductionist methodology 还原论方法学
reductionist point of view 还原论视点
redundant aspect 冗长方面
redundant code 冗余代码
redundant state 冗余状态
redundant system 冗余系统
referability 可参考性
reference 参考；基准；引用
reference architecture 参考架构
reference data 参考数据
reference input 参考输入
reference model 参考模型
reference value 参考值
reference variable 参考变量
reference version 参考版本
referenced variable 参考变量
referent 指示物
referent model 参照模型
referent of an assumption 假设的参照物
referential constraint 参考约束
referential ontology 参考本体
referential value 参考值

refined mesh   细化网格
refinement   细分；精化
refinement technique   细分技术
reflected attribute   反射的属性
reflected object   反射的物体
reflective   反射的
reflective knowledge processing   反射式知识处理
reflective simulation   反射式仿真
reflexive emergence   反射涌现
refractory period   不应期；顽固期；耐火期
regenerative simulation   再生仿真
regime   政权；政体
region   区域
region-based   基于区域的
register   注册；寄存器
register error   寄存器错误
registration error   注册错误
registry error   注册表错误
regressand   回归方程中的从属变量
regressed variable   回归变量
regression   回归；退化
regression analysis   回归分析
regression metamodel   回归元模型
regression metamodeling technique   回归元建模技术
regression model   回归模型
regression technique   回归技术
regression testing   回归测试
regressor   回归量
regula falsi   试位法
regular error   常规错误；正则错误
regular simulation   常规仿真
regular tessellation   规则曲面细分

rehearsal   排练，预演
reinitialization discontinuity   重新初始化不连续性
rejection   拒绝
rejection error   拒绝错误
rejection probability   拒绝概率
related   相关的
related model   相关模型
related simulation   相关仿真
relation   关系
relational assumption   关联性假设
relational concept   关系概念
relational database   关系数据库
relational dictionary   关系词典
relational model   关系模型
relationally equivalent system   关系等价系统
relationship   关系
relative   相对的
relative error   相对误差
relative error control   相对误差控制
relative truth   相对真理
relaxation   松弛
relaxation algorithm   松弛算法
relaxation method   松弛法
relevance   关联
relevant   相关的
relevant data   相关数据
relevant information   相关信息
relevant input   相关输入
relevant model   相关模型
relevant processing   相关处理
relevant stimulus   相关激励
relevant variable   相关变量
reliabilism   可靠主义

reliabilist 可靠者
reliabilist epistemology 可靠性认识论
reliability 可靠性
reliability analysis 可靠性分析
reliability evaluation 可靠性评价
reliability issue 可靠性问题
reliability model 可靠性模型
reliability modeling 可靠性建模
reliability prediction 可靠性预测
reliable 可靠的
reliable approximation 可靠的逼近
reliable model 可靠的模型
reliable model repository 可靠的模型库
reliable product 可靠的产品
reliable service 可靠的服务
reliable simulation 可靠的仿真
reliable simulation model 可信的仿真模型
reliance 信任
remaining fault 剩余故障
remote error 远程错误
remote simulation 远程仿真
renaming error 重命名错误
repeatability 可重复性
repeatable 可重复的
repeatable environment 可重复环境
repeated game 重复博弈
repetitive aspects of modeling 建模的重复方面
repetitive behavior 重复的行为
replacement 替换
replacement technology 置换技术
replacement value 置换值
replica 复制品；副本

replicability 可复制性
replicability-aware 可复制性感知
replicability-aware M&S 可复制性感知建模与仿真
replicability-aware modeling 可复制性建模
replicable experimental result 可复制的实验结果
replicable model 可复制模型
replicable result 可复制结果
replicate (v.) 复制
replicated experiment 复制的实验
replicated model 复制的模型
replication 复制
replication analysis 复制分析
replication experiment 复制实验
replicative experiment 复制性实验
replicative model validity 复制性模型有效性
replicative simulation 复制性仿真
replicative validation 复制验证
replicative validity 复制有效性
replicatively valid 复制有效的
report 报告
repository 库；栈
representation 表示（法），表征
representation consistency 表示一致性
representation error 表示错误
representation technology 表示技术
representational accuracy 表征精度
representational model 表征模型
representational reality 表征现实
representational requirements 代表性需求

representative output 典型输出
reproducibility 再现性
reproducible 可重现的；可重复的
reproducible computational experiment 可重复计算实验
reproducible emergence 可重现涌现
reproducible experiment 可重复实验
reproducible research 可重复研究
reproducible simulation 可重复仿真
repurposed game 新目标博弈
request 请求
require (v.) 要求；需要
required 必需的
required condition 必需条件
requirement 需求
requirement analysis 需求分析
requirement analysis phase 需求分析阶段
requirement decomposition 需求分解
requirement error 需求错误
requirement specification phase 需求说明阶段
requirements 需求；要求
requirements compliance 需求合规
requirements credibility 需求可信性
requirements determination 需求测定
requirements document 需求文档
requirements management 需求管理
requirements modeling 需求建模
requirements quality 需求质量
requirements specification 需求规范
requirements traceability matrix 需求可追溯性矩阵
requirements validation 需求验证
Rescorla-Wagner model 雷斯科拉-瓦格纳模型
research 研究；考证
research environment 研究环境
research game 博弈研究
research model 研究模型
model research 模型研究
research reproducibility 研究重现性
research variable 研究变量
research, development and acquisition 研究、开发与采办
resemblance 相似
resemblance relation 相似关系
resemblance value 相似值
residual equation 残差方程
residual error 残差
residual standard error 剩余标准差
residual variable 残差变量
resilient design 弹性设计
resilient system 弹性系统
resolution 分辨率；决议
resolution and validation management 解决与验证管理
resolution and validation management tool 解决与验证管理工具
resolution error 分辨率误差
resolution level 分辨率水平
resolution management 决议管理
resolution of understanding 理解的决议
resolution refinement 决议细化
resource 资源
resource constraint 资源约束
resource description layer 资源描述层
resource driver 资源驱动器
resource error 资源错误

resource model 资源模型
resource repository 资源库
resource use efficiency 资源利用率
responding variable 响应量
response 响应
response characteristic 响应特性
response curve 响应曲线
response diagram 响应图
response function 响应函数
response space 响应空间
response surface 响应面，反应面
response surface design 响应面设计
response surface method 响应面法
response variable 响应变量
restart (v.) 重启动；重新开始
restart point 重启动点
restart request 重启动请求
restart routine 重启动例程
restore (v.) 修复；还原
restriction 限制，约束
result 结果，结论
result caching 结果缓存
result investigation 结果调查
result validation 结果验证
resultant 作为结果而发生的；结果
resultant coupling 结果上的耦合
resultant model 结果模型
resulting system 结果系统
retained data 保留的数据
retina display 视网膜显示屏
retina-display game 视网膜显示游戏
retraction 回缩；撤回
retrieval 检索
retrieving error 检索错误
retrocausal 回溯原因的

retrocausality 回溯因果关系
retrocausation 回溯因果关系
retrodiction 追溯性
retrodocumentation 回溯存档
retrosimulate (v.) 回溯仿真
retrosimulated 回溯仿真的
retrosimulating 回溯仿真
retrosimulation 回溯仿真
retrospective simulation 可追溯仿真
return on investment 投资回报
reusability 可重用性
reusability framework 可重用性框架
reusable 可重用的
reusable component 可重用的组件
reusable federation 可重用的联邦
reusable model 可重用的模型
reusable simulation component 可重用的仿真组件
reusable simulation library 可重用仿真库
reusable simulation model 可重用仿真模型
reuse 重用
reuse (v.) 重用
reuse of simulation asset 仿真资产重用
revalidation 重验证
reverification 重校核
reverse engineering 逆向工程
reverse simulation 逆向仿真
reverse time causality 逆时因果关系
review 评审；检查
review (v.) 评审；检查
review technique 评审技术
revised data 修改的数据

rho-DEVS ρ-DEVS
richly structured model 结构完全的模型
right eigenvector 右特征向量
right model matrix 权限模型矩阵
rigid-body simulation 刚体仿真
rigid game 标准博弈；死板式博弈
risk 风险
risk assessment 风险评估
risk calculation 风险计算
risk impact 风险影响
risk level 风险级别
risk management 风险管理
risk model 风险模型
risk prediction 风险预测
risk reduction 风险降低
risk simulation 风险仿真
risk variable 风险变量
robust 鲁棒的
robust control 鲁棒控制
robust decision making 鲁棒决策
robust design 鲁棒设计
robust model 鲁棒模型
robust optimization 鲁棒优化
robust simulation runtime library 鲁棒仿真运行库
robustness 鲁棒性
robustness control 鲁棒控制
role-based experience 基于角色的体验
role-play simulation 角色扮演仿真
role-playing 角色扮演
role-playing game 角色扮演游戏
role playing simulation 角色扮演仿真
rollback-based discrete simulation 基于回滚的离散仿真
rollback-based parallel discrete-event simulation 基于回滚的并行离散事件仿真
rollback-based simulation 基于回滚的仿真
root cause 根本原因
root coordinator 根协调器
root mean square error 均方根误差
rotational variable 旋转变量
round-off error 舍入误差
rounded value 舍入值
rounding error 舍入误差
routine 常规；例程
routine knowledge processing 常规的知识处理
routing 路由；路径
routing space 路由空间；路径空间
RT-DEVS 实时离散事件系统规范
rule 规则
rule base 规则库
rule-based 基于规则的
rule-based information exchange 基于规则的信息交换
rule-based model 基于规则的模型
rule-based simulation 基于规则的仿真
rule-based system 基于规则的系统
rule-based system embedded simulation 基于规则的系统嵌入仿真
rule defuzzification 规则去模糊化
rule engine 规则引擎
rule fuzzification 规则模糊化
rule markup language 规则标记语言
rule model 规则模型

run 运行
run (v.) 运行；执行
run-control variable 运行控制变量
run length 运行长度
run phase 运行阶段
Runge-Kutta algorithm 龙格-库塔算法
Runge-Kutta integration 龙格-库塔积分
running 正在运行的
runtime 运行时间的；运行时
runtime activity 运行时活动
runtime coupling 运行时耦合
runtime data 运行时数据
runtime environment 运行时环境
runtime executive 运行时执行程序
runtime extensibility 运行时可扩展性
runtime federate 运行时联邦成员
runtime federate composition 运行时联邦成员组合
runtime federate discovery 运行时联邦成员发现
runtime federate instantiation 运行时联邦成员实例化
runtime federate interoperation 运行时联邦成员互操作性
runtime infrastructure 运行时基础设施
runtime infrastructure-message 运行时基础设施消息
runtime interface 运行时接口
runtime linked data 运行时连接的数据
runtime model qualification 运行时模型鉴定
runtime model recommendation 运行时模型推荐
runtime model replacement 运行时模型替换
runtime model switching 运行时模型切换
runtime model update 运行时模型更新
runtime output 运行时输出
runtime recommendation 运行时推荐
runtime simulation linking 运行时仿真连接
runtime simulation reconfiguration 运行时仿真重置
runtime simulation update 运行时仿真更新
runtime update 运行时更新
runtime updating 运行时更新

Ss

safety 安全性
safety analysis 安全性分析
safety certification 安全认证
safety component system 安全组件系统
safety-critical system 安全关键系统
safety layer 安全层
safety mechanism 安全机制
safety model 安全模型
safety model repository 安全模型库
safety prediction 安全预测
safety property 安全属性
safety system 安全系统
safety testing 安全测试

sample 样本
sample and hold 取样与保持
sampled-data control system 采样数据控制系统
sampled-data system 采样数据系统
sampling 采样
sampling distribution 采样分布
sampling error 采样误差
sandbox game 沙箱游戏
sandbox-style simulation 沙箱式仿真
satisficing 满意的
satisficing behavior 令人满意的行为
scalability 可扩展性
scalability constraint 可扩展性约束
scalability of model transformation 模型转换的可扩展性
scalable 可扩展的
scalable analysis 可扩展分析
scalable data 可扩展数据
scalable model 可扩展模型
scalable modeling 可扩展建模
scalable simulation 可扩展仿真
scalable solution 可扩展的解决方案
scalar variable 标量变量
scale 缩放；尺度；规模
scale-free complex interactive system 无标度复杂交互系统
scale-free topology 无标度拓扑
scale invariance 尺度不变性
scale-invariant behavior 尺度不变的行为
scale linking 尺度连接
scale model 缩比模型
scaled 成比例的
scaled conjugate gradient algorithm 标度共轭梯度算法
scaled-down replica 缩小的副本
scaled real-time simulation 缩放的实时仿真
scaled-up replica 放大的副本
scaled wallclock time 缩放的挂钟时间
scaling 缩放
scaling law 缩放规则
scan 扫描；浏览
scan (v.) 扫描；浏览
scan activity 扫描活动
scanned 已扫描的
scanning 扫描
scenario 情景；想定
scenario adaptivity 情景自适应性
scenario-based 基于情景的
scenario-based design 基于情景的设计
scenario-based game management 基于情景的对策管理
scenario-based learning 基于情景的学习
scenario-based model synthesis 基于情景的模型综合
scenario-based virtual environment 基于想定的虚拟环境
scenario development 想定开发
scenario evolution 情景演变
scenario generation 情景生成；想定生成
scenario instantiation 想定实例化
scenario model 想定模型
scenario modularity 想定模块性
scenario simulation 情景仿真

scenario space   情景空间；想定空间
scenario specification   想定规范
scenario testing   想定测试
schedular   时间表的；调度器
schedule (v.)   计划；调度
scheduled   预定的
scheduled activity   预订的活动
scheduled event   计划事件
scheduled message   预订的消息
scheduler   调度器；调度程序
scheduling   调度；计划
scheduling an event   计划事件
scheduling mechanism   调度机制
schema   架构；模式；计划
schema mapping   模式映射
schemata   图式；纲要
schematic type variable   图式类型变量
schematic variable   图式变量
scheme   方案；机制；计划
scholarly knowledge   学术知识
science   科学
scientific error   科学错误
scientific knowledge   科学知识
scientific simulation   科学性仿真
scope   范围；作用域
scope creep   范围蠕变
scope of a model   模型的范围
scope of an assumption   假设的范围
scope of applicability   适用范围
scope of understanding   理解力范畴
scope of usability   可用性范围
scope of validity   有效性范围
screen   屏幕
screen readability   屏幕可读性
screening   筛查

script   脚本
script-based   基于脚本的
scripted   脚本的
scripted game   脚本游戏
scripted role-playing game   脚本角色扮演游戏
scripted scenario   脚本化想定
scripted testing   脚本测试
scripting   脚本编写
scripting language   脚本语言
seamless   无缝的
search   查询；搜索
search model   搜索模型
search space   搜索空间
second degree simulation   二级仿真
second derivative model   二阶倒数模型
second-order accurate   二阶精度
second-order cybernetics   二阶控制论
second-order effect   二阶效应
second-order emergence   二阶涌现
second order synergy   二阶协同
section   段落；部分
sectioning   分割
secure component system   安全组件系统
secure model repository   安全模型库
secure system   安全系统；可靠系统
security   安全性
security analysis   安全性分析
security certification   安全认证
security-critical system   安全关键系统
security in MAS   多智能体系统安全性
security layer   安全层

security mechanism 安全机制
security model 安全模型
security prediction 安全预测
security testing 安全测试
see-through head-mounted display 透视式头盔显示器
seed 种子；根源
seed selection 种子选择
segment 片段；分割
segmenting time series 分段时间序列
selected action 选定的动作
selected output 选择的输出
selection 选择；筛选
selection criterion 筛选准则
selection phase 筛选阶段
selective attention 选择性注意
selective documentation 选择文档
selector 选择器
self 自身
self-adaptable system 自适应系统
self-adaptation 自适应
self-adaptive 自适应的
self-adaptive simulation 自适应仿真
self-adaptive system 自适应系统
self-archiving model 自存档模型
self-aware system 自我意识系统
self-configuration system 自配置系统
self-configuring system 自配置系统
self-driven 自驱动的
self-driven input testing 自驱动输入测试
self-driven model 自驱动模型
self-driven simulation 自驱动仿真
self-evolution 自进化
self-excitation 自激励
self-healing system 自修复系统
self-healing technique 自修复技术
self-learning simulation 自学习仿真
self-managing 自我管理
self-managing system 自管理系统
self-model 自我模型
self-monitoring system 自监控系统
self-optimized behavior 自优化行为
self-optimizing system 自优化系统
self-organization 自组织
self-organized 自组织的
self-organized intelligence 自组织智能
self-organizing 自组织的
self-organizing model 自组织模型
self-organizing simulation 自组织仿真
self-organizing system 自组织系统
self-organizing system simulation 自组织系统仿真
self-protective system 自保护系统
self-reconfiguration 自重置
self-regulating simulation 自调节仿真
self-repair system 自修理模型
self-replicating system simulation 自复制系统仿真
self-similar 自相似的
self-similar process 自相似过程
self-similarity 自相似性
self-simulation 自仿真
self-stabilization 自稳定
self-stabilizing system simulation 自稳定系统仿真
self-test input 自测试输入
self-tuning system 自校正系统

semantic agent   语义智能体；语义代理
semantic analysis   语义分析
semantic assessment   语义评估
semantic-based   基于语义的
semantic-based tool   基于语义的工具
semantic communication   语义通信
semantic compatibility   语义的相容性
semantic composability   语义的可组合性
semantic constraint   语义约束
semantic context   语义上下文
semantic data model   语义数据模型
semantic-directed   语义导向的
semantic discovery   语义发现
semantic domain   语义论域
semantic emergence   语义涌现
semantic emergent behavior   语义突现行为
semantic entity   语义实体
semantic error   语义错误
semantic evaluation   语义评价
semantic gap   语义间隙
semantic graph   语义图
semantic improvement   语义改进
semantic interoperability   语义互操作性
semantic interoperability level   语义互操作水平
semantic layer   语义层
semantic level   语义层次
semantic markup   语义标记
semantic memory model   语义记忆模型
semantic metadata   语义元数据
semantic model   语义模型
semantic model assessment   语义模型评估
semantic model comparison   语义模型比较
semantic model differencing   语义模型差分
semantic model evaluation   语义模型评价
semantic modeling   语义建模
semantic network   语义网络
semantic operability   语义可操作性
semantic-pragmatic model   语义-语用模型
semantic scheme   语义机制
semantic structure   语义结构
semantic theory of truth   语义真理论
semantic understanding   语义理解
semantic value   语义值
semantic web   语义网
semantic web for legacy transformation   用于遗留转换的语义网
semantic web for model transformation   用于模型转换的语义网
semantically augmented   语义增强的
semantically augmented metadata   语义增强元数据
semantically rich metadata   语义丰富的元数据
semantics   语义学
semantics-based composition   基于语义的组合
semantics of language   语言的语义
semantics of model transformation   模型转换的语义

semantics of modeling language 建模语言的语义
semianalytic algorithm 半解析算法
semiautomatic code generation 半自动代码生成
semiautomatic semicode generation 半自动半代码生成
semichaotic 半混沌
semichaotic state 半混沌状态
semicode generation 半代码生成
semiempirical model 半经验模型
semiformal 半正式的
semiformal definition 半正式定义
semiformal language 半正式语言
semi-implicit algorithm 半隐式算法
semi-implicit method 半隐式方法
semi-implicit trapezoidal formula 半隐式梯形公式
semilinear defuzzification 半线性去模糊化
semi-Markov model 半马尔可夫模型
semi-Markov process 半马尔可夫过程
semiotic mental simulation 符号心理仿真
semiotic simulation 符号仿真
sensed input 感应输入
sensed variable 感应变量
sensitive 敏感的
sensitive data 敏感数据
sensitivity 灵敏度
sensitivity analysis 灵敏度分析
sensitivity analysis technique 敏感度分析技术
sensitivity error 灵敏度误差
sensitivity model 灵敏度模型
sensitivity simulation 灵敏度仿真
sensitivity study 灵敏度研究
sensor 传感器
sensor-based system 基于传感器的系统
sensor data 传感器数据
sensor error 传感器误差
sensor fusion 传感器融合
sensor input 传感器输入
sensor system 传感器系统
sensory 感官的
sensory data 感官数据
sensory data conversion 感官数据转换
sensory experience 感官经验
sensory input 感官输入
sensory interface 感官接口
sensory model 感官模型
sensory perception 感官知觉
sensory touch 感官触摸
sentential variable 句子变量
sentinel event 前哨事件；标志性事件
separate error 独立的错误；分离误差
separate task 单独任务
separation of concerns 关注点分离
separation of model and experimental frame 模型与试验框架分离
separation of variables 变量分离
sequence 序列
sequence diagram 序列图，时序图
sequence error 序列误差
sequenced 序列式
sequencing 排序
sequencing error 排序错误

sequential 连续的；顺序的
sequential algorithm 顺序算法；序贯算法
sequential behavior 顺序行为
sequential bifurcation 顺序分叉
sequential execution 顺序执行
sequential game 序贯博弈
sequential Gaussian simulation 序贯高斯仿真
sequential mapping 顺序映射
sequential multimodel 连续多模型
sequential processing 顺序处理
sequential simulation 顺序仿真
sequential simulator 顺序仿真器
sequential SL 顺序仿真语言
sequential state 顺序状态
sequential update 顺序更新
serial connection 串行连接
serial model 串行模型
serial simulation 串行仿真
serious game 严肃游戏
serious game initiative 严肃游戏倡议
serious gamer 严肃玩家
serious simulation 严肃仿真
server 服务器
server error 服务器错误
server requirements 服务器需求
server simulator 服务器仿真器
service 服务
service availability 服务可用性
service-based 基于服务的
service-based simulation 基于服务的仿真
service-based simulation system approach 基于服务的仿真系统方法
service bus 服务总线
service discipline 服务规则
service encapsulated 服务封装的
service encapsulation 服务封装
service modeling 服务建模
service-oriented 面向服务的
service-oriented architecture 面向服务的架构
service-oriented architecture approach 面向服务的架构方法
service-oriented technology 面向服务的技术
service performance 服务性能
service provider 服务提供方
service quality 服务质量
service reliability 服务可靠性
service requirements 服务需求
service specification 服务规范
service specification requirements 服务规范需求
service view 服务视图
servitization 服务化
set-theoretic model 集合论模型
set theory 集合论
set-up 组织；建立；设置
set-up error 设置误差
set value 设定值
shallow model 浅层模型
shape modeling 外形建模
shape simulation 形状仿真
shareability 可共享性
shared conceptualization 共享概念化
shared data 共享数据
shared environment 共享环境

shared event 共享事件
shared experience 共享经验
shared immersive experience 共享沉浸体验
shared model 共享模型
shared secret data random variable 共享机密数据随机变量
shared state 共享状态
shared state event 共享状态事件
shared variable 共享变量
shareware 共享软件
sharing 共享
shell 外壳
shift error 移位误差
shift operator 移位算子
shooter game 射击游戏
short-lived ontology 短时本体
short-lived state 短时状态
short-term predictability 短期可预测性
short time-scale variable 短时间尺度变量
sidereal time 恒星时
signal 信号
significance term 重要性术语
significant data 重要数据
significant event 重要事件
significant feature 重要特征
similar 相似的
similar model 相似模型
similarity 相似性
similarity criterion 相似性准则
similarity degree 相似度
similarity measure 相似度测量
similarity method 相似法

similarity relation 相似性关系
similarity style 相似类型，相似风格
similarity theory 相似性理论
similarity transformation 相似变换
similitude 相像
simple behavior 简单行为
simple emergence 简单涌现
simple emergent behavior 简单应急行为
simple failure 简单故障
simple game 简单博弈
simple Kriging 简单克里格
simple output 简单输出
simple variable 简单变量
simplex algorithm 单纯形算法
simplicity 简单；朴素
simplifiable model 可简化的模型
simplification 简化
simplification error 简化误差
simplification methodology 简化方法学
simplification procedure 简化过程
simplified model 简化的模型
simplify (v.) 简化
simplifying abstraction 简化抽象
simplifying assumption 简化假设
simulacra 幻影；模拟物
simulacre 幻影；模拟物
simulacrum 幻影
simuland 模拟对象
simulant 模拟的
simulatable 可仿真的
simulatable model 可仿真模型
simulate (v.) 仿真
simulated 仿真的

simulated annealing 模拟退火
simulated annealing method 模拟退火算法
simulated attack 仿真攻击
simulated behavior 仿真行为
simulated data 仿真数据
simulated data acceptability 仿真数据的可接受性
simulated entity 仿真实体
simulated environment 仿真环境
simulated event horizon 仿真事件视界
simulated evolution 仿真进化
simulated input 仿真输入
simulated learning platform 仿真学习平台
simulated model 仿真模型
simulated mulfunction 模拟的故障
simulated object 仿真对象
simulated pseudoobservation 模拟的伪观察
simulated random sampling 仿真随机采样
simulated real-time 仿真实时,实时仿真的
simulated reality 仿真现实
simulated representation 仿真的表示（法）
simulated retrocausality 仿真的回溯因果关系
simulated system 仿真系统
simulated time 仿真时间
simulated view 仿真视图
simulated warfare 仿真对抗
simulated world 仿真世界
simulating program 仿真程序
simulating state 模拟状态
simulation 仿真
simulation accuracy 仿真精度
simulation adaptation 仿真适配
simulation-aided 仿真辅助的
simulation algorithm 仿真算法
simulation analysis 仿真分析
simulation analytics 仿真分析
simulation analytics tool 仿真分析工具
simulation animation 仿真动画
simulation annealing process 模拟退火过程
simulation application 仿真应用
simulation application federate 仿真应用联邦成员
simulation architecture 仿真架构
simulation argument 仿真论证
simulation as a service 仿真即服务
simulation asset 仿真资产
simulation asset management 仿真资产管理
simulation association 仿真关联
simulation-augmented reality 仿真增强现实
simulation availability 仿真可用性
simulation-based 基于仿真的
simulation-based acquisition 基于仿真的采办
simulation-based algorithm 基于仿真的算法
simulation-based analysis 基于仿真的分析
simulation-based assessment 基于仿

真的评估
simulation-based assessment tool 基于仿真的评估工具
simulation-based augmented reality 基于仿真的增强现实
simulation-based calibration 基于仿真的标定
simulation-based collaborative engineering 基于仿真的协同工程
simulation-based communication approach 基于仿真的通信方法
simulation-based computational discovery 基于仿真的计算发现
simulation-based confidence 基于仿真的置信度
simulation-based control 基于仿真的控制
simulation-based control graph 基于仿真的控制图
simulation-based data mining 基于仿真的数据挖掘
simulation-based decision making 基于仿真的决策
simulation-based decision support 基于仿真的决策支持
simulation-based design 基于仿真的设计
simulation-based design approach 基于仿真的设计方法
simulation-based design methodology 基于仿真的设计方法学
simulation-based design optimization 基于仿真的设计优化
simulation-based diagnosis 基于仿真的诊断
simulation-based discipline 基于仿真的学科
simulation-based discovery 基于仿真的发现
simulation-based distributed training 基于仿真的分布式训练
simulation-based education 基于仿真的教育
simulation-based education system 基于仿真的教育系统
simulation-based education technique 基于仿真的教育技术
simulation-based engineering 基于仿真的工程
simulation-based enterprise 基于仿真的企业
simulation-based entertainment 基于仿真的娱乐
simulation-based environment 基于仿真的环境
simulation-based estimation 基于仿真的评估
simulation-based evaluation 基于仿真的评价
simulation-based experience 基于仿真的经验
simulation-based experiment 基于仿真的实验
simulation-based experimentation 基于仿真的实验
simulation-based failure analysis 基于仿真的故障分析
simulation-based field 基于仿真的区域
simulation-based forecasting 基于仿

真的预测

simulation-based game 基于仿真的博弈

simulation-based genetic algorithm 基于仿真的遗传算法

simulation-based individual training 基于仿真的个体训练

simulation-based inference 基于仿真的推理

simulation-based initiative 基于仿真的起始

simulation-based learning 基于仿真的学习

simulation-based learning system 基于仿真的学习系统

simulation-based management 基于仿真的管理

simulation-based maximization 基于仿真的最大化

simulation-based method 基于仿真的方法

simulation-based methodology 基于仿真的方法学

simulation-based minimization 基于仿真的最小化

simulation-based modeling 基于仿真的建模

simulation-based operational support 基于仿真的运行支持

simulation-based optimal design 基于仿真的优化设计

simulation-based optimization 基于仿真的优化

simulation-based ordinal optimization 基于仿真的序优化

simulation-based performance analysis 基于仿真的性能分析

simulation-based performance evaluation 基于仿真的性能评估

simulation-based planning 基于仿真的规划

simulation-based prediction 基于仿真的预估

simulation-based predictive display 基于仿真的预测显示

simulation-based problem solving 基于仿真的问题求解

simulation-based problem solving environment 基于仿真的问题求解环境

simulation-based proof-of-concept 基于仿真的概念证明

simulation-based prototype 基于仿真的原型

simulation-based prototyping 基于仿真的原型设计

simulation-based real-time prediction 基于仿真的实时预测

simulation-based reliability assessment 基于仿真的可靠性评估

simulation-based research 基于仿真的研究

simulation-based robust design 基于仿真的鲁棒设计

simulation-based scheduling 基于仿真的调度

simulation-based science 基于仿真的科学

simulation-based security 基于仿真的安全

simulation-based serious game 基于仿真的严肃游戏
simulation-based strategy game 基于仿真的战略博弈
simulation-based system 基于仿真的系统
simulation-based system acquisition 基于仿真的系统采办
simulation-based systems analysis 基于仿真的系统分析
simulation-based systems engineering 基于仿真的系统工程
simulation-based teaching 基于仿真的教学
simulation-based teamwork training 基于仿真的团队训练
simulation-based tool 基于仿真的工具
simulation-based training 基于仿真的训练
simulation-based training system 基于仿真的训练系统
simulation-based understanding 基于仿真的理解
simulation-based validation 基于仿真的验证
simulation-based virtual environment 基于仿真的虚拟环境
simulation benchmark 仿真基准
simulation branching 仿真分支
simulation brokering 仿真代理
simulation business 仿真业务
simulation business practice 仿真商业实践
simulation-C4ISR interoperability C4ISR仿真的互操作性
simulation capability 仿真能力
simulation certification 仿真认证
simulation challenge 仿真挑战
simulation clock 仿真时钟
simulation code 仿真代码
simulation coercion 仿真胁迫
simulation complexity 仿真复杂性
simulation component 仿真组件
simulation component interface 仿真组件接口
simulation component reuse methodology 仿真组件重用方法学
simulation composability 仿真可组合性
simulation composition 仿真组合
simulation comprehensibility 仿真可理解性
simulation concept 仿真概念
simulation confederation 仿真联盟
simulation configuration 仿真组态
simulation consultant 仿真恒量
simulation control 仿真控制
simulation control program 仿真控制程序
simulation controller 模拟控制器
simulation coordination 仿真协调
simulation correctness 仿真正确性
simulation credibility 仿真可信度
simulation curriculum 仿真课程
simulation customer 仿真客户
simulation cycle 仿真周期
simulation data 仿真数据
simulation data management 仿真数据管理

simulation delivery 仿真交付
simulation deployment 仿真部署
simulation design 仿真设计
simulation design methodology 仿真设计方法学
simulation developer 仿真开发者
simulation development 仿真开发
simulation development environment 仿真开发环境
simulation development life cycle 仿真开发生命周期
simulation development program 仿真开发程序
simulation documentation 仿真文档
simulation domain 仿真域
simulation domain requirements 仿真领域需求
simulation-driven 仿真驱动
simulation-driven education 仿真驱动的教育
simulation-driven experimentation 仿真驱动的实验
simulation-driven optimization 仿真驱动的优化
simulation-driven training 仿真驱动的训练
simulation effort 仿真尝试
simulation embedded within expert system 嵌入专家系统中的仿真
simulation embedded within linear programming 嵌入线性规划中的仿真
simulation embedded within optimization 嵌入优化中的仿真
simulation embedded within rule-based system 嵌入基于规则的系统中的仿真
simulation engine 仿真引擎
simulation engineering 仿真工程
simulation-enhanced 增强型仿真
simulation enterprise 仿真企业
simulation entity 仿真实体
simulation environment 仿真环境
simulation epistemology 仿真认识论
simulation error 仿真误差
simulation error assessment 仿真误差评估
simulation ethics 仿真道德规范
simulation evolution 仿真进化
simulation execution 仿真执行
simulation execution environment 仿真执行环境
simulation execution speed 仿真执行速率
simulation execution synchronization 仿真的执行同步
simulation exercise 仿真演练
simulation experiment 仿真实验
simulation experiment description 仿真实验描述
simulation experiment description mark-up language 仿真实验描述标记语言
simulation experiment reproducibility 仿真实验可复现性
simulation expert 仿真专家
simulation exposure 模拟曝光
simulation fallacy 仿真谬论
simulation federation 仿真联合
simulation fidelity 仿真保真度

simulation fitness 模拟适应度
simulation for acquisition 采办仿真
simulation for decision making 决策仿真
simulation for decision support 决策支持仿真
simulation for education 教育仿真
simulation for system analysis 系统分析的仿真
simulation for system design 仿真设计的系统
simulation for systems engineering life cycle 系统工程生命周期的仿真
simulation for training 用于训练的仿真
simulation framework 仿真框架
simulation game 仿真博弈
simulation game design 仿真游戏设计
simulation game development 仿真游戏开发
simulation game management 仿真游戏管理
simulation gaming 仿真博弈
simulation gaming application 仿真博弈应用
simulation gaming process 仿真博弈过程
simulation gaming project 仿真博弈工程
simulation gaming software 仿真博弈软件
simulation gaming tool 仿真博弈工具
simulation-generated 仿真生成
simulation granularity 仿真粒度
simulation grid 仿真网格
simulation grid architecture 仿真网格架构
simulation GUI 仿真图形用户界面
simulation heuristic 仿真启发式
simulation horizon 仿真视界
simulation horizon requirement 仿真视界需求
simulation hypothesis 仿真假设
simulation implementation 仿真实现
simulation in entertainment 娱乐仿真
simulation in entertainment game 娱乐游戏仿真
simulation infrastructure 仿真基础设施
simulation instance 仿真实例
simulation interaction 仿真交互
simulation interface 仿真接口
simulation interface issue 仿真接口问题
simulation interface system 仿真接口系统
simulation interoperability 仿真互操作性
simulation interoperability standard 仿真互操作标准
simulation investment 仿真投资
simulation knowledge 仿真知识
simulation knowledge acquisition 仿真知识获取
simulation knowledge sharing 仿真知识共享
simulation language 仿真语言
simulation language processor 仿真语言处理器
simulation latency 仿真时延

simulation layer 仿真层
simulation library 仿真库
simulation life cycle 仿真生命周期
simulation linkage 仿真连接
simulation literacy 仿真素养
simulation literature 仿真文献
simulation machine 仿真机
simulation maintenance program 仿真维护程序
simulation management 仿真管理
simulation management capability 仿真管理能力
simulation manager 仿真管理器
simulation mark up language 仿真标记语言
simulation market 仿真市场
simulation mash-up 仿真混合
simulation mechanism 仿真机制
simulation metamodel 仿真元模型
simulation metamodeling 仿真元建模
simulation method 仿真方法
simulation methodology 仿真方法学
simulation middleware 仿真中间件
simulation mode 仿真模式
simulation mode control 仿真模式控制
simulation model 仿真模型
simulation model assessment 仿真模型评估
simulation model design quality 仿真模型设计质量
simulation model interoperability 仿真模型互操作
simulation-model response 仿真模型响应
simulation modeling 仿真建模
simulation-modeling infrastructure 仿真建模的基础设施
simulation modernization 仿真现代化
simulation modularity 仿真模块化
simulation monitoring 仿真监控
simulation node 仿真节点
simulation object model 仿真对象模型
simulation on-demand 按需仿真
simulation on high-performance computer 高性能计算机上仿真
simulation ontology 仿真本体
simulation operation 仿真操作
simulation operator 仿真操作者
simulation optimization 仿真优化
simulation optimization methodology 仿真优化方法学
simulation optimization problem 仿真优化问题
simulation organization 仿真组织
simulation output 仿真输出
simulation output-data-analysis federate 仿真输出数据分析联邦
simulation package 仿真包
simulation paradigm 仿真范式
simulation parameter 仿真参数
simulation participant 仿真参与者
simulation phase 仿真阶段
simulation plan 仿真计划
simulation platform 仿真平台
simulation platform degree of freedom 仿真平台自由度
simulation plug-in 仿真插件
simulation policy group 仿真策略组

simulation portal 仿真入口
simulation problem 仿真问题
simulation process 仿真过程
simulation processor 仿真处理器
simulation program 仿真程序
simulation program generator 仿真程序生成器
simulation program management 仿真程序管理
simulation programming 仿真程序设计
simulation programming language 仿真编程语言
simulation project 仿真工程
simulation project management 仿真工程管理
simulation project report 仿真项目报告
simulation proof 仿真证明
simulation protocol 仿真协议
simulation proxy 仿真代理
simulation quality assurance 仿真质量保证
simulation reconfiguration 仿真重置
simulation reference markup language 仿真参考标记语言
simulation-related 仿真相关的
simulation relation 仿真关系
simulation reliability 仿真可靠性
simulation repository 仿真库
simulation reproducibility 仿真再现性
simulation requirement 仿真需求
simulation resolution 仿真分辨率
simulation resource 仿真资源
simulation response 仿真响应
simulation response function 仿真响应函数
simulation response surface 仿真响应面
simulation result 仿真结果
simulation results comprehensibility 仿真结果解释
simulation results quality 仿真结果质量
simulation reusability 仿真可重用性
simulation reuse 仿真重用
simulation robustness 仿真鲁棒性
simulation rule 仿真规则
simulation run 仿真运行
simulation run control 仿真运行控制
simulation run length 仿真运行长度
simulation run monitoring 仿真运行监控
simulation runtime library 仿真运行库
simulation runtime monitoring 仿真运行监控
simulation scenario 仿真想定
simulation semantics 仿真语义学
simulation sequence 仿真序列
simulation server 仿真服务器
simulation service 仿真服务
simulation service provider 仿真服务提供方
simulation skill 仿真技能
simulation society 仿真协会
simulation software 仿真软件
simulation software assessment 仿真软件评估
simulation software development 仿真软件开发

simulation-specific 特定仿真
simulation specification 仿真规范
simulation specification environment 仿真规范环境
simulation specification language 仿真规范语言
simulation specification repository 仿真规范库
simulation stability 仿真稳定性
simulation stakeholder 仿真参与者
simulation standard 仿真标准
simulation status 仿真状态
simulation strategy 仿真策略
simulation structure 仿真结构
simulation study 仿真研究
simulation study acceptability 仿真学习可接受性
simulation study monitoring 仿真研究监控
simulation support 仿真支持
simulation support entity 仿真支撑实体
simulation support tool 仿真支撑工具
simulation supported 仿真支持的
simulation-supported game 仿真支持的博弈
simulation-supported war game 仿真支持的战争博弈
simulation system 仿真系统
simulation systems engineer 仿真系统工程师
simulation systems engineering 仿真系统工程
simulation systems engineering body of knowledge 仿真系统工程知识体系
simulation task 仿真任务
simulation team 仿真团队
simulation technique 仿真技术
simulation technology 仿真技术
simulation term 仿真术语
simulation terminology 仿真术语
simulation theory 仿真理论
simulation time 仿真时间
simulation time step 仿真时间步长
simulation tool 仿真工具
simulation topology 仿真拓扑
simulation trainer development 仿真训练器开发
simulation update 仿真更新
simulation update time 仿真更新时间
simulation updating 仿真更新
simulation use 仿真使用
simulation user 仿真用户
simulation user issue 仿真用户问题
simulation utility 仿真效用
simulation utility type 仿真应用类型
simulation validation 仿真验证
simulation visualization federate 仿真可视化联邦成员
simulation warm up period 仿真预热阶段
simulation warm up time 仿真预热时间
simulation web service 仿真网络服务
simulation with multistage model 多阶段模型的仿真
simulation within optimization 优化中的仿真
simulation, test, and evaluation process 仿真、测试与评估过程

simulational context 仿真情境
simulationist 仿真专家
simulative 仿真的
simulative design 仿真的设计
simulative design approach 仿真设计方法
simulative design environment 仿真设计环境
simulative environment 仿真环境
simulative problem solving environment 仿真问题求解环境
simulative solution technique 仿真求解技术
simulator 仿真器
simulator-based training 基于仿真器的训练
simulator coordination 仿真器协调
simulator design 仿真器设计
simulator middleware 仿真器中间件
simulator parameter 模拟器参数
simulator program 仿真器程序
simulator server 仿真器服务器
simulism 仿真主义
simultaneous 同步
simultaneous attribute update 同步属性更新
simultaneous control 同步控制
simultaneous event 同时发生的事件
simultaneous interval technique 同步间隔技术
simultaneous simple failure 同时发生的简单故障
simultaneous simulation 同步仿真
simultaneous update 同步更新
simware 仿真套件

single-aspect model 单方面模型
single-aspect multimodel 单方面多模型
single-aspect simulation 单方面仿真
single-aspect system 单方面系统
single-component simulation 单部件仿真
single-error 单项误差
single-event 单个事件
single-event effect 单个事件影响
single-input 单输入
single-input multi-output model 单输入多输出模型
single-input multi-output system 单输入多输出系统
single-input single-output model 单输入单输出模型
single-input single-output system 单输入单输出系统
single-input system 单输入系统
single-message 单个消息
single-output 单输出
single paradigm 单一范式
single-paradigm simulation system 单范式仿真系统
single-player game 单人游戏
single-player gaming technique 单人游戏（博弈）技术
single-point object 单点目标
single-pole 单极的
single-processor simulation 单处理器仿真
single-run experiment 单次运行实验
single-run simulation study 单次运行仿真研究

single-simulation time step 单次仿真时间步长
single-space variable 单空间变量
single-step algorithm 单步算法
single-step formula 单步公式
single-step integration method 单步积分法
single-transition 单一变换
single-transition subnet 单转换子网
single-valued function 单值函数
single-vision understanding 单目视觉理解
single-zero-crossing 单零交叉
singleton global variable 单个全局变量
singleton variable 单个变量
singular coupling 奇点耦合
singular input 奇点输入
singular point 奇点
singularity 奇点；奇异性
situated activity 情景活动
situated immersion 情境沉浸
situation 态势；情景
situation-action 情境-行动
situation-action rule 情境-行动规则
situation awareness 情境感知；态势感知
situation model 态势模型
situational attribute 情境属性
situational awareness 情境感知；态势感知
situational complexity 情境的复杂性
situational variable 情境变量
situative interaction 情景交互
six degree of freedom 六自由度

size 尺寸
skeleton-driven simulation 骨架驱动的仿真
sketch-based simulation 基于草图的仿真
skill 技能
skill enhancing 技能增强
SL (Simulation Language) 仿真语言
slack variable 松弛变量
slicing 切片
slot 槽
slow time 标准时间；减缓时间
slow time constant 慢时间常数
slower than real-time system 欠实时系统
slower than real-time time 欠实时时间
small-scale system 小尺度系统
smart agent 智能代理
smart experiment agent 智能实验代理
smart phone activated simulation 智能电话激活仿真
smart system 智能系统
smooth 平滑；平缓
smooth ($v.$) 平滑；平缓；消除
smooth behavior 平滑行为
smooth data 平滑数据
smoothed 平滑的
smoothing 平滑的
smoothing parameter 平滑参数
smoothness simulation 平滑度仿真
social game 社会博弈
social holon 社会合弄
social impact game 社会影响博弈
social media 社会媒体
social media environment 社会媒体

环境
social media game 社会媒体游戏
social network 社会网络
societal complexity 社会复杂性
socio-cultural knowledge 社会文化知识
sociodynamic model 社会动力学模型
soft-body simulation 柔体仿真
soft computing 软计算
soft computing simulation 软计算仿真
soft constraint 软约束
soft coupling 软耦合
soft data 软数据
soft error 软错误
soft input 软输入
soft output 软输出
soft real-time constraint 软实时约束
soft sensor input 软传感器输入
soft simulation 软仿真
soft system 软系统
soft system thinking 软系统思考
software 软件
software agent 软件代理
software architecture 软件架构
software-based 基于软件的
software-based approach 基于软件的方法
software-based continuous system simulation 基于软件的连续系统仿真
software-based discrete-event simulation 基于软件的离散事件仿真
software-based simulation 基于软件的仿真
software certification 软件认证

software code 软件代码
software compatible 软件兼容的
software component 软件组件
software composition 软件组成
software design error 软件设计错误
software development 软件开发
software documentation 软件文档
software engineering 软件工程
software engineering process 软件工程过程
software environment 软件环境
software error 软件错误
software implementation 软件实现
software-in-the-loop 软件在环
software-in-the-loop simulation 软件在回路仿真
software integrity 软件完整性
software-intensive 软件密集型
software-intensive system 软件密集系统
software methodology 软件方法学
software model 模型软件
software modeler 软件建模器
software modeling 软件建模
software modeling language 软件建模语言
software project life cycle 软件项目生命周期
software requirement 软件需求
software resource 软件资源
software simulator 软件仿真器
software socket interface technique 软件套接字接口技术
software specification 软件规范
software systems engineering 软件系

统工程
solicited model   要求的模型
solution   解决方案
solution documentation   解决方案文档
solution error   求解错误
solution scalability   解决方案的延展性
solution set   解集合
solution space   解空间
solution-specific   特定解
solution stability   解的稳定性
solution technique   求解技术
solvability   可解性
solvability issue   可解性问题
solver   求解器
solving   求解
sonar input   声呐输入
sophism   诡辩（法）；似是而非的论点
sorter error   分类器错误
sound value   公允价值，稳定价值
source   源，来源；源点；信源
source code   源代码
source data   源数据
source documentation   源文件
source domain   源域
source of cognition   认知源
source of input   输入源
source simulation program   源仿真程序
source SL   源仿真语言
source system   源系统
sourcing   采购；获得
SP (Schedule-Preserving) DEVS   调度保持的离散事件系统规范
space   空间

space discretization   空间离散化
space leak   空间泄露
space tessellation   空间曲面细分
space-time continuum   时空连续体
span   跨度；范围；量程
span error   量程误差
sparse connection   稀疏连接
sparse data   稀疏数据
sparse linear system solver   稀疏线性系统求解器
spatial   空间的
spatial assumption   空间假设
spatial characteristic   空间特性
spatial condition   空间条件
spatial consistency   空间一致性
spatial data modeling   空间数据建模
spatial derivative   空间导数
spatial discretization   空间离散化
spatial dynamics   空间动力学
spatial error   空间误差
spatial microsimulation model   空间微观仿真模型
spatial model   空间模型
spatial modeling   空间建模
spatial relation   空间关系
spatial resolution   空间分辨率
spatial similarity   空间相似性
spatial similarity degree   空间相似度
spatial similarity relation   空间相似关系
spatial simulation   空间仿真
spatial structure   空间结构
spatial variable   空间变量
spatially correlated error   空间相关误差

spatially distributed system 空间分布式系统
spatially emergent behavior 空间突现行为
spatiotemporal analytical method 时空分析方法
spatiotemporal event 时空事件
spatiotemporal method 时空方法
spatiotemporal model 时空模型
spatiotemporal simulation 时空仿真
special input testing 特定输入测试
special-purpose gaming hardware 专用游戏硬件
special-purpose simulation hardware 专用仿真硬件
special purpose SL 专用仿真语言
special value 特殊值
speciality model 特性模型
specific 特定的；明确的；专门的
specific data 具体数据；特定数据
specific model 特定模型
specific system 特定系统
specific value 特定值
specification 规范；规格说明；技术规范
specification-based 基于规范的
specification condition 规范条件
specification environment 规范环境
specification error 规范误差
specification formalism 规范形式
specification language 规范语言
specification maintenance 规范维护说明
specification method 规范方法
specification of experimental condition 实验条件规范
specification of simulation task 仿真任务规范
specification phase 规格说明阶段
specification repository 规范库
specification validation 规格验证
specification verification 规格校验
specified error 给定误差
specified model 指定模型
specify (v.) 描述；指定
spectral algorithm 频谱算法
spectral analysis 频谱分析
spectral analysis technique 频谱分析技术
spectrum of simulators 仿真器频谱
spectrum of trainers 训练器频谱
speculative data 推测数据
speech compression 语音压缩
speech input 语音输入
speech markup language 语音标记语言
speed 速率
speedup 加速
split (v.) 拆分；分离
split event 分离的事件
sponsor 赞助商；发起人
spooky emergence 幽灵般的涌现
spooky emergent behavior 幽灵般的突现行为
sports game 体育博弈
spread parameter 扩散参数
spreadsheet simulation 电子表格仿真
spurious behavior 虚假行为
stability 稳定性
stability condition 稳定性条件

stability criterion 稳定性判据
stability domain 稳定域
stability mechanism 稳定机制
stability property 稳定特性
stability region 稳定区域
stabilization 稳定化
stabilized variable 稳定变量
stabilized variable model 固定变量模型
stable 稳定的；不变的
stable behavior 稳定的行为
stable clustering state 稳定聚合状态
stable discretization scheme 稳定的离散化机制
stable environment 稳定的环境
stable error 稳态误差
stable integration algorithm 稳定积分算法
stable limit cycle 稳定极限环
stable linear time-invariant system 稳定线性时不变系统
stable model 稳定的模型
stable region 稳定范围
stable result 稳定结果
stable solution 稳态解
stable state 稳定状态
stable system 稳定系统
stage 级；阶段；步骤
staged composition 阶段的组合
staging 级间分离，分级
stakeholder 参与者；利益相关者
stakeholder requirements 参与者需求
stamp 标记；特征；类型
stamp coupling 标记耦合
stand-alone simulation 单机仿真

standard 标准
standard assumption 标准假设
standard-based 基于标准的
standard-based language 基于标准的语言
standard error 标准差
standard for acceptability 合格标准
standard input 标准输入
standard view 标准视图
standardization 标准化；标定
standardized value 标准化值
standardized variable 标准化变量
standards testing 标准测试
start firing phase 开始启动阶段
start place 起始点
start-up interpolation 启动插值
start-up period 启动周期
starting condition 启动条件
starting model 启动模型，起始模型
starting state 启动状态
state 状态
state-based 基于状态的
state-based description 基于状态的描述
state-based formal method 基于状态的形式化方法
state-based model 基于状态的模型
state-based stability 基于状态的稳定性
state-based system description 基于状态的系统描述
state-based system model 基于状态的系统模型
state change 状态变化
state changing 状态改变

state chart 状态表
state chart diagram 状态图
state control 状态控制
state coupling 状态耦合
state density 状态密度
state dependent 状态依赖
state-dependent coupling 相关依赖的耦合
state-dependent event 状态依赖的事件
state-determined 状态确定的
state diagram 状态图
state equation 状态方程
state event 状态事件
state event detection 状态事件检测
state event localization 状态事件定位
state feedback 状态反馈
state flow 状态流
state graph 状态图
state history vector 状态历史向量
state independent 状态独立
state-independent coupling 状态独立的耦合
state machine 状态机
state maintaining simulation 状态保持仿真
state matrix 状态矩阵
state model 状态模型
state quantization 状态量化
state space 状态空间
state-space description 状态空间描述
state-space form 状态空间形式
state-space graph 状态空间图
state-space model 状态空间模型
state-space notation 状态空间表示法
state superposition 状态叠加
state trajectory 状态轨迹
state transformation 状态转换
state transition 状态转换
state transition analysis 状态转移分析
state transition diagram 状态迁移图
state-transition function 状态转移函数
state-transition machine 状态转换机
state-transition matrix 状态转移矩阵
state-transition model 状态转换模型
state-transition system 状态转移系统
state-triggered 状态触发的
state-triggered control 状态触发的控制
state value 状态值
state value set 状态值集合
state variable 状态变量
state-variable derivative discontinuity 状态变量导数不连续性
state-variable discontinuity 状态变量不连续性
state-variable identification 状态变量辨识
state-variable quantization 状态变量量化
state vector 状态向量
stateflow 状态流
statement 语句；声明
statement testing 语句测试
statement variable 语句变量
static 静态的
static analysis 静态分析
static assumption 静态假设
static behavior 静态的行为

static check 静态的检查
static coupling 静态耦合
static data 静态数据
static decoupling 静态解耦
static error 静态误差
static game 静态博弈
static microsimulation model 静态微观仿真模型
static model 静态模型
static model documentation 静态模型文档
static model structure 静态模型结构
static modeling formalism 静态建模形式化
static network 静态网络
static output 静态输出
static program check 静态程序检查
static reality 静态实境
static relationship 静态关系
static simulation 静态仿真
static structure 静态结构
static-structure model 静态结构模型
static-structure multimodel 静态结构多模型
static technique 静态技术
static value 静态值
static variable 静态变量
static VV&T technique 静态校核、验证与测试技术
statically coupled multimodel 静态耦合的多模型
statically scoped variable 静态作用域变量
station 站；台
stationary 不变的；稳定的；静止的
stationary process 平稳过程
stationary variable 平稳变量
statistical 统计的；统计学的
statistical analysis 统计分析
statistical confirmation 统计确认
statistical data compression 统计数据压缩
statistical decision 统计决策
statistical design of simulation experiments 仿真实验的统计设计
statistical experiment 统计试验
statistical methodology 统计方法学
statistical model 统计模型
statistical model checking 统计模型检验
statistical modeling 统计建模
statistical sampling 统计采样
statistical significance testing 统计显著性检验
statistical simulation 统计仿真
statistical technique 统计技术
statistical test 统计检验
statistical testing 统计检验
statistical validation 统计验证
statistical validation technique 统计验证技术
statistical validity 统计有效性
statistical variable 统计变量
statistics 统计；统计学
status 状态；情况；身份
steady 稳定的；稳态的；平衡的
steady behavior 稳态行为
steady state 稳态，定常态
steady-state error 稳态误差
steady-state model 稳态模型

steady-state period 稳态周期
steady-state phase 稳态相位；稳态阶段
steady-state simulation 稳态仿真
steady-state technique 稳态技术
stealth viewer 隐形观察者
step 步；阶
step-by-step 步进的；逐步的
step-change 阶跃变化
step function 阶跃函数
step input 阶跃输入
step-size adjustment 步长调整
step-size control 步长控制
step-size control algorithm 步长控制算法
step-size controlled algorithm 步长可控的算法
step-size 步长
stiff 刚性的；不易移动的
stiff differential equation 刚性微分方程
stiff discontinuous model 刚性不连续的模型
stiff dynamic system 刚性动力学系统
stiff model 刚性模型
stiff simulation 刚性仿真
stiff system 刚性系统
stiff system integration algorithm 刚性系统积分算法
stiff system simulation 刚性系统仿真
stiffly stable 刚性稳定的
stiffly stable algorithm 刚性稳定算法
stiffly stable implicit algorithm 刚性稳定隐式算法
stiffly stable step-size control 刚性稳定步长控制
stiffness 刚度；硬挺度
stiffness matrix 刚度矩阵
stiffness ratio 刚度比
stigmergic 群体激发工作现象的，共识主动性的
stigmergic behavior 群体激发工作的行为
stigmergic collaboration 群体激发工作式协作
stigmergic communication 群体激发工作式通信
stigmergic coordination 群体激发工作式协调
stigmergic interaction 群体激发工作式交互
stigmergic system 群体激发工作式系统
stigmergy 群体激发工作
stigmergy model 群体激发工作行为
stigmergy system 群体激发工作系统
stimulant 刺激物；激励的
stimulate (v.) 激励；有助于；促进
stimulation 刺激；激励；兴奋
stimulator 激励器；刺激物
stimulus 刺激；激励源
stochastic 随机的；推测的
stochastic agent-based model 基于随机代理的模型
stochastic analysis 随机分析
stochastic approximation method 随机近似法
stochastic automaton 随机的自动机
stochastic dependence 随机关系
stochastic differential equation 随机微

分方程

stochastic differential equation model 随机微分方程模型

stochastic differential game 随机微分博弈

stochastic error 随机误差

stochastic event 随机事件

stochastic game 随机博弈

stochastic input process 随机输入过程

stochastic Kriging 随机克里格法

stochastic matrix 随机矩阵

stochastic model 随机模型

stochastic modeling 随机建模

stochastic noise 随机噪声

stochastic output process 随机输出过程

stochastic Petri net 随机佩特里网

stochastic process 随机过程

stochastic simulation 随机仿真

stochastic-simulation-based 基于随机仿真的

stochastic simulation model 随机仿真模型

stochastic simulation optimization 随机仿真优化

stochastic system 随机系统

stochastic timed-transition 随机时间变迁

stochastic transition 随机转换

stochastic value 随机值

stochastic variability 随机可变性

stochastic variable 随机变量

stock and flow variable 库存的流量变量

stock variable 库存变量

stopping criterion 停止准则

storage 存储器；贮存

stored energy 存储的能量

strange attractor 奇怪吸引子

strategic-decision simulation 战略决策仿真

strategic game 战略博弈

strategic management knowledge 战略管理知识

strategic simulation 战略仿真

strategic variable 战略变量

strategic view 战略观点

strategy 策略，战略

strategy game 战略博弈

strategy simulation 战略仿真

strategy video game 战略电子游戏

stratified sampling 分层采样

stress testing 压力测试；强度测试

strict validity 严格有效性

string variable 字符串变量

strong anticipation 强烈的期待

strong biosimilarity 强生物相似性

strong bisimilarity 强互相似

strong bisimulation 强互仿真

strong emergence 强涌现

strong emergent behavior 强涌现行为

strong simulation 强健仿真

strong-strong simulation 强强仿真

strong two-way coupling 强双向耦合

strong two-way coupling simulation 强双向耦合仿真

strongly anticipatory system 强预期系统

strongly coupled algorithm 强耦合算法

strongly NP-hard problem 强 NP 难题
structural analysis 结构分析
structural assessment 构造评估
structural behavior 结构性能
structural-change 结构变化
structural comparison 结构比较
structural complexity 结构复杂性
structural condition 结构条件
structural coupling 结构耦合
structural discontinuity 结构不连续性
structural domain 结构论域
structural emergence 结构涌现
structural equation model 结构方程模型
structural equation modeling 结构方程建模
structural factor 结构因素
structural inference 结构推论
structural model 结构模型
structural model comparison 结构模型比较
structural model fidelity 构造模型保真度
structural model validity 结构模型有效性
structural property 结构属性
structural similarity 结构相似性
structural simulation 构造仿真
structural singularity 结构奇异性
structural singularity elimination 结构奇异性消除
structural SL 结构仿真语言
structural stability 结构稳定性
structural technique 结构化技术
structural testing 结构化测试
structural uncertainty 结构不确定性
structural understanding 结构化理解
structural validation 结构验证
structural validity 结构有效性
structurally coupled 结构耦合的
structurally singular model 结构奇异模型
structurally singular system 结构奇异系统
structurally valid model 结构有效的模型
structure 结构；组织
structure analysis 结构分析
structure assessment 构造评估
structure credibility 结构可信度
structure digraph 结构有向图
structure identification 结构辨识
structure identification methodology 结构辨识方法学
structure incidence matrix 结构关联矩阵
structure modeling 结构建模
structure simulation 结构仿真
structure-transition function 结构转换函数
structured 结构化的
structured data 结构化数据
structured documentation 结构化文档
structured entity 结构化实体
structured information 结构化信息
structured methodology 结构化方法学
structured modeling 结构化建模
structured reducibility 结构化可还原性

structured set 结构化集合
structured simulation knowledge 结构化仿真知识
structured system 结构化系统
structured walkthrough 结构化预演
study 研究
style 格式；类型；风格；式样
style guide 风格指南
stylized model 程式化模型
stylus input 手写笔输入
subactivity 子活动
subclass 子类
subclass coupling 子类耦合
subclassing 子类化
subcomponent 子组件，子部件；子分量；次要成分
subcontext 子上下文
subcontract management 分包管理
subcoupling 局部连接；子耦合
subevent 子事件
subformalism 子形式
subframe 副帧；副框架
subgame 子博弈
subgoal 子目标
subject area 主题范围
subject matter expert 主题专家
subject-related knowledge 主题相关知识
subject variable 主题变量
subjective data 主观数据
subjective error 主观错误；主观误差
subjective experience 主观经验
subjective metric 主观度量
subjective reality 主观现实
subjective validation 主观验证

subjective value 主观价值
subliminal stimulus 阈下刺激
submodel 子模型
submodel activation 子模型激活
submodel deactivation 子模型失活
submodel replacement 子模型替换
submodel testing 子模型测试
submodel validity 子模型有效性
submodularity 子模块性；子调制性
submodule 子模块
subnet 子网
subnet membership function 子网隶属度函数
subscripted variable 下标变量
substance 物质；实质
substantive interoperability 实质互操作性
substitutability 可替代性；互换性；替代度
substitutable 可替换的
substitution error 替代误差
substructure 子结构
subsumption relation 包含关系
subsystem execution 子系统执行
subsystem extrapolation 子系统外推
subsystem goal 子系统目标
subtreshold stimulus 次级、限值一下的刺激
subvariable 子变量
succession of events 事件的连续性
successor 后继站；后继
successor model 后继模型
successor relation 后继关系
successor simulation 后继仿真
successor simulation study 后继仿真

研究

suitability 适用性

suitability of a language for modeling 建模语言的适用性

suitability of a language for semantic modeling 语义建模语言的适用性

suitability of a paradigm for modeling 建模范式的适用性

suitability of a paradigm for semantic modeling 语义建模范式的适用性

suitability of modeling methodology 建模方法学的适用性

suitable 适合的；相适应的

suitable model 适合的模型；相配的模型

suitable simulation 合适的仿真

suite 套件；组

superadditive trust game 超可加信任博弈

superbasic variable 超基变量

superclass 超类

superclassing 超类化

supercomputer 超级计算机

superficial model 表面的模型

superintelligence 超级智能

superposed state 叠加的状态

superposition principle 叠加原理

superposition state 叠加状态

supersystem 超级系统

supplementary maintenance 辅助的维护

supplementary model maintenance 辅助的模型维护

supplementary variable 附加的变量；辅助变量

supply chain modeling 供应链建模

support 支持，支撑

support software 支撑软件

support tool 支撑工具

supportability 可支持性；保障性

supported 支撑的，支持的

supporting domain 支撑论域

suppressor variable 抑制变量

supremum 上确界

surface 表面；曲面

surface model 表面模型；曲面模型

surface modeling 表面建模；曲面建模

surplus variable 剩余变量

surrogate-based 基于代理的

surrogate-based model 基于代理的模型

surrogate model 代理模型；替代模型

survivability simulation 生存性仿真

sustainable 可持续的

sustainable emergent behavior 可持续的应急行为

sustainable group 可持续的群体

sustainable simulation 可持续的仿真

swap (v.) 交换；调动

swapping 对换

swapping method 交换方法

swarm behavior 群体行为

swarm intelligence 群体智能

swarm model 群体模型

swarm simulation 群体仿真

switch 切换；开关

switchable understanding 可切换的理解

switching 切换；交换；开合；通断

switching model 切换模型

syllogism 三段论
symbiotic behavior 共生行为
symbiotic simulation 共生仿真
symbiotic simulation-based training 基于共生仿真的训练
symbolic algorithm 符号算法
symbolic analysis 符号分析
symbolic approach 符号方法
symbolic computing 符号计算
symbolic debugging 符号调试
symbolic DEVS 符号离散事件系统规范
symbolic differential equation solver 符号微分方程求解器
symbolic differentiation 符号微分法
symbolic entity 符号实体
symbolic evaluation 符号评价
symbolic execution 符号化执行
symbolic index reduction algorithm 符号索引约简算法
symbolic knowledge 符号化知识
symbolic model 符号模型
symbolic model processing 符号模型处理
symbolic preprocessing 符号预处理
symbolic processing 符号处理
symbolic processor 符号处理器
symbolic regression 符号回归
symbolic simulation 符号仿真
symbolic technique 符号技术
symmetric game 对称博弈
symmetric simulation 对称仿真
symmetrical distribution 对称分布
symmetrical function 对称函数
symmetrical wargame 对称的军事演习
symmetry model 对称性模型
synchronization 同步
synchronization mechanism 同步机制
synchronization point 同步点
synchronization protocol 同步协议
synchronized 同步化的
synchronized collaborative modeling 同步协同建模
synchronized execution 同步执行
synchronized input 同步输入
synchronized modeling 同步建模
synchronous 同步；同步的；同时的
synchronous analysis 同步分析
synchronous input 同步输入
synchronous model 同步的模型
synchronous state 同步的状态
synchronous transmission 同步传输
synergetic scenario 协同想定
synergetics 协同学
synergy 协同
synoptic model 天气模型
syntactic assessment 语法的评估
syntactic composability 句法可组合性
syntactic conformance 句法一致性
syntactic emergent behavior 句法突现行为
syntactic error 语法错误
syntactic evaluation 句法评价
syntactic interoperability 句法互操作性
syntactic interoperability level 句法互操作性水平
syntactic level 语法水平

syntactic validation 语法验证
syntactic variability 语法可变性
syntactical error 语法错误
syntactical interoperability level 语法互操作水平
syntax 句法；语法
syntax analysis 语法分析
syntax check 语法检查
syntax error 语法错误
syntax of model component 模型组件的语法
syntax of modeling language 建模语言语法
synthesis 综合
synthesizable scenario 可合成想定
synthesized model 综合模型
synthetic 合成的；综合的；人造的
synthetic analog 合成模拟
synthetic battlefield 合成战场
synthetic battlespace 合成战场空间
synthetic component 合成组件
synthetic environment 合成环境
synthetic-environment based 基于合成环境的
synthetic-environment-based acquisition 基于合成环境的采办
synthetic environment content 合成环境内容
synthetic environment data 合成环境数据
synthetic environment modeling 综合环境建模
synthetic input 合成输入
synthetic model 综合模型
synthetic modeling 综合建模

synthetic reality 合成现实
synthetic space 合成空间
synthetic theater of war 战争综合演示室
system 系统
system analysis 系统分析
system approach 系统方法
system architecture 系统架构
system architecture specification 系统架构规范
system assessment 系统评估
system behavior 系统行为
system boundary 系统边界
system characteristic 系统特性
system classification 系统分类
system complexity 系统复杂性
system connectivity 系统连通性
system constraint 系统约束
system control 系统控制
system coupling 系统耦合
system coupling recipe 系统耦合方法
system description 系统描述
system design 系统设计
system design process 系统设计过程
system design theory 系统设计理论
system development 系统开发
system development process 系统开发过程
system development system 系统开发系统
system diagnostics 系统诊断学
system dynamics 系统动力学
system dynamics-based modeling 基于系统动力学的建模
system dynamics modeling 系统动力

学建模

system entity　系统实体
system entity structure　系统实体结构
system error　系统性误差
system event　系统事件
system failure　系统故障
system goal　系统目标
system identification　系统辨识
system in transition　系统转换中
system integration　系统集成
system integration service　系统集成服务
system integrity　系统完整性
system interaction　系统交互
system layout　系统配置
system level　系统层
system-level metric　系统级度量
system-level misbehavior　系统级不当行为
system-level performance　系统性能
system-level property　系统级性能
system-level simulation　系统级仿真
system-level soft error　系统级软错误
system-level specification　系统层规范
system life cycle　系统生命周期
system model　系统模型
system modeling　系统建模
system modeling language　系统建模语言
system morphism　系统同型
system of interest　收益系统
system of systems　体系，系统的系统，复合系统，复杂大系统
system of systems approach　体系方法
system of systems ecology　体系生态学

system of systems emergent behavior　体系应急行为
system of systems level　体系级
system of systems model　体系模型
system of systems model instance　体系模型实例
system of systems simulation　体系仿真
system of systems SL　体系仿真语言
system of systems taxonomy　体系分类
system on a chip　片上系统
system plasticity　系统可塑性
system property　系统属性
system quality characteristic　系统质量特性
system quality prediction　系统质量预测
system representation　系统表示（法）
system response　系统响应
system safety　系统安全性
system safety standard　系统安全标准
system simulation　系统仿真
system simulation theory　系统仿真理论
system-specific　特定系统的
system-specific library　特定系统库
system specification　系统规范
system stability　系统稳定性
system state　系统状态
system test requirements　系统测试需求
system test system　系统测试系统
system-theory-based continuous system simulation　基于系统理论的连续系

统仿真

system-theory-based discrete-event simulation 基于系统理论的离散事件仿真

system-theory-based simulation 基于系统理论的仿真

system thinking 系统思维

system under investigation 系统调查中

system variable 系统变量

system verification 系统校核

system view 系统视图

systematic 系统的；系统化的；体系的

systematic dictionary 系统词典

systematic error 系统性误差

systematic experimentation 系统化实验

systematic sampling 系统采样

systematic simulation 系统化仿真

systematic specification 系统化规范

systemic 系统的

systemic activity 系统活动

systemic cause 系统原因

systemic error 系统性误差

systemic model 系统模型

systemic solution 系统解

systemness 系统性

systems 系统；体制

system's behavior 系统行为

systems biology markup language 系统生物学标记语言

systems engineering 系统工程

systems engineering for simulation life cycle 面向仿真生命周期的系统工程

systems engineering life cycle 系统工程生命周期

systems engineering system 系统工程化的系统

system's observed behavior 系统的观测行为

systems theory 系统理论

systems-theory-based 基于系统理论的

systems-theory-based approach 基于系统理论的方法

systems-theory-based continuous system simulation 基于系统理论的连续系统仿真

systems-theory-based discrete-event simulation 基于系统理论的离散事件仿真

systems-theory-based M&S 基于系统理论的建模与仿真

systems-theory-based simulation 基于系统理论的仿真

Tt

t-simulation t仿真

table error 表格错误

table look up 表格查询，表格查阅

table model 模型表格

tableau 表；场面，局面；画面

tableau Butcher 表格剪切

tabu search 禁忌搜索

tabu search method 禁忌搜索方法

tabula rasa understanding 白板的理解

tabular 列表的，表格式的；按表格计算的

tabular model 表格式的模型
tacit hypothesis 隐性假设
tacit knowledge 隐性知识
tactical decision simulation 战术决策仿真
tactical simulation 战术仿真
tactics training 战术训练
tactile 触觉的；能触知的
tactile characteristic 触觉特性
tactile display 触觉显示
tactile feedback 触觉反馈
tactile feedback hardware 触觉反馈硬件
tactile I/O 触觉输入/输出
tactile input 触觉输入
tactile interface 触觉界面
tactile interfacing 触觉交互
tactile signal 触觉信号
tagged value 标记值
tailor (v.) 剪裁；制作
tailoring 剪裁
tandem simulation 串联仿真
tangible interaction 可触知的互动
tangible value 有形的价值
target control model 目标控制的模型
target data 目标数据
target domain 目标域
target model 目标模型
target registration error 目标配准误差
target SL 目标仿真语言
target state 目标状态
target system 目标系统
targeted coupling 定向耦合
Tarjan algorithm 塔尔简算法
task 任务

task level 任务级别
task-oriented 面向任务的
task-oriented agent 面向任务的智能体；面向任务的代理
task-oriented help 面向任务的帮助
task planner 任务规划器
task sharing 任务共享
task specific interface 任务特定接口
task switching 任务切换
taxonomy 分类；分类学
taxonomy tree 分类树
Taylor series 泰勒级数
teaching 教学
team 团队
team-based 基于团队的
team-based training 基于团队的训练
team-oriented 面向团队的
team-oriented design 面向团队的设计
team-oriented multidisciplinary design 面向团队的多学科设计
team support tool 团队支持工具
tearing algorithm 撕裂算法
tearing variable 撕裂变量
technical data 技术数据
technical infrastructure 技术的基础设施
technical interoperability 技术互操作性
technical interoperability level 技术互操作级别
technical requirements 技术需求
technical risk 技术风险
technical simulation 技术仿真；工艺仿真
technical validity 技术有效性

technical view 科技视角
technique 技术；方法；工艺
technological attribute 技术属性
technological model attribute 技术模型属性
technological singularity 技术奇点
technologically obsolete model 技术上过时的模型
technology 技术；工艺
technology analysis 技术分析
technology-dependent model 依赖于技术的模型
technology-enhanced 技术增强的
technology-enhanced simulation 技术增强型仿真
technology-independent model 技术独立的模型
technology requirements 技术需求
teleimmersion 远程沉浸
teleogenetic system 远生系统
teleogenic model 远源模型
teleological model 目的论模型
teleological system simulation 目的论系统仿真
teleology 目的论
teleonomic model 目的性的模型
teleonomic system simulation 目的性系统仿真
teleozetic simulation 远传仿真
template 模板，样板，样型
template-based 基于模板的
template-based documentation 基于模板的文档
template-based modeling 基于模板的建模
temporal 暂时的；时间的
temporal assumption 时间假设
temporal behavior 暂时的行为
temporal boundary condition 时间边界条件
temporal condition 时间条件
temporal coupling 时间耦合
temporal error 时间误差
temporal model 时序模型
temporal relation 时序关系
temporal specification 时序规范
temporal state 暂态
temporal structure 时序结构
temporal uncertainty 时间不确定性
temporal variable 临时变量
temporary variable 中间变量；临时变量
TENA (Test and Training Enabling Architecture) 测试和训练使能架构
tendency error 倾向误差
Terahertz technology 太赫兹技术
terascale computer 万亿级计算机
terascale computing 万亿级计算
terascale simulation 万亿级仿真
term 期限；术语；条款；学期
terminal 终端
terminal section 末段
terminal value 终值
terminating simulation 终止仿真
termination condition 终止条件
termination criterion 终止准则，终止判据
terminology 术语学；术语，专有名词
tessellation 曲面细分
test 测试；试验；检验

test and evaluation 测试和评价
test and evaluation master plan 测试和评估主计划
test and evaluation simulation 测试和评估仿真
test and training enabling architecture 试验和训练使能架构
test-based 基于测试的
test-based confidence 基于测试建立的信任
test management 测试管理
test methodology 测试方法学
test model 测试模型
test observance 试验观测
test qualification 试验鉴定
test qualification technique 试验鉴定技术
test tool 测试工具
testability 可测试性；可检验性
testable 可测试的
testable consequence 可测试的结果
testable emergent behavior 可测试的应急行为
testbed 试验床，试验台
testing 测试
testing and evaluation 测试和评价
testing data 测试数据
testing data set 测试数据集合
testing methodology 测试方法学
testing phase 测试阶段
testing technique 测试技术
tests execution 测试执行
text-based 基于文本的
text-based screen design 基于文本的屏幕设计
text-to-model transformation 文字-模型转换
textual specification 文本规范
texture simulation 纹理仿真
theater of war 战争演示厅
thematic dictionary 专题词典
theorem 定理，法则；命题；理论
theoretical 理论的
theoretical confirmation 理论的证实
theoretical data 理论数据
theoretical framework 理论框架
theoretical model 理论模型
theoretical modeling framework 理论建模框架
theoretical study 理论研究
theoretical validity 理论有效性
theoretical value 理论值
theory 理论；原理；学说
theory-based 基于理论的
theory centric 以理论为中心
theory construction 理论构建
theory of truth 真理论
theory validation 理论验证
thermal field problem 温度场问题
thermodynamic variable 热力学变量
thesis 论文；论点
thinking 思维；思考；思想
third degree simulation 三维仿真
third-order accurate 三阶精度
third-order overimplicit Adams formula 三阶超隐式亚当斯公式
third-order Runge-Kutta integration 三阶龙格-库塔积分
third-person shooter game 第三人称射击游戏

third variable 第三变量
thought controlled simulation 思维控制仿真
thought experience 思维体验
thought experiment simulation 思维实验仿真
thought simulation 思维仿真
threading game-technology game 线程化游戏技术制作的游戏
threading technology 线程技术
threat representation 威胁表征
threshold 阈；阈值，门限
threshold model 阈值模型
threshold value 阈值
throttled time-warp simulation 扼制时间翘曲的仿真
through variable 直通变量
tie branching 联络分支
tie-breaking function 决胜函数
tie-breaking rule 打破和局规则
tight coupling 紧耦合
tightly coupled 紧耦合的
tightly coupled model 紧耦合的模型
tightly coupled system 紧耦合的系统
time 时间
time advance 时间推进
time advance function 时间推进函数
time advance grant 时间推进准许
time advance mechanism 时间推进机制
time advance request 时间推进请求
time advancement 时间推进
time alignment 时间对准
time aspect 时间方面
time axis 时间轴

time base 时基
time-base accuracy 时基精度
time-base expansion 基于时间的扩展
time-base generator 时基发生器
time-based 基于时间的
time-based signal 基于时间的信号
time consistency 时间一致性
time constant 时间常数
time constrained 时间约束的
time constraint 时间约束
time-consuming activity 耗时的活动
time delay 时延
time dependent 时间相关
time-dependent constraint 时间相关的约束
time-dependent coupling 时间相关的耦合
time-dependent covariate 时间相关的协变量
time-dependent event 时间相关事件
time discretization 时间离散化
time domain 时域
time-domain design 时域设计
time-domain simulation 时域仿真
time-driven 时间驱动
time-driven simulation 时间驱动的仿真
time-event 时间事件
time-event system 时间事件系统
time flow 时间流
time flow mechanism 时间流机制
time granularity 时间粒度
time independent 时间独立的
time-independent constraint 时间独立的约束

time-independent covariate 时间独立的协变量
time-indexed 时间索引的
time-indexed data 时间索引数据
time interval 时间间隔
time-interval simulation 时间间隔仿真
time invariance 时不变
time invariant 时不变的
time-invariant continuous system 时不变连续系统
time-invariant coupling 时不变耦合
time-invariant model 时不变模型
time-invariant system 定常系统，时不变系统
time management 时间管理
time management algorithm 时间管理算法
time-management data 时间管理数据
time-management module 时间管理模块
time-management transparency 时间管理透明度
time period 时间周期
time persistent statistic 持续时间统计
time resolution 时间分辨率
time scale 时间尺度
time-scale variable 时间尺度变量
time series 时间序列
time series analysis 时序分析
time-series data 时序数据
time series time 时序时间
time-series validity 时间序列有效性
time set 时间集合
time-shared input/output system 分时输入/输出系统
time-slice 时间片段
time-slice simulation 时间片仿真
time slicing 时间分片
time-slicing simulation 时间分片的仿真
time slot 时间槽；时隙
time-stamp 时间标记，时间戳
time-stamp of an event 事件的时间戳
time-stamp order 时间戳顺序
time-stamped message 时间戳消息
time step 时间步长
time step model 时间步长模型
time-stepped execution 时间步进式执行
time stepping 时间步进
time-stepping simulation 时间步进仿真
time-synchronized 时钟同步
time trajectory 时间轨迹
time-triggered 时间触发的
time-triggered control 时间触发的控制
time value 时间值
time variable 时间变量
time-variant system 时变系统
time-varying coupling 时变耦合
time-varying model 时变模型
time-varying system simulation 时变系统仿真
time warp 时间翘曲
time-warp algorithm 时间翘曲算法
time-warp mechanism 时间翘曲机理
time-warp simulation 时间翘曲仿真
timed automata 时间自动机

timed automaton 时间自动机
timed-event system 定时事件系统
timed formalism 定时形式
timed input/output automaton 定时的输入/输出自动机
timed Petri net 时间佩特里网
timed system 定时系统
timed transition 时间变迁；定时转换
timed transition system 时间变迁系统
timing simulation 时序仿真
token 标记；令牌；权标
token knowledge attribute 标记知识属性
token numeric attribute 标记数值属性
token string attribute 标记字符串属性
token value 标记值
tolerance 公差；容忍
tolerance interval 容许区间
tolerance on comparison 容差比较
tolerant 容忍的，耐受的
tool 工具
tool error 工具误差
tool integration 工具集成
tool integration framework 工具集成框架
tool interoperability 工具互操作性
tool support 工具支持
top-down construction 自上向下构建
top-down design 从上至下设计
top-down emergence 自上而下涌现
top-down testing 自上而下测试
top-down understanding 自上而下理解
topic 主题；专题；话题
topic model 主题模型；话题模型
topographic model 地形学模型
topological 拓扑的；拓扑学的
topological coupling 拓扑耦合
topological interaction 拓扑交互
topological modeling 拓扑建模
topological relation 拓扑关系
topological shape modeling 拓扑形状建模
topological transformation 拓扑变换
topologically coupled 拓扑耦合的
topology 拓扑；拓扑学
total immersion gaming 全沉浸式游戏
total state 全部状态
total system behavior 全系统行为
totally ordered set 全有序集合
touch input 触摸输入
touch sensory interface 触摸式界面
trace 跟踪；痕迹
trace (v.) 跟踪
trace analysis 踪迹分析
trace-based 基于轨迹的
trace-based modeling 基于轨迹的建模
trace data 跟踪数据
trace-driven 轨迹驱动的
trace-driven input testing 轨迹驱动的输入测试
trace-driven model 轨迹驱动的模型
trace-driven simulation 踪迹驱动的仿真
trace-driven simulator 轨迹驱动的仿真器
traceability 可追溯性；溯源性；跟踪能力

traceability assessment 可追溯性评估
traceability matrix 可追溯性矩阵
traceable 可追溯的；可追踪的；起源于
tracing 追踪
tracking error 跟踪误差
tractability 易处理，驯良；精细性；溯源性
tractable 易于驾驭的；易处理的
tractable equation 易处理方程
tractable model 易处理模型
tractable simulation 可跟踪的仿真
trade-off algorithm 折中算法
trade-off requirements 权衡需求
traditional design approach 传统设计方法
traditional reality 传统现实
train as you operate 操作训练
trainer 训练器；教练机
training 培训；训练；教练；练习
training analysis 训练分析
training confederation 训练联盟
training data 训练数据
training data-set 训练数据集合
training device 训练装置
training effectiveness 训练效能
training effectiveness evaluation 训练效能评价
training environment 训练环境
training game 训练游戏
training on the real-system 实际系统上的训练
training platform 训练平台
training simulation 训练仿真
training simulation game 训练仿真游戏
training simulator 训练仿真器
training staff 训练人员
training system 训练系统
training task analysis 训练任务分析
training technology 训练技术
trajectory 迹线，轨迹，轨道
trajectory behavior 轨迹行为
trajectory comparison 轨迹比较
trajectory of a variable 变量的轨迹
trajectory planner 弹道规划器
trajectory simulation 弹道仿真
trajectory stability 轨道稳定性
transaction-based 基于事物的
transaction-based Petri net 基于事务的佩特里网
transcription error 转录错误
transdisciplinary collaboration 跨学科协同
transdisciplinary modeling 跨学科建模
transducer 传感器，换能器；转换器
transfer 转运；转移
transfer (v.) 转运；转移
transfer error 转移误差
transfer function simulation 传递函数仿真
transfer value 转移值
transferability 可移动性；可传递性；可转移性
transform (v.) 转换
transformation 变换；转化
transformation algorithm 变换算法
transformation chain 变换链
transformation-driven 变换驱动的

transformation-driven replication 变换驱动的复制
transformation engine 变换引擎
transformation for validation 用于验证的转换
transformation language 转换语言
transformation paradigm 转换范式
transformation strategy 转换策略
transformational accuracy 转换精度
transient 过渡过程，暂态，瞬态
transient behavior 瞬时行为
transient elimination 瞬变消除
transient phase 过渡阶段
transient sensitivity analysis 瞬态灵敏度分析
transient state 瞬态；暂态
transition 过渡；转换；变迁
transition condition 过渡条件
transition enabling 转换使能
transition-enabling firing 转换使能触发
transition function 转换函数
transition operation 转换操作
transition output 转换输出
transition selection rule 跃迁选择规则
transition system 转换系统
transition variable 过渡变量
transitivity 传递性，转移性
transitory state 暂态
translate ($v.$) 翻译；转化
translation error 翻译错误
translation function 转换函数
translation methodology 翻译方法学
translational variable 平移变量
translator 翻译程序

transmission 传输；透射；传动；传动装置；变速器
transmission error 传输误差
transmit ($v.$) 传输；发送
transparency 透明；透明性，透明度
transparent cognitive model 透明认知模型
transparent model 透明模型
transparent reality simulation 透明现实仿真
transport 运输，传送
transporter model 转运体模型
transpose 调换，变换；转置
transposition error 换位误差
trapezoidal formula 梯形公式
tree 树
tree-based modeling 基于树的建模
tree formalism 树形式
tree model 树模型
trend 趋势，倾向
trend verification 趋势校核
trends in simulation 仿真趋势
trial-and-error 试错，反复试验
trial-and-error method 试错法，尝试法
trial-level model 审级模型
trial simulation 试验仿真
triangular distribution 三角分布
trigger 触发器；触发
trigger behavior 触发行为
trigger input 触发输入
triggered 触发的
triggering phase 触发阶段
triggering state 触发状态
trip generation 行程生成
triple pole 三极的

Trojan simulator 木马仿真器
true 真实的；准确的；正直的；正确的
true global time 准确的世界时间
true knowledge 真知
true value 真值
truncated dependent variable 截断因变量
truncated distribution 截断分布
truncated normal distribution 截断正态分布
truncated random variable 截断随机量
truncated variable 截断变量
truncation 舍去，截断
truncation error 截断误差
trust game 信任博弈
trust game experiment 信任博弈实验
trust management language 信任管理语言
trustworthiness 信誉，可信度
trustworthy simulation 可靠的仿真
truth 真值；真理；真实
truth value 真值，真假值
tunable parameter 可调参数
tuner error 调谐器误差
Turing machine-like model 图灵机样模型
Turing machine model 图灵机模型
Turing machine simulation 图灵机仿真
Turing model 图灵模型
Turing test 图灵测试
turn-based game 回合制博弈
turning bands conditional simulation 转向带条件仿真
turning bands simulation 转向带仿真
two-action coordination game 双动作协调博弈
two-action pure coordination game 双动作纯协调博弈
two-dimensional network 二维网络
two-directional causality 双向因果关系
two-directional input 双向输入
two-level factorial design 两级析因设计
two-level simulation 两级仿真
two-level simulation methodology 两级仿真方法学
two person game 双人博弈
two-person hypergame 二人超对策
two person zero-sum game 双人零和博弈
two sided game 双边博弈
two-stage transformation 两阶段变换
two-stage understanding 两阶段理解
two-way coupling 双向耦合
type 类型，型
type 1 fuzzy system model 第一类模糊系统模型
type 2 fuzzy system model 第二类模糊系统模型
type I error Ⅰ型错误
type I error probability Ⅰ型错误概率
type Ⅱ error Ⅱ型错误
type Ⅱ error probability Ⅱ型错误概率
type Ⅲ error Ⅲ型错误
type Ⅲ error probability Ⅲ型错误概率
type of error 错误类型

type variable 类型变量
typed variable 类型化变量
types of fallacies in logic 逻辑错误类型
types of knowledge processing perspective 知识处理视角类型
typical error 典型错误
typing error 打字错误

Uu

ubiquitous computing simulation 泛在计算仿真
ubiquitous computing simulator 泛在计算仿真器
ubiquitous game 泛在博弈
ubiquitous simulation 泛在仿真
ubiquitous simulator 泛在仿真器
ubiquitous system 泛在系统
ultrascale simulation 超大规模仿真
unaccounted assumption 未说明的假设
unacknowledged error 未确认的错误
unambiguity 明确性，无歧义性
unambiguous 明确的，清晰的
unambiguous input 明确的输入
unambiguous model 明确的模型
unambiguous representation 无歧义表示（法）
unambiguous rule 明确的规则
unanticipated event 未预料到的事件
unbalanced error 不平衡误差
unbiased analysis 无偏估计
unbiased data 无偏数据
unbiased error 无偏误差
unbiased estimator 无偏估值器
unbiased sample 无偏样本
unbound variable 无约束变量
unbounded variable 无界变量
unbundling 非绑定
uncanny valley 诡异谷
uncertain 不确定的
uncertain data 不确定数据
uncertain information 不确定信息
uncertainty 不确定性，不定度，不确定度
uncertainty in design 设计不确定性
uncertainty in modeling 建模中的不确定性
uncertainty modeling 不确定性建模
uncertainty quantification 不确定性量化
uncertainty simulation 不确定性仿真
unchallenged assumption 无异议的假设
uncomplicated 不复杂的
unconditional simulation 无条件的仿真
unconditionally stable 无条件稳定
unconnected configuration 非连接配置
unconstrained simulation 无约束仿真
uncontrollable component 不可控的组件
uncontrollable factor 不可控因素
uncontrollable variable 不可控变量
unconventional input 非常规输入
uncorrelated error 不相关误差
uncorrelated random variable 无关随机变量

uncouple (v.)　解耦
uncoupled　解耦合的
uncoupled simulation　解耦仿真
uncoupling　解耦
undefined error　未定义误差
undefined variable　未定义变量
understand (v.)　理解；以为
understandability　可理解性，易懂性
understanding　理解，谅解
understanding process　理解过程
understanding system　理解系统
understanding theory　理解理论
understood reality　理解现实
undesirable system　不合需要的系统
undetected error　漏检错误
undiscovered error　未发现的错误
unessential state event　非本质的状态事件
unexpected behavior　意外的行为
unexpected input　意外的输入
unexplained behavior　无法解释的行为
unexplained emergent behavior　无法解释的突现行为
unfeasible state　非可行状态
unicast　单一传播
unidentified state　未经确认的状态，未识别状态
unification error　合一错误
unified discrete and continuous simulation　统一的离散合连续仿真
unified framework　统一框架
unified metamodel　统一的元模型
unified metric　统一的度量标准
unified modeling approach　统一的建模方法
unified modeling language　统一的建模语言
unified multimodeling methodology　统一的多重建模方法学
unified simulation　统一仿真
uniform bisimulation equivalence　一致互仿真等价
uniform distribution　均匀分布
uniform input　一致输入
uniform modeling　统一建模
uniform modeling language　统一建模语言
uniform random number　均匀分布的随机数
uniform random variable　均匀随机变量
uniform system　统一系统
uniformization simulation　归一化仿真
uniformized event　归一化事件
unifying framework　统一框架
unifying modeling framework　统一建模框架
unimportant state event　不重要的状态事件
uninitialized variable　未初始化变量
uninstaller error　卸载程序错误
unintended consequence　意外后果
unintended emerging behavior　意外出现的行为
unintended outcome　意外结果
unintended behavior　意外行为
unintentional state　无意识状态
uninteresting variable　无趣变量
unique random variable　单一随机变量

uniqueness of DEVS representation 离散事件系统规范表示的唯一性
uniqueness of representation 表示的唯一性
uniqueness of solution 解的唯一性
unit 单元，单位；装置
unit conversion 单位换算
univariate 单变量的
universal coordination 世界坐标
universal function approximator 通用函数逼近器
universal property 通用属性
universal time 世界时
universal Turing machine simulation 通用图灵机仿真
universality of DEVS representation 离散事件系统规范表示的通用性
universality of representation 表示的通用性
unjustified assumption 不合理假设
unjustified reliance 不合理信任
unknown error 未知错误
unknown reality 未知现实
unknown state 未知状态
unknown value 未知值
unknown variable 未知变量
unmodeled dynamics 未建模动力学
unmodifiable property 不可改变的特性
unobserved variable 未观察到的变量
unplanned context 计划外情境
unpredictable behavior 不可预见的行为
unpredictable data 不可预见的数据
unpredictable emergence 不可预见的突现
unpredictable emergent behavior 不可预见的突现行为
unpredicted data 未预测数据
unqualified variable 未限定变量
unquantifiable relationship 无法量化的关系
unrecognized assumption 未公认的假设
unrecoverable error 不可恢复的错误
unreferenced variable 无参照的变量
unscheduled event 非调度事件
unset value 未设置的值
unsimulatable 不可仿真的
unsolicited model 未经授权的模型
unspecified error 未指明的错误
unstable 不稳定的
unstable error 不稳定误差
unstable model 不稳定模型
unstable result 不稳定结果
unstable solution 非稳定解
unstable state 非稳态
unstructured data 非结构化数据
unstructured grid 非结构化网格
unsuitable simulation 不适合的仿真
unsustainable emergent behavior 不可持续的应急行为
untestable emergent behavior 不可检测的突现行为
untimed 不计时的，不限期的
untimed discrete-event system model 不计时的离散事件系统模型
untimed formalism 不计时形式
untimed functional model 非时变功能模型

untimed model　非时变模型
ununiformized event　非均匀化事件
ununiformized rate　非均匀化速率
unvariable　恒定的，不变化的
unverified understanding　未经核实的理解
up-to-date model　最新的模型
updatable continuous-change model　可更新的连续变化模型
updatable continuous model　可更新的连续模型
updatable discrete-change model　可更新的离散变化模型
updatable discrete model　可更新的离散模型
updatable event model　可更新的事件模型
updatable memoryless model　可更新的无记忆模型
updatable model　可更新的模型
updatable piece-wise continuous model　可更新的分段连续模型
updatable process model　可更新的过程模型
update　更新；修正
update (v.)　更新；修正
update methodology　更新方法学
update region　更新区域
updated data　更新的数据
updated model　更新的模型
updating　更新
upper bound　上界
upper bound on the time stamp　时间戳上界
upper limit　上限

upper merged ontology　上层合并本体
upper ontology　上层本体
upward causation　向上因果
usability　可用性
usability requirements　可用性需求
usability testing　可用性测试
usability tool　可用性工具
usable　可用的
usage error　使用错误
usage mining　使用挖掘
use　使用；用途；用法
use case diagram　用例图
use of M&S　建模与仿真（M&S）的使用
use of simulator in training　仿真器在训练中的应用
used　使用；习惯的；旧的
used metamodel　使用的元模型
usefulness　有用；有效性；有用性
useless data　无用数据
user　用户，使用者
user acceptance　用户验收
user behavior　用户行为
user behavior model　用户行为模型
user-centered　以用户为中心
user-centered analysis　用户为中心的分析
user-centered design　以用户为中心的设计
user-centered experimentation　以用户中心的实验
user-created　用户创造的
user documentation　用户文档
user domain　用户域
user domain requirements　用户领域

需求
user error 用户错误
user input 用户输入
user interface 用户接口
user interface analysis 用户界面分析
user interface design 用户界面设计
user interface error 用户界面错误
user interface testing 用户接口测试，用户界面测试
user modeling 用户建模
user support 用户支持
user/system interface 用户/系统接口
user's conceptual model 用户概念模型
user's mental model 用户心理模型
user's model 用户的模型
utilitarian simulation 效用仿真
utility 实用程序；效用；可用性
utility-based 基于效用的
utility-based system 基于效用的系统
utility model 效用模型
utility program 实用程序

Vv

V&V (Verification and Validation) 校核与验证
vague 含糊的，不清楚的
vagueness 含糊，暧昧
valid 有效的；正当的
valid composition 有效的组成
valid model 有效的模型
valid simplification 有效简化
valid understanding 有效理解
validatability 可验证性

validatability of data 数据的可验证性
validate (v.) 验证，证实；确认
validated model 经过验证的模型
validated system 经过验证的系统
validation 确认；验证
validation agent 验证代理
validation algorithm 验证算法
validation assessment 验证评估
validation check 验证检查
validation criterion 验证准则
validation data 验证数据
validation data requirements 验证数据需求
validation error 验证错误
validation management 确认管理
validation method 验证方法
validation metric 验证度量
validation of conceptual model 概念模型验证
validation operator 验证操作者
validation plan 验证计划
validation requirement 验证需求
validation system 验证系统
validation technique 验证技术
validity 有效性；效度
validity checking 有效性检查
validity condition 有效性条件
validity measure 有效性测量
validity metric 有效性度量
validity of optimality 最优性的有效性
validity of robustness 鲁棒有效性
validity period 有效期
validity range 有效范围
value 价值；数值；评价；评估
value-free decision 价值无关的决策

value of simulation 仿真值
value system 价值系统
variability 可变性，易变性；变异性
variable 变量；可变的
variable-component-model coupling 可变组件模型的耦合
variable coupling 变量耦合
variable coupling modeling 变量耦合建模
variable derivability 变量可微性
variable error 可变误差
variable fidelity simulation 变保真度仿真
variable input 可变的输入
variable-oriented 面向变量的
variable-oriented model 面向变量的模型
variable parameter system 可变参数系统
variable resolution modeling 变分辨率建模
variable resolution simulation 变分辨率仿真
variable-step integration algorithm 变步长积分算法
variable-step method 变步长法
variable structure 变结构
variable-structure automaton 变结构自动机
variable-structure coupled model 变结构耦合模型
variable-structure coupling model 变结构耦合模型
variable-structure model 变结构模型
variable-structure modeling 变结构建模
variable-structure multimodel 变结构多模型
variable-structure system 变结构系统
variable time-step 可变的时间步长
variable topology simulation 变拓扑结构的仿真
variable transformation 变量转换
variable type 变量类型
variable validity 变量有效性
variance 变度；方差
variance estimator 方差估计器
variance reduction 方差缩减
variance reduction technique 方差减缩技术
variant 变体；变量；变异的
variant model 变异模型
variate 变量，变值
variation 变分；变差
vector 向量；矢量
vector-based cellular model 基于向量的元胞模型
vector-valued variable 向量值变量
velocity error 速度误差
veracity 真实性；准确性；精确性
verbal model 言语模型
verbal specification 语言规范
verbalized data 语言化数据
verifiability 置信度；可证实性，可验证性
verifiable 可检验的；可核实的
verifiable subclass 可核实的子类
verification 验证；检定；确认；证实；校核
verification agent 校核代理

verification algorithm 校核算法
verification and validation 校核与验证
verification and validation paradigm 校核和验证范式
verification and validation plan 校核和验证计划
verification and validation proponent 校核与验证提议者
verification assessment 校核评估
verification criterion 校核准则
verification error 校核误差
verification metric 校核性度量
verification of conceptual model 概念模型校核
verification operator 校核操作者
verification plan 校核计划
verification requirement 校核需求
verification system 校核系统
verification technique 校核技术
verification, validation and accreditation 校核、验证与确认
verification, validation and accreditation process 校核、验证与确认过程
verification, validation and certification 校核、验证与确认
verified 检验，鉴定；验证；核实；校核
verified model 经过校核的模型
verified system 经过校核的系统
verified understanding 已核实的协定
verifier 验证者；检验者；核实者
verify (v.) 检验，鉴定；验证；核实；校核
verisimilar 逼真的

verisimilitude 逼真性
verity 真实性；真实；真理；事实
versatile approach 通用方法
version 版本
version control 版本控制
versioning 版本控制
vertex model 顶点模型
very high-level language 超高级的语言
very large eddy simulation 超大涡仿真
very large simulation 超大型仿真
vetted model 经过审核的模型
vibrotactile display 震动触觉演示
vicarious experience 替代经验
vicarious experiment 替代实验
video game 视频游戏
video game simulation 视频游戏仿真
video gaming 视频游戏
video input 视频输入
view 视图
viewer 观察者；观察器；取景器
virtual 虚拟；虚拟的
virtual actor 虚拟演播室
virtual analog computer 虚拟模拟计算机
virtual battlespace 虚拟战场空间
virtual clock 虚拟时钟
virtual collaboration 虚拟协同
virtual community 虚拟社区
virtual computer 虚拟计算机
virtual content 虚拟内容
virtual device error 虚拟设备错误
virtual entity 虚拟实体
virtual environment 虚拟环境

virtual experience 虚拟体验
virtual experiment 虚拟实验
virtual holonomic constraint 虚拟完整约束
virtual human 虚拟人
virtual human markup language 虚拟人体标记语言
virtual image 虚拟图像
virtual knowledge 虚拟知识
virtual laboratory 虚拟实验室
virtual modeling display 虚拟建模显示
virtual network 虚拟网络
virtual network device error 虚拟网络设备错误
virtual player 虚拟参赛者
virtual presence 虚拟呈现
virtual prototype 虚拟样机
virtual prototyping 虚拟样机
virtual reality 虚拟现实
virtual reality modeling language 虚拟现实建模语言
virtual reality simulation 虚拟现实仿真
virtual reality simulation methodology 虚拟现实仿真方法学
virtual reality technology 虚拟现实技术
virtual reality tool 虚拟现实工具
virtual representation 虚拟表达
virtual research environment 虚拟研究环境
virtual retina display 虚拟视网膜显示器
virtual sensor input 虚拟传感器输入
virtual simulation 虚拟仿真
virtual simulation environment 虚拟仿真环境
virtual simulator 虚拟仿真器
virtual state 虚拟状态
virtual system-of-systems 虚拟系统体系
virtual system simulation 虚拟系统仿真
virtual time 虚拟时间
virtual time simulation 虚拟时间仿真
virtual training 虚拟训练
virtual training environment 虚拟训练环境
virtual training simulation 虚拟训练仿真
virtual training system 虚拟训练系统
virtual value 虚拟价值；有效值
virtual world 虚拟世界
virtuality 虚拟性
virtualization 虚拟化
virtualization simulation 虚拟化仿真
virtualize (v.) 虚拟化
virtualized reality 虚拟化现实
virtualized system 虚拟化系统
vision 视觉
vision input 视觉输入
visual 视觉的
visual characteristic 视觉特性
visual environment 视觉环境；可视化环境
visual I/O 视觉输入/输出
visual input 视觉输入
visual interactive environment 可视化交互环境

visual interactive modeling 可视化交互建模
visual interactive simulation 可视化交互仿真
visual interactive simulation environment 可视化交互仿真环境
visual model 视觉模型
visual model specification 视觉模型规范
visual modeling 可视化建模
visual modeling language 可视化建模语言
visual modeling technique 视觉建模技术
visual monitoring 视觉监控
visual perception 视觉感知
visual programming environment 可视化编程环境
visual sensation 视觉感受
visual simulation 可视化仿真
visual simulator 视景仿真器
visual specification 可视化规范
visual stealth 视觉隐身
visual system 视觉系统；视景系统
visual system specification 视觉系统规范
visualization 可视化
visualization capability 可视化能力
visualization of output 输出可视化
visualization parameter 可视化参数
visualization software 可视化软件
visualization technique 可视化技术
visualization technology 可视化技术
visualize (v.) 形象，形象化；可视化
visualized 直观的；具体的；形象化的；可视化的
vocabulary 词汇
voice input 语音输入
volatile variable 易失变量
volatility 波动性；变更率；不稳定性；易失性
Von Neuman model 冯-诺依曼模型
Voronoi boundary 沃罗诺伊边界
Voronoi cell 沃罗诺伊单元
Voronoi decomposition 沃罗诺伊分解
Voronoi diagram 沃罗诺伊图
Voronoi tessellation 沃罗诺伊曲面细分
VV&A (Validation, Verification and Accreditation) 校核、验证与确认
VV&A data cost 校核、验证与确认数据成本
VV&A documentation 校核、验证与确认文档
VV&A-related 校核、验证与确认相关的
VV&A-related issue 校核、验证与确认相关的问题
VV&C (Verification, Validation and Certification) 校核、验证与认证
VV&T (Verification, Validation and Testing) 校核、验证和测试
VV&T technique 校核、验证与测试技术

Ww

Wagner model 瓦格纳模型
waiting model 等待模型
walkthrough 预演

wall clock time　挂钟时间
war　战争
war game　战争博弈
war gaming　战争游戏
war simulation　战争模拟
warez　盗版软件
warez game　盗版软件游戏
warfare　战争；冲突；交战
warfare simulation　作战仿真
warfare training　作战训练
wargame　军事演习；作战对抗推演
wargaming　战争博弈；作战对抗推演
warm up　加热的；预热的；熟练的
warm up (v.)　加温，加热；准备工作
warm up period　暖机时段；发射前的准备阶段
warm up time　暖机时间，预热时间
warp　翘曲；弯曲；变形
wave equation　波动方程
weak anticipation　弱预期
weak bisimulation　弱互仿真
weak bisimulation equivalence　弱互仿真等价
weak classical simulation　弱经典仿真
weak emergence　弱涌现
weak emergent behavior　弱涌现行为
weak simulation　弱仿真
weakly anticipatory system　预期不充分系统，弱预期系统
wearable computer　可穿戴式计算机
wearable computer-based simulation　基于可穿戴式计算机的仿真
wearable simulation　可穿戴仿真
web-accessible　网络可访问的
web-accessible ontology　万维网可访问本体
web-based　基于万维网的
web-based business game　基于万维网的商业博弈
web-based computing　基于万维网的计算
web-based game　网络游戏
web-based infrastructure　基于万维网的基础设施
web-based learning　基于万维网的学习
web-based modeling　基于万维网的建模
web-based modeling system　基于万维网的建模系统
web-based multi-user simulation　基于万维网的多用户仿真
web-based simulation　基于万维网的仿真
web-based simulation game　基于万维网的仿真博弈
web-based standard　基于万维网的标准
web-based training　基于万维网的训练
web-centric　以万维网为中心
web-centric simulation　以万维网为中心的仿真
web-enabled　万维网使能的
web-enabled emerging trend　万维网使能的新趋势
web-enabled M&S application　基于万维网使能的建模与仿真应用程序
web-enabled simulation　万维网使能的仿真

web-enabled technique 万维网使能技术
web language 万维网语言
web service 万维网服务
web-service-based simulation 基于万维网服务的仿真
Weibull random variable 威布尔随机变量
weighted average 加权平均
weighted value 加权值
well-posed problem 适定问题
wet-lab data 湿实验室数据
wet laboratory 湿式实验室
what-if analysis 假设分析
what-if study 假设研究
white box 白箱，白盒
white box approach 白箱法，白盒法
white box approach in modeling 建模中的白箱法，建模中的白盒法
white box model 白箱模型，白盒模型
white box testing 白箱测试，白盒测试
white box validation 白箱验证，白盒验证
wicked problem 恶劣问题；奇特问题
width 宽度
willful error 故意错误
wireless input 无线电输入
within subject variable 组内变量
workflow modeling 工作流建模
working memory 工作记忆；工作存储器
working phase 作业阶段
world 世界
world model 世界模型
world view 世界观
world view of a formalism 形式主义的世界观
wrap 包
wrap (v.) 打包
wrapped 包装的
wrapper 封装器，包装器
wrapping 封装，包装
wrapping technique 包装技术
write error 写入错误
write protect error 写保护错误
writing error 书写错误，写入错误
wrong 错误的
wrong model 错误的模型
wrong self-model 错误的自模型

Yy

yoked 共轭的；束缚的；联动的
yoked relation 共轭的关系
yoked simulation 联动仿真
yoked variable 共轭变量；联动变量

Zz

Z-transform Z 变换
zero-crossing 过零；零交点
zero-crossing function 过零函数
zero error 零位误差
zero-scale error 零刻度误差
zero-sum game 零和博弈
zero-sum simulation 零和仿真
zero-sum trust game 零和信任博弈
zero-variance simulation 零方差仿真

参 考 文 献

李伯虎, Ören T, 赵沁平, 等. 2012. 汉英-英汉建模与仿真术语集. 北京: 科学出版社.

全国科学技术名词审定委员会. http://www.cnctst.cn/.

清华大学外语系《英汉科学技术词典》编写组. 1991. 英汉科学技术词典. 北京: 国防工业出版社.

中国仿真学会. 2018. 建模与仿真技术词典. 北京: 科学出版社.

章含. 2007. 简明英汉计算机词典. 上海: 上海科学技术出版社.

张霄军. 2009. 英汉·汉英航空航天词汇手册. 上海: 上海外语教育出版社.

Ören T. 2014. The richness of modeling and simulation and an index of its body of knowledge// Obaidat M, Filipe J, Kacprzyk J, et al. Simulation and Modeling Methodologies, Technologies and Applications: Advances in Intelligent Systems and Computing, Vol 256. Cham: Springer.

Ören T, et al. 2006. Modeling and Simulation Dictionary: English-French-Turkish. Marseilles: Université d'Aix-Marseille.